Design and Synthesis of Organic Molecules as Antineoplastic Agents

Design and Synthesis of Organic Molecules as Antineoplastic Agents

Editors

Carla Boga
Gabriele Micheletti

MDPI • Basel • Beijing • Wuhan • Barcelona • Belgrade • Manchester • Tokyo • Cluj • Tianjin

Editors
Carla Boga
Department of Industrial
Chemistry "Toso Montanari",
Alma Mater Studiorum -
Università di Bologna
Italy

Gabriele Micheletti
Department of Industrial
Chemistry "Toso Montanari",
Alma Mater Studiorum -
Università di Bologna
Italy

Editorial Office
MDPI
St. Alban-Anlage 66
4052 Basel, Switzerland

This is a reprint of articles from the Special Issue published online in the open access journal *Molecules* (ISSN 1420-3049) (available at: https://www.mdpi.com/journal/molecules/special_issues/synthesis_antineoplastic_agents).

For citation purposes, cite each article independently as indicated on the article page online and as indicated below:

LastName, A.A.; LastName, B.B.; LastName, C.C. Article Title. *Journal Name* **Year**, *Article Number*, Page Range.

ISBN 978-3-03936-666-8 (Hbk)
ISBN 978-3-03936-667-5 (PDF)

Cover image courtesy of Carla Boga.

Contents

About the Editors

Carla Boga graduated with 110/110 "summa cum laude" in Chemistry at the University of Bologna in 1983. She then completed a Ph.D. with a thesis entitled "Chiral ligands from D-mannitol and their use in asymmetric synthesis". She won fellowship applications with chemical companies (Recordati, Proter, Tecnofarmaci), carrying out research on the Fries reaction, the synthesis of anthraquinonic systems, and novel synthetic derivatives of beta-lactam. In 1992, she took up a postdoctoral fellowship in the Department of Biochemistry at the University of Bologna, where she worked for two years on the mechanisms of controlling tumoral growth. In 1995, she became Assistant Professor in the Faculty of Industrial Chemistry (Department of Organic Chemistry "A. Mangini"), then in the Department of Industrial Chemistry ("Toso Montanari") at the University of Bologna. As of 2020, she is Associate Professor at the University of Bologna. She is a member of the American Chemical Society and the Italian Chemical Society. She is author of more than 130 articles in international scientific journals, as well as one patent and three book chapters. Her teaching activity is mainly in the field of organic and bioorganic chemistry. She has supervised or co-supervised 6 Ph.D. theses and more than 65 dissertations for bachelor's and master's degrees, mainly in the field of industrial chemistry.

Gabriele Micheletti graduated with 110/110 in Industrial Chemistry at the University of Bologna in 2006, where he later completed a Ph.D. in Chemical Science, in 2011. His thesis was entitled "Abiotic and prebiotic phosphorus chemistry". From 2011 to 2018, he took up a postdoctoral fellowship at the Department of Industrial Chemistry, University of Bologna, where he worked on organic synthesis, organic mechanisms reaction, and the mechanisms of controlling tumoral growth. In 2017, 2018, and 2020, he is Adjunct Professor in the Department of Chemistry and Geology at the University of Modena and Reggio Emilia. From 2019 to 2020, he is working as a collaborator in the Department of Industrial Chemistry at the University of Bologna. He is author of 42 articles in international scientific journals. His teaching activity is mainly in the field of organic and bioorganic chemistry. He has supervised 2 Ph.D. theses and 26 dissertations for bachelor's and master's degrees, mainly in Industrial Chemistry.

Editorial

Design and Synthesis of Organic Molecules as Antineoplastic Agents

Carla Boga * and Gabriele Micheletti *

Department of Industrial Chemistry "Toso Montanari", Alma Mater Studiorum - Università di Bologna, Viale del Risorgimento 4, 40136 Bologna, Italy
* Correspondence: carla.boga@unibo.it (C.B.); gabriele.micheletti3@unibo.it (G.M.)

Received: 30 April 2020; Accepted: 9 May 2020; Published: 18 June 2020

The fight against cancer is one of the most challenging tasks currently for lots of researchers in many fields, such as pharmaceuticals, medicine, and chemicals. In this context, synthetic organic chemistry plays a key role, representing a flexible tool that uses variations of functional groups or motifs within organic architecture; it permits the disclosure and modulation of structure and activity relationships for designing new molecules or structural hybrids once a molecular target has been identified.

A query for "anticancer" on Web of Science [1] returns more than 25,000 records just for the years of 2018 and 2019, more than half of them concerning chemical and biochemical subjects (Figure 1).

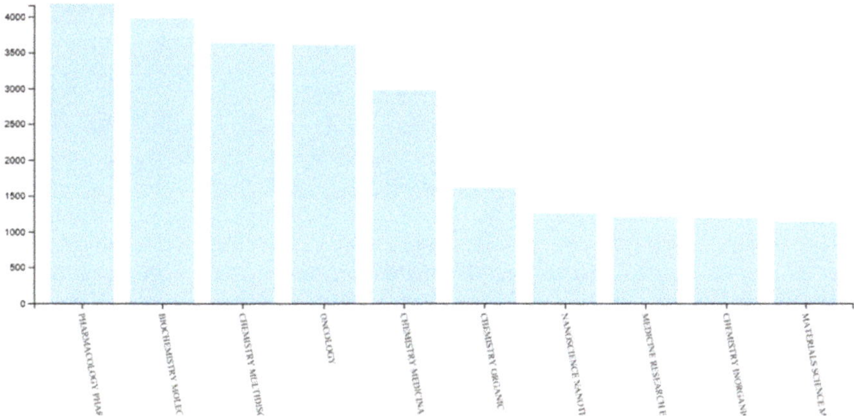

Figure 1. Results analysis for Web of Science query "anticancer" within the publication years of 2018 and 2019 [1].

Due to the cutting-edge nature of the topic, we recognized the need to propose a Special Issue on "Design and Synthesis of Organic Molecules as Antineoplastic Agents" in the Molecules journal, with the aim of offering an appropriate opportunity to scientists to share their more relevant results within the scientific community. This volume constitutes a collection of papers for the Special Issue dedicated to the topic.

The articles exhibit a cross-field character, focusing both synthetic aspects and biological activity of a wide variety of organic molecules, which have been often designed with the support of in silico studies. Molecular hybridization approaches have been used in several cases as a successful multi-target strategy for the design and development of novel antitumor agents.

The articles included in the collection globally show about 180 newly synthesized organic species, and more than 200 have been tested as anticancer agents.

In most cases [2–11], the structural skeleton of these molecules contain a heterocyclic scaffold, and biological investigations were supported by docking calculations to identify and predict the possible molecular targets.

Thus, 1,2,3-triazole was used as a linker between 4-hydroxycoumarin and benzoyl-substituted arylamines to obtain novel derivatives whose biological activity was evaluated against human breast cancer MDA-MB-231 cells [2]. Novel pyrazole and benzofuropyrazole derivatives were prepared from resorcinol and tested as cell growth inhibitors in leukemia K562, lung tumor A549, and breast tumor MCF-7 cells [3]. Effects of the novel class of pyrido[3′,2′:4,5]furo(thieno)[3,2-*d*]pyrimidin-8-amines were evaluated on the methylation of DNA of murine sarcoma S-180 cells, while a docking analysis and an assessment of the anti-proliferative activity of the most active compounds were performed on both cancer (HeLa) and normal (Vero) cells [4]. Geronikaki et al. [5] report an exhaustive study on 4,5-diaryl 3(2H)furanones, ranging from synthesis to molecular docking, including a complete evaluation of the anti-inflammatory and COX-1/2 inhibitory action and the influence on cancer growth.

Both in vitro and in vivo anticancer activity for a library of 2-(het)arylpyrrolidine-1-carboxamides, synthesized by a modular approach based on the intramolecular cyclization/Mannich-type reaction of *N*-(4,4-diethoxybutyl)ureas, was tested on a M-Hela cell line [6]. Furthermore, the ability to suppress bacterial biofilm growth of some compounds bearing a benzofuroxan moiety was evaluated. A study by Mancini and Defant was based on the synthesis of new aminoquinone hybrid molecules containing covalently linked pharmacophoric units, present individually in compounds acting as inhibitors of the cancer protein targets tubulin, human topoisomerase II, and ROCK1 [7]. Docking calculation of complexes with each protein allowed the selection of some molecules to be subjected to screening on a panel of 60 human cancer cell lines. In the context of molecular hybridization, a synthesis of molecules containing azaheteroaromatics bound to azelaic acid fragments has been planned, owing to their analogous structure with some histone deacetylase inhibitors [8]. Cell lines included in the evaluation of toxicity profiles were: malignant U2OS, HT29, PC3, IGROV1, and normal human adult fibroblast HDFa cells. Biological tests on U2OS cells suggested a post-transcriptional modification of both H2/H3 and H4 histones, and an in silico investigation revealed a plausible interaction with HDAC7. Janecky et al. [9] prepared a library of 3-methylidene chroman-4-ones via the Horner–Wadsworth–Emmons methodology and made an accurate structural analysis of the products and related intermediates. Anti-proliferative activity of the compounds against leukemia and breast cancer cell lines (HL-60, NALM-6, and MCF-7) was studied, revealing that one of them promoted caspase-mediated apoptosis. Bisphenol units containing tetrahydrothiepine and dihydrothiophene derivatives as estrogen receptor α (ERα) modulators have been designed, synthesized, and investigated as therapeutic candidates for breast cancer drugs by Ohta et al. [10]. Cardiovascular activity of some nitro-substituted sulfur-containing heterocycles (a thiopyran S,S-dioxide and 3-aryl-4-nitrobenzothiochromans S,S-dioxide species) was assessed by ex-vivo studies. At the same time, molecular drug resistance (MDR) reverting effect was evaluated for selected compounds by using tumor cell lines (breast MDA-MB-453, neuroblastoma SHSY5Y, ovarian carcinoma A2780, and multidrug-resistant A2780/DX3 cells) [11]. Dihydrobetulinic-, ursolic- and oleanolic-acid derivatives have been synthesized by Spivak et al. [12], and their cytotoxicity was tested on five different human tumor cell lines (Jurkat, K562, U937, HEK, and HeLa cells) and compared with tests carried out on normal human fibroblasts. The effect of the substituent in position nine and of the functionality of methyl esterification on the biological activity of 9-hydroxystearic acid (9-HSA), an endogenous cellular lipid with anti-proliferative and selective activity against cancer cells, was evaluated [13]. The study indicated the importance, in position nine, of groups able to make hydrogen bonding with the molecular target, and the preservation of the biological activity even if the 9-HSA carboxy group is esterified.

In exploiting the effect on cancer proliferation, one may note that the reported contributions reflect a wide variety of molecular building blocks, sometimes with a natural origin. The synthesized compounds have been tested on more than 70 human cancer cell lines, besides control cells. It has been evidenced that the strategic function of docking calculations can support both synthetic design and to disclose or predict appropriate molecular targets.

The Guest Editors thank all the authors for their contributions to this Special Issue, all the reviewers for their efforts in evaluating the submitted articles, and the editorial staff of Molecules, especially the Assistant Editor of the journal Katie Zhang for gently encouraging the realization of this Special Issue.

Funding: This research received no external funding.

Conflicts of Interest: The authors declare no conflict of interest.

References and Note

1. Citation Report graphic is derived from Clarivate Web of Science, Copyright Clarivate 2020. All rights reserved.
2. An, R.; Hou, Z.; Li, J.; Yu, H.; Mou, Y.; Guo, C. Design, Synthesis and Biological Evaluation of Novel 4-Substituted Coumarin Derivatives as Antitumor Agents. *Molecules* **2018**, *23*, 2281. [CrossRef] [PubMed]
3. Cui, Y.; Tang, L.; Zhang, C.; Liu, Z. Synthesis of Novel Pyrazole Derivatives and Their Tumor Cell Growth Inhibitory Activity. *Molecules* **2019**, *24*, 279. [CrossRef]
4. Sirakanyan, S.; Spinelli, D.; Geronikaki, A.; Hakobyan, E.; Sahakyan, H.; Arabyan, E.; Zakaryan, H.; Nersesyan, L.; Aharonyan, A.; Danielyan, I.; et al. Synthesis, Antitumor Activity, and Docking Analysis of New Pyrido[3′,2′:4,5]furo(thieno)[3,2-d]pyrimidin-8-amines. *Molecules* **2019**, *24*, 3952. [CrossRef]
5. Semenok, D.; Medvedev, J.; Giassafaki, L.; Lavdas, I.; Vizirianakis, I.; Eleftheriou, P.; Gavalas, A.; Petrou, A.; Geronikaki, A. 4,5-Diaryl 3(2H)Furanones: Anti-Inflammatory Activity and Influence on Cancer Growth. *Molecules* **2019**, *24*, 1751. [CrossRef] [PubMed]
6. Smolobochkin, A.; Gazizov, A.; Sazykina, M.; Akylbekov, N.; Chugunova, E.; Sazykin, I.; Gildebrant, A.; Voronina, J.; Burilov, A.; Karchava, S.; et al. Synthesis of Novel 2-(Het)arylpyrrolidine Derivatives and Evaluation of Their Anticancer and Anti-Biofilm Activity. *Molecules* **2019**, *24*, 3086. [CrossRef] [PubMed]
7. Defant, A.; Mancini, I. Design, Synthesis and Cancer Cell Growth Inhibition Evaluation of New Aminoquinone Hybrid Molecules. *Molecules* **2019**, *24*, 2224. [CrossRef] [PubMed]
8. Micheletti, G.; Calonghi, N.; Farruggia, G.; Strocchi, E.; Palmacci, V.; Telese, D.; Bordoni, S.; Frisco, G.; Boga, C. Synthesis of Novel Structural Hybrids between Aza-Heterocycles and Azelaic Acid Moiety with a Specific Activity on Osteosarcoma Cells. *Molecules* **2020**, *25*, 404. [CrossRef] [PubMed]
9. Kędzia, J.; Bartosik, T.; Drogosz, J.; Janecka, A.; Krajewska, U.; Janecki, T. Synthesis and Cytotoxic Evaluation of 3-Methylidenechroman-4-ones. *Molecules* **2019**, *24*, 1868. [CrossRef] [PubMed]
10. Ohta, K.; Kaise, A.; Taguchi, F.; Aoto, S.; Ogawa, T.; Endo, Y. Design and Synthesis of Novel Breast Cancer Therapeutic Drug Candidates Based upon the Hydrophobic Feedback Approach of Antiestrogens. *Molecules* **2019**, *24*, 3966. [CrossRef] [PubMed]
11. Micucci, M.; Viale, M.; Chiarini, A.; Spinelli, D.; Frosini, M.; Tavani, C.; Maccagno, M.; Bianchi, L.; Gangemi, R.; Budriesi, R. 3-Aryl-4-nitrobenzothiochromans S,S-dioxide: From Calcium-Channel Modulators Properties to Multidrug-Resistance Reverting Activity. *Molecules* **2020**, *25*, 1056. [CrossRef] [PubMed]
12. Spivak, A.; Khalitova, R.; Nedopekina, D.; Dzhemileva, L.; Yunusbaeva, M.; Odinokov, V.; D'yakonov, V.; Dzhemilev, U. Synthesis and Evaluation of Anticancer Activities of Novel C-28 Guanidine-Functionalized Triterpene Acid Derivatives. *Molecules* **2018**, *23*, 3000. [CrossRef] [PubMed]
13. Calonghi, N.; Boga, C.; Telese, D.; Bordoni, S.; Sartor, G.; Torsello, C.; Micheletti, G. Synthesis of 9-Hydroxystearic Acid Derivatives and Their Anti-proliferative Activity on HT 29 Cancer Cells. *Molecules* **2019**, *24*, 3714. [CrossRef] [PubMed]

Article

Design, Synthesis and Biological Evaluation of Novel 4-Substituted Coumarin Derivatives as Antitumor Agents

Ran An [1,†], **Zhuang Hou** [1,†], **Jian-Teng Li** [1], **Hao-Nan Yu** [1], **Yan-Hua Mou** [2,*] and **Chun Guo** [1,*]

[1] School of Pharmaceutical Engineering, Shenyang Pharmaceutical University, Shenyang 110016, China; bear2015a@163.com (R.A.); houzhuang8@sina.com (Z.H.); jtli2014@outlook.com (J.-T.L.); yhna380@hotmail.com (H.-N.Y.)

[2] Department of Pharmacology, Shenyang Pharmaceutical University, Shenyang 110016, China

[*] Correspondence: mu_hua_jj@sina.com (Y.-H.M.); chunguo@syphu.edu.cn (C.G.); Tel.: +86-24-4352-0226 (C.G.)

[†] These authors contributed equally to this work.

Received: 17 August 2018; Accepted: 4 September 2018; Published: 6 September 2018

Abstract: Herein, fifteen new compounds containing coumarin, 1,2,3-triazole and benzoyl-substituted arylamine moieties were designed, synthesized and tested in vitro for their anticancer activity. The results showed that all tested compounds had moderate antiproliferative activity against MDA-MB-231, a human breast cancer cell line, under both normoxic and hypoxic conditions. Furthermore, the 4-substituted coumarin linked with benzoyl 3,4-dimethoxyaniline through 1,2,3-triazole (compound **5e**) displayed the most prominent antiproliferative activities with an IC_{50} value of 0.03 µM, about 5000 times stronger than 4-hydroxycoumarin (IC_{50} > 100 µM) and 20 times stronger than doxorubicin (IC_{50} = 0.60 µM). Meanwhile, almost all compounds revealed general enhancement of proliferation-inhibiting activity under hypoxia, contrasted with normoxia. A docking analysis showed that compound **5e** had potential to inhibit carbonic anhydrase IX (CA IX).

Keywords: anticancer; coumarin; hypoxia; 1,2,3-triazole

---

1. Introduction

Coumarin, a significant scaffold of both natural and synthetic origin, displays versatile pharmacological properties that include antibacterial [1], antioxidant [2], anticoagulant [3], anti-Alzheimer [4], anti-HIV [5], antimicrobial [6,7] and anticancer activities. Their antitumor effects are widely reported to be related to the inhibition of the cellular proliferation through binding to different targets and diverse pharmacological mechanisms.

For example, the coumarins attached an iodinated aromatic ring (Figure 1A), initially identified by Basanagouda et al. as a potential anti-cancer agent, exerted an anti-proliferative effect in MDA-MB-231 human adenocarcinoma mammary gland and A-549 human lung carcinoma [8]. Similarly, the coumarin linked 6-methylpyridine (Figure 1B) was reported to show potent inhibition of 17β-hydroxysteroid dehydrogenase type 3 (17β-HSD3) with an IC_{50} value of 0.0015 µM [9]. In addition, the hybrid of 1,2,3-triazole and 4-subsitituted coumarin (Figure 1C) had an IC_{50} value of 0.52 µM against A-549 cells and induces G2/M phase cell cycle arrest [10]. Interestingly, the supuran group revealed that 4-substituted coumarins (Figure 1D) are very effective against transmembrane, tumor-associated isoforms carbonic anhydrase IX (CA IX) [11] with activity in the submicromolar range [12]. From the precedents mentioned above, 4-substituted coumarin derivatives are thus excellent leads for designing antitumor agents.

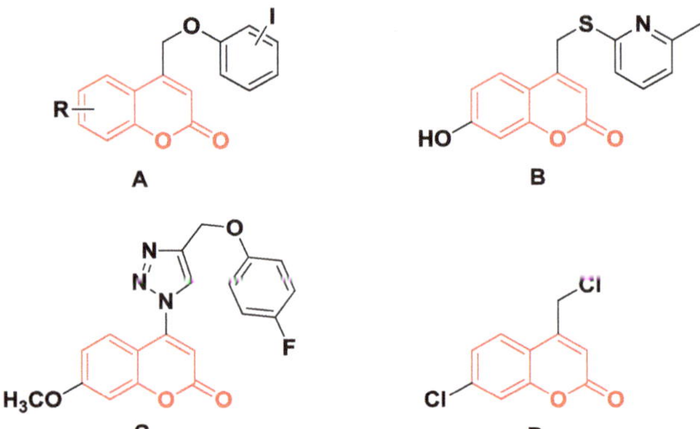

Figure 1. Structures of some 4-substituted coumarins. (**A**) the coumarins attached an iodinated aromatic ring. (**B**) the coumarin linked 6-methylpyridine. (**C**) the hybrid of 1,2,3-triazole and 4-subsitituted coumarin. (**D**) 4-substituted coumarins reported by supuran group.

Benzanilide moieties are importantly active groups in anticancer agents and introduction of a benzanilide is considered as an efficient method to improve the activity of compounds. For instance, the Su group introduced electron-donating- or withdrawing group-substituted benzamides into compounds (Figure 2A) and some derivatives exhibited good growth inhibitory activity against SK-BR-3 breast cancer cells at low nanomolar concentrations [13]. Analogously, Yang et al. focused on the modification of the aniline in benzanilide, and identified the compound (Figure 2B) with an IC_{50} value of 2.57 μM, which inhibited HepG2 cell proliferation more effectively than sorafenib (IC_{50} = 9.61 μM) [14].

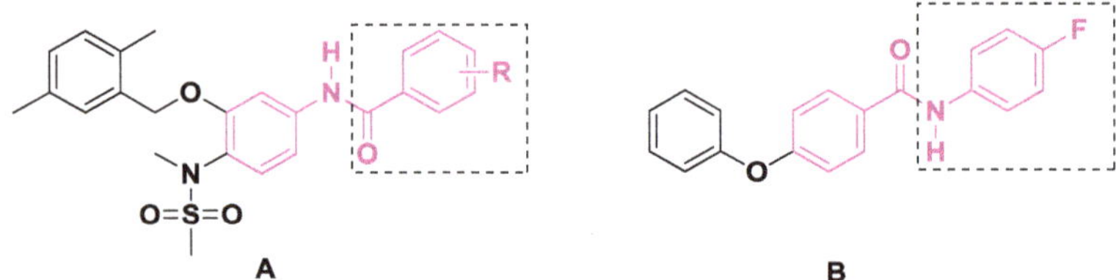

Figure 2. Structures of some benzanilide. (**A**) compounds reported by Su group. (**B**) compounds reported by Yang et al.

Based on the molecular hybridization strategy and mentioned above, we combined a pharmacophore (coumarin) which can inhibit proliferation of cancer cells through varied mechanisms with another anticancer pharmacophore (benzanilide) which can increase the diversity of compounds to quickly screen target compounds with good anticancer activities. Herein, we referred to Pingaew's work [15] and designed the compounds comprising three core structural elements (Figure 3): (i) a (4-substituted) coumarin moiety as a scaffold, (ii) a 1,2,3-triazole moiety as a biocompatible, covalent linker [16] and (iii) a benzoyl-substituted arylamine group as a variable group.

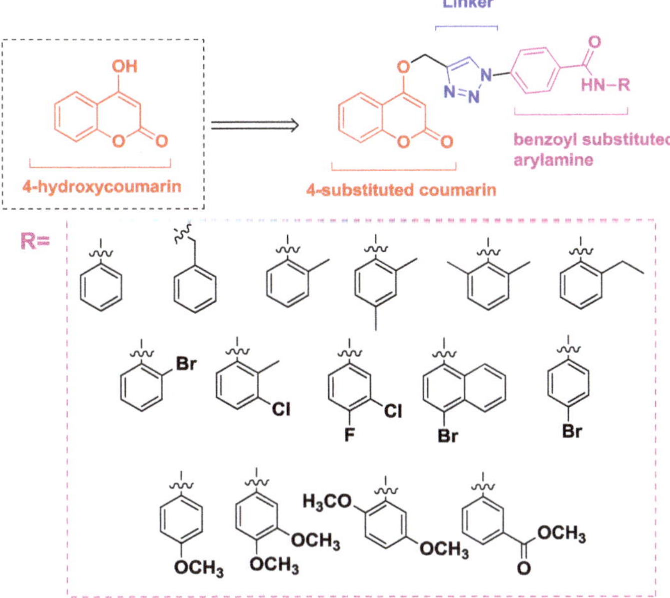

Figure 3. Rationale design of the title compounds.

2. Results and Discussion

2.1. Chemistry

The synthesis of target compounds **5a–5o** was presented at Scheme 1. First, intermediate **1** was obtained by substitution of propargyl bromide in 4-hydroxycoumarin at room temperature [17], and **2** was synthesized by the azidation of 4-aminobenzoic acid at the presence of $NaNO_2/NaN_3$ in H_2O. Next, **1** and **2** were treated with CuI in dichloromethane at room temperature via click chemistry to give compound **3** and then a chlorination reaction with $SOCl_2$ was conducted to obtain intermediate **4**. Last, **4** was reacted with the corresponding substituted arylamines to obtain target compounds **5a–5o**.

Scheme 1. Synthesis of compounds **5a–5o**. *Reagents and conditions*: (**a**) Propargyl bromide, K_2CO_3, DMF, r.t., 4 h; (**b**) $NaNO_2$, HCl, 0.5 h, NaN_3, 0.5 h; (**c**) CuI, Et_3N, DCM, r.t., 3 h; (**d**) $SOCl_2$, 1.5 h; (**e**) RNH_2, DCM, 2 h.

2.2. Biological Evaluation

The antiproliferative activities of new compounds against MDA-MB-231 cells (a kind of breast cancer cells) was evaluated by the 3-(4,5-dimethyl-2-thiazolyl)-2,5-diphenyl-2-*H*-tetrazolium bromide (MTT) assay. Specifically, we aimed to identify compounds that inhibit proliferation more powerfully under hypoxic conditions, in line with the physiological response to hypoxia in increased CA IX activity in this cell line, so both normoxic and hypoxic conditions were evaluated [18]. Doxorubicin (DOX), cisplatin(*cis*-Pt) and 4-hydroxycoumarin were selected as positive reference drugs.

The in vitro antiproliferative activities are summarized in Table 1. From the structure activity relationships of the target compounds the following was concluded:

(i)   All compounds showed better inhibition than that of 4-hydroxycoumarin and **5e** had the best antiproliferative activity of this series of compounds, even better inhibition than that of DOX and *cis*-Pt under hypoxia;

(ii)   The IC_{50} of most compounds under hypoxic conditions were lower than that under normoxic conditions;

(iii)   The $IC_{50, \text{normoxia}}/IC_{50, \text{hypoxia}}$ of DOX and *cis*-Pt were lower than that of **5b**, **5e**, **5h**, **5m**, **5o**;

(iv)   The presence of a 4-substitued phenyl in the R group enhances the antiproliferative potential and antiproliferative activity increases remarkably when a 3,4-substitued phenyl is present in the R group.

2.3. Molecular Docking

In the last years, several approaches have reported that the coumarins are truly CA IX-selective inhibitors [12,19,20]. The coumarin moiety is the scaffold of our newly synthesized compounds and we wanted to predict the binding mode of the most antiproliferative **5e** of this series of compounds into the binding site of carbonic anhydrase IX, so we decided to carry out a molecular docking study [21]. We knew that coumarin moiety was hydrolyzed within the CA IX active site [12,22]. Meanwhile, according to our previous work [23], as shown in Figure 4(d), the hydrolyzed compound **5e** was used as the ligand of this docking analysis. The molecular docking results (Figure 4(b,c)) showed that the carboxyl group in 2-hydroxycinnamic acid moiety engaged hydrogen bonds with the hydroxyl group of Thr199. The oxygen atom of enol ether acted as acceptor receiving two H-bonds from the backbone NH of Thr199 and Thr200. Meanwhile, the nitrogen atom of triazole moiety formed a H-bond with His64 and the carbonyl group of the benzoyl arylamine moiety formed H-bond with Gln67. In addition, the triazole moiety displayed a π-π stacking with His94, and the aromatic benzene ring of the coumarin group showed lipophilic interactions with Leu98 and Val121, respectively. As shown in Figure 4a,b, the coumarin moiety will generally adopt a conformation to interact with the hydrophobic half (red part) of the CA IX active site cavity; similarly, the triazole and the benzoyl aniline moieties will generally interact with the hydrophilic half (blue part). In addition, the binding energy is −6.57 kcal/mol. On the basis of the docking results, it was found that compound **5e** had the potential to inhibit CA IX, which will be the probable anticancer activity mechanism of these derivatives.

Table 1. IC_{50} of the compounds against MDA-MB-231 cell.

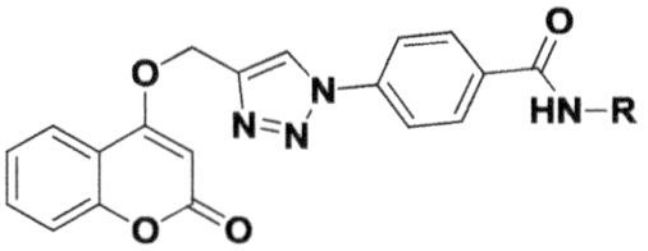

Compd.	R	IC$_{50}$ (Mm) [a]		IC$_{50, normoxia}$/IC$_{50, hypoxia}$ [b]
		Hypoxia	Normoxic	
5a	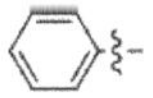 (phenyl)	23.47	24.41	1.04
5b	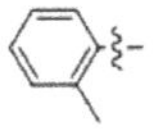 (2-methylphenyl)	8.14	108.72	13.36
5c	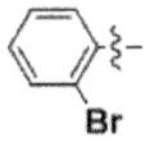 (2-bromophenyl)	75.21	73.77	0.98
5d	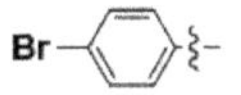 (4-bromophenyl)	6.72	6.78	1.01
5e	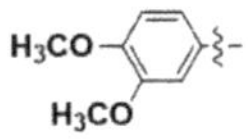 (3,4-dimethoxyphenyl)	0.03	1.34	46.31
5f	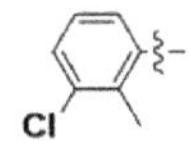 (2-methyl-3-chlorophenyl)	73.82	91.61	1.24
5g	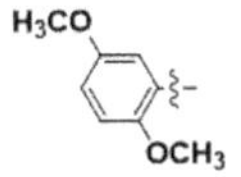 (2,5-dimethoxyphenyl)	53.98	62.79	1.16
5h	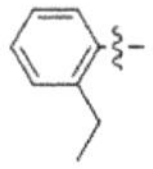 (2-ethylphenyl)	3.44	18.45	5.36
5i	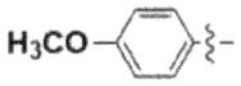 (4-methoxyphenyl)	20.35	20.98	1.03
5j	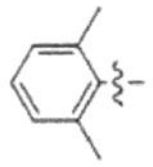 (2,3-dimethylphenyl)	12.87	12.28	0.95
5k	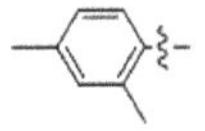 (3,4-dimethylphenyl)	8.70	10.86	1.25
5l	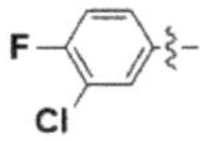 (3-fluoro-5-chlorophenyl)	1.30	7.03	5.39
5m	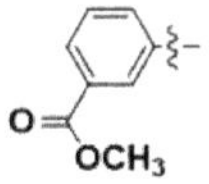 (2-methoxycarbonylphenyl)	0.25	5.06	20.46
5n	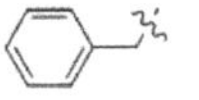 (benzyl)	34.82	39.58	1.14
5o	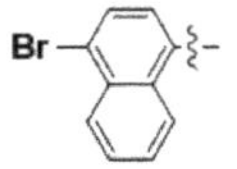 (bromonaphthyl)	9.42	16.76	1.78
DOX		0.60	1.07	1.79
cis-Pt		4.68	7.87	1.68
4-hydroxycoumarin		>100	>100	

[a] Values are the average of three independent experiments. Relative errors are generally within 5–10%. [b] A higher IC$_{50, normoxia}$/IC$_{50, hypoxia}$ indicates more antiproliferative activity in hypoxic conditions.

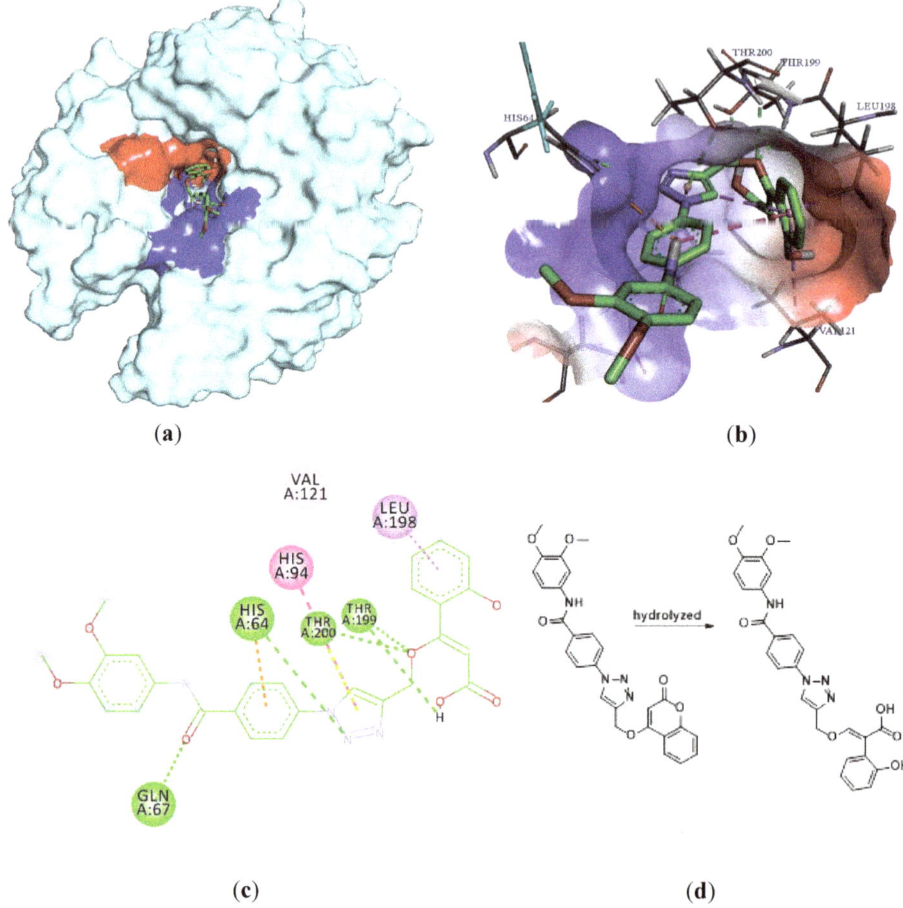

Figure 4. Interaction diagrams of the selected docked conformations for hydrolyzed compound **5e** inside the active site of CA IX enzyme. (**a**) The surface representation of binding pocket has been shown at the top of the figure. (**b**) 3D ligand interactions diagram. (**c**) 2D ligand interactions diagram. (**d**) The ligand of this docking analysis.

3. Materials and Methods

3.1. Chemistry

Column chromatography was carried out on the 200–300 mesh silica gel (Qingdao Haiyang Chemical Co. Ltd., Qingdao, Shandong, China). Analytical thin-layer chromatography (TLC) was performed on silica gel precoated GF254 plates (Qingdao Haiyang Chemical Co. Ltd.). ^{1}H-NMR and ^{13}C-NMR spectra were recorded on an AV-400 spectrometer (Bruker Bioscience, Billerica, MA, USA), with tetramethylsilane as an internal standard. ^{1}H and ^{13}C-NMR spectra of these compounds are available in the Supplementary Materials. ESI-MS spectra were obtained on an Agilent ESI-QTOF instrument. High resolution mass spectra (HRMS) were measured with an Agilent Accurate-Mass Q-TOF 6530 (Agilent, Santa Clara, CA, USA) in ESI mode and are available in the Supplementary Materials. Melting points were determined using a X-4 microscope melting point apparatus (Beijing Tech Instrument Co., Ltd., Beijing, China) without calibration.

3.1.1. 4-(Prop-2-ynyloxy)-2*H*-chromen-2-one (**1**)

4-Hydroxycoumarin (20 g, 0.12 mol) was dissolved in DMF (100 mL), and K_2CO_3 (2 eq) was added. Propargyl bromide (1.5 eq) was then added under nitrogen. The reaction mixture was kept stirring at room temperature. After the completion of reaction (as monitored by TLC), the reaction mixture was poured onto crushed ice and set aside for some time. Then it was filtered and dried to obtain the desired product. The crude product was used without further purification. Yield 63%; m.p. 155–156 °C ^{1}H-NMR (DMSO-d_6, 600 MHz, δ, TMS = 0): 7.82–7.75 (m, 1H), 7.71–7.64 (m, 1H), 7.44–7.39 (m, 1H), 7.40 7.34 (m, 1H), 5.96 (d, J = 1.3 Hz, 1H), 5.11 (dd, J = 2.4, 1.1 Hz, 2H), 3.83 (t, J = 2.4 Hz, 1H) ESI-MS [M − H]$^-$: (m/z) Calcd. for $C_{12}H_7O_3$: 199.0. Found: 199.1.

3.1.2. 4-Azidobenzoic acid (**2**)

4-Aminobenzoic acid (11.2 g, 0.08 mol) and 3 M HCl (250 mL) were added to a 500 mL three-necked flask, and then cooled to 0 °C. $NaNO_2$ (aq, 6.8g (0.098 mol)/50 mL) was added dropwise to the cooled mixture while the temperature was kept between 0 and 5 °C. After stirring for 30 min, NaN_3 (aq, 7.6g (0.12 mol)/50 mL) was added dropwise to the cooled mixture. The reaction mixture was stirred for 30 min at 0 °C and 1 h at room temperature. Then it was filtered, washed with water and dried to obtain the desired product. The crude product was used without further purification. Yield 90%; m.p. 178–180 °C. ^{1}H-NMR (DMSO-d_6, 600 MHz, δ, TMS = 0):12.97 (s, 1H), 8.01–7.90 (m, 2H), 7.26–7.15 (m, 2H). ESI-MS [M − H]$^-$: (m/z) Calcd. for $C_7H_4N_3O_2$: 162.0 Found: 162.0.

3.1.3. 4-(4-(((2-Oxo-2*H*-chromen-4-yl)oxy)methyl)-1*H*-1,2,3-triazol-1-yl)benzoic acid (**3**)

In a 250 mL flask, compounds **1** (10 g, 0.05 mol) and **2** (9 g, 0.055 mol) were added to DCM (150 mL) at room temperature. To this mixture was added CuI 1g (5 mmol), followed by trimethylamine 2 g (0.02 mol) under an argon atmosphere. After the completion of reaction (as evidenced by TLC), the mixture was washed with 1 M HCl and evaporated. The residue was directly used in the next step without further purification. Yield 90%; m.p. 218–220 °C. ^{1}H-NMR (DMSO-d_6, 600 MHz, δ, TMS = 0): 9.18 (s, 1H), 8.15 (d, J = 8.2 Hz, 2H), 8.07 (d, J = 8.2 Hz, 2H), 7.84 (d, J = 7.9 Hz, 1H), 7.66 (t, J = 7.7 Hz, 1H), 7.42 (d, J = 8.3 Hz, 1H), 7.35 (t, J = 7.6 Hz, 1H), 6.21 (s, 1H), 5.54 (s, 2H). ESI-MS [M − H]$^-$: (m/z) Calcd. for $C_{19}H_{12}N_3O_5$: 362.1 Found: 362.1.

3.1.4. Synthesis of Compound **4**

In a 100 mL flask, compound **3** (1 g, 2.8 mmol) were added to anhydrous DCM (50 mL), and cooled to 0 °C. Then, sulfoxide chloride (15 mL) was added dropwise to the cooled mixture under stirring. After added, the reaction mixture was stirred for 2 h at room temperature. Solvent and excess sulfoxide chloride were evaporated. The residue was directly used in the next step without further purification.

3.1.5. General Procedure for the Synthesis of Compound **5a–5e**

In a 50 mL flask, the appropriate arylamine (1 mmol) was added to anhydrous DCM (10 mL), and cooled to 0 °C. Then, compound **4** (0.2 g, 0.5 mmol) mixed with anhydrous DCM (4 mL) was added dropwise to the cooled mixture under stirring. After stirring for 30 min, the reaction mixture was stirred for 4 h more at room temperature. After the completion of reaction (as evidenced by TLC), the mixture was evaporated. The residue was finally purified by column chromatography (DCM:MeOH = 50:1) to obtain the desired products.

4-(4-(((2-Oxo-2H-chromen-4-yl)oxy)methyl)-1H-1,2,3-triazol-1-yl)-N-arylbenzamide (**5a**): Yield 78%, m.p. 239–42 °C. ^{1}H-NMR (DMSO-d_6, 600 MHz, δ, TMS = 0): 10.40 (s, 1H), 9.22 (s, 1H), 8.21 (d, J = 8.6 Hz, 2H), 8.15 (d, J = 8.7 Hz, 2H), 7.85 (dd, J = 8.0, 1.6 Hz, 1H), 7.80 (d, J = 7.3 Hz, 2H), 7.67 (ddd, J = 8.6, 7.3, 1.7 Hz, 1H), 7.42 (dd, J = 8.3, 1.0 Hz, 1H), 7.40–33 (m, 3H), 7.13 (tt, J = 7.3, 1.2 Hz, 1H), 6.23 (s, 1H), 5.56 (s, 2H). ^{13}C-NMR (DMSO-d_6, 150 MHz, δ, TMS = 0): 164.46, 161.67, 152.91, 142.68, 139.07, 138.59,

135.01, 133.01, 129.63, 128.79, 124.37, 124.02, 123.63, 123.17, 120.58, 119.95, 116.60, 115.16, 91.63, 62.90. ESI-HRMS [M − H]⁻: (*m/z*) Calcd. for $C_{25}H_{17}N_4O_4$: 437.1250. Found: 437.1271.

4-(4-(((2-Oxo-2H-chromen-4-yl)oxy)methyl)-1H-1,2,3-triazol-1-yl)-N-(o-tolyl)benzamide (**5b**): Yield 72%, m.p. 248–250 °C. ¹H-NMR (DMSO-d_6, 600 MHz, δ, TMS = 0): 10.06 (s, 1H), 9.22 (s, 1H), 8.22 (d, *J* = 8.5 Hz, 2H), 8.15 (d, *J* = 8.7 Hz, 2H), 7.86 (dd, *J* = 8.0, 1.6 Hz, 1H), 7.67 (ddd, *J* = 8.6, 7.3, 1.6 Hz, 1H), 7.43 (dd, *J* = 8.3, 1.0 Hz, 1H), 7.39–7.34 (m, 2H), 7.32–7.27 (m, 1H), 7.24 (td, *J* = 7.6, 1.7 Hz, 1H), 7.19 (td, *J* = 7.4, 1.4 Hz, 1H), 6.23 (s, 1H), 5.56 (s, 2H), 2.26 (s, 3H). ¹³C-NMR (DMSO-d_6, 150 MHz, δ, TMS = 0): 164.44, 164.25, 161.65, 152.89, 142.68, 138.58, 136.29, 134.58, 133.93, 132.99, 130.48, 129.56, 126.78, 126.30, 126.17, 124.35, 123.60, 123.16, 119.97, 116.58, 115.15, 91.62, 62.89, 18.03. ESI-HRMS [M − H]⁻: (*m/z*) Calcd. for $C_{26}H_{19}N_4O_4$: 451.1406. Found: 451.1415.

N-(2-Bromophenyl)-4-(4-(((2-oxo-2H-chromen-4-yl)oxy)methyl)-1H-1,2,3-triazol-1-yl)benzamide (**5c**): Yield 55%, m.p. 285–287 °C. ¹H-NMR (DMSO-d_6, 600 MHz, δ, TMS = 0): 10.54 (s, 1H), 9.22 (s, 1H), 8.20 (d, *J* = 8.4 Hz, 2H), 8.16 (d, *J* = 8.4 Hz, 2H), 8.13 (d, *J* = 2.3 Hz, 1H), 7.88–7.82 (m, 1H), 7.78 (d, *J* = 7.8 Hz, 1H), 7.67 (t, *J* = 7.9 Hz, 1H), 7.43 (d, *J* = 8.3 Hz, 1H), 7.34 (dt, *J* = 21.7, 7.6 Hz, 3H), 6.22 (s, 1H), 5.56 (s, 2H). ¹³C-NMR (DMSO-d_6, 150 MHz, δ, TMS = 0): 164.48, 164.39, 161.70, 152.92, 142.73, 138.82, 136.44, 134.07, 133.04, 132.88, 129.64, 129.28, 128.38, 128.32, 124.39, 123.65, 123.19, 120.96, 120.11, 116.62, 115.17, 91.64, 62.91. ESI-HRMS [M − H]⁻: (*m/z*) Calcd. for $C_{25}H_{16}N_4O_4Br$: 517.0334. Found: 517.0327.

N-(3-Bromophenyl)-4-(4-(((2-oxo-2H-chromen-4-yl)oxy)methyl)-1H-1,2,3-triazol-1-yl)benzamide (**5d**): Yield 58%, m.p. 289–292 °C. ¹H-NMR (DMSO-d_6, 600 MHz, δ, TMS = 0): 10.56 (s, 1H), 9.23 (s, 1H), 8.21 (d, *J* = 8.7 Hz, 2H), 8.16 (d, *J* = 8.7 Hz, 2H), 8.13 (d, *J* = 2.1 Hz, 1H), 7.86 (dd, *J* = 7.9, 1.6 Hz, 1H), 7.79 (dt, *J* = 7.9, 1.6 Hz, 1H), 7.68 (ddd, *J* = 8.7, 7.3, 1.6 Hz, 1H), 7.43 (d, *J* = 8.3 Hz, 1H), 7.40–7.30 (m, 3H), 6.23 (s, 1H), 5.56 (s, 2H). ¹³C-NMR (DMSO-d_6, 150 MHz, δ, TMS = 0): 164.62, 164.38, 161.59, 152.82, 142.62, 140.62, 138.69, 134.48, 132.94, 130.75, 129.63, 126.52, 124.30, 123.57, 123.09, 122.70, 121.47, 119.92, 119.15, 116.53, 115.07, 91.55, 62.81. ESI-HRMS [M − H]⁻: (*m/z*) Calcd. for $C_{25}H_{16}N_4O_4Br$: 517.0334. Found: 517.0327.

N-(3,4-Dimethoxyphenyl)-4-(4-(((2-oxo-2H-chromen-4-yl)oxy)methyl)-1H-1,2,3-triazol-1-yl)benzamide (**5e**): Yield 80%; m.p. 283–286 °C. ¹H-NMR (DMSO-d_6, 600 MHz, δ, TMS = 0): 10.26 (s, 1H), 9.21 (s, 1H), 8.20 (d, 2H), 8.14 (d, 2H), 7.85 (dd, *J* = 8.0, 1.6 Hz, 1H), 7.67 (ddd, *J* = 8.6, 7.3, 1.7 Hz, 1H), 7.49 (d, *J* = 2.4 Hz, 1H), 7.42 (dd, *J* = 8.4, 1.0 Hz, 1H), 7.36 (td, *J* = 8.4, 1.9 Hz, 2H), 6.95 (d, *J* = 8.7 Hz, 1H), 6.22 (s, 1H), 5.56 (s, 2H), 3.77 (s, 3H), 3.75 (s, 3H). ¹³C-NMR (DMSO-d_6, 150 MHz, δ, TMS = 0): 164.45, 163.96, 161.65, 152.89, 148.52, 145.40, 142.66, 138.48, 135.06, 132.54, 129.46, 124.36, 123.61, 123.16, 119.93, 116.59, 115.15, 112.55, 111.93, 105.62, 91.62, 62.90, 55.80, 55.51. ESI-HRMS [M − H]⁻: (*m/z*) Calcd. for $C_{27}H_{21}N_4O_6$: 497.1461. Found: 497.1475.

N-(3-Chloro-2-methylphenyl)-4-(4-(((2-oxo-2H-chromen-4-yl)oxy)methyl)-1H-1,2,3-triazol-1-yl)benzamide (**5f**): Yield 61%; m.p. 270–273 °C. ¹H-NMR (DMSO-d_6, 600 MHz, δ, TMS = 0): 10.31 (s, 1H), 9.22 (s, 1H), 8.23 (d, *J* = 8.6 Hz, 2H), 8.16 (d, *J* = 8.7 Hz, 2H), 7.85 (dd, *J* = 7.9, 1.6 Hz, 1H), 7.67 (ddd, *J* = 8.6, 7.3, 1.6 Hz, 1H), 7.43 (dd, *J* = 8.3, 1.0 Hz, 1H), 7.39 (dd, *J* = 8.0, 1.3 Hz, 1H), 7.38–7.33 (m, 2H), 7.28 (t, *J* = 7.9 Hz, 1H), 6.23 (s, 1H), 5.56 (s, 2H), 2.27 (s, 3H). ¹³C-NMR (DMSO-d_6, 150 MHz, δ, TMS = 0): 164.49, 164.45, 161.65, 152.90, 142.70, 138.73, 137.95, 134.17, 133.95, 133.00, 132.40, 129.65, 127.07, 126.06, 124.35, 123.61, 123.16, 120.00, 116.59, 115.15, 91.62, 62.89, 15.50. ESI-HRMS [M − H]⁻: (*m/z*) Calcd. for $C_{26}H_{18}N_4O_4Cl$: 485.1017. Found: 485.1016.

N-(2,5-Dimethoxyphenyl)-4-(4-(((2-oxo-2H-chromen-4-yl)oxy)methyl)-1H-1,2,3-triazol-1-yl)benzamide (**5g**): Yield 74%; m.p. 267–269 °C. ¹H-NMR (DMSO-d_6, 600 MHz, δ, TMS = 0): 9.61 (s, 1H), 9.21 (s, 1H), 8.19 (d, *J* = 8.2 Hz, 2H), 8.13 (d, *J* = 8.3 Hz, 2H), 7.85 (d, *J* = 7.9 Hz, 1H), 7.67 (t, *J* = 7.9 Hz, 1H), 7.49 (d, *J* = 3.1 Hz, 1H), 7.43 (d, *J* = 8.3 Hz, 1H), 7.36 (t, *J* = 7.6 Hz, 1H), 7.03 (d, *J* = 8.9 Hz, 1H), 6.77 (dd, *J* = 9.0, 3.2 Hz, 1H), 6.22 (s, 1H), 5.56 (s, 2H), 3.80 (s, 3H), 3.73 (s, 3H). ¹³C-NMR (DMSO-d_6, 150 MHz, δ, TMS = 0): 164.43, 164.03, 161.64, 152.92, 152.88, 145.71, 142.68, 138.62, 134.52, 132.98, 129.48, 127.44, 124.34, 123.57,

123.16, 120.00, 116.57, 115.14, 112.27, 110.77, 110.07, 91.61, 62.89, 56.32, 55.54. ESI-HRMS [M − H]$^-$: (*m*/*z*) Calcd. for $C_{27}H_{21}N_4O_6$: 497.1461. Found: 497.1473.

N-(2-Ethylphenyl)-4-(4-(((2-oxo-2H-chromen-4-yl)oxy)methyl)-1H-1,2,3-triazol-1-yl)benzamide (**5h**): Yield 70%; m.p. 244–246 °C. ^{1}H-NMR (DMSO-d_6, 600 MHz, δ, TMS = 0): 10.07 (s, 1H), 9.21 (s, 1H), 8.22 (d, *J* = 8.6 Hz, 2H), 8.15 (d, *J* = 8.6 Hz, 2H), 7.86 (dd, *J* = 8.0, 1.6 Hz, 1H), 7.67 (ddd, *J* = 8.6, 7.3, 1.6 Hz, 1H), 7.43 (dd, *J* = 8.3, 1.0 Hz, 1H), 7.36 (ddd, *J* = 8.1, 7.3, 1.1 Hz, 1H), 7.34–7.29 (m, 2H), 7.29–7.22 (m, 2H), 6.23 (s, 1H), 5.56 (s, 2H), 2.65 (q, *J* = 7.6 Hz, 2H), 1.15 (t, *J* = 7.6 Hz, 3H). ^{13}C-NMR (DMSO-d_6, 150 MHz, δ, TMS = 0): 164.65, 164.45, 161.65, 152.89, 142.68, 140.02, 138.59, 135.65, 134.56, 132.98, 129.51, 128.66, 127.74, 126.81, 126.20, 124.33, 123.59, 123.15, 120.01, 116.57, 115.14, 91.62, 62.89, 24.10, 14.28. ESI-HRMS [M − H]$^-$: (*m*/*z*) Calcd. for $C_{27}H_{21}N_4O_4$: 465.1563. Found: 465.1567.

N-(4-Methoxyphenyl)-4-(4-(((2-oxo-2H-chromen-4-yl)oxy)methyl)-1H-1,2,3-triazol-1-yl)benzamide (**5i**): Yield 70%; m.p. 283–287 °C. ^{1}H-NMR (DMSO-d_6, 600 MHz, δ, TMS = 0): 10.32 (s, 1H), 9.22 (s, 1H), 8.20 (d, *J* = 8.4 Hz, 2H), 8.14 (d, *J* = 8.4 Hz, 2H), 7.85 (dd, *J* = 8.0, 1.6 Hz, 1H), 7.68 (d, *J* = 6.1 Hz, 2H), 7.43 (d, *J* = 8.3 Hz, 1H), 7.36 (t, *J* = 7.6 Hz, 1H), 7.18 (d, *J* = 8.2 Hz, 2H), 6.75 (d, *J* = 8.6 Hz, 1H), 6.23 (s, 1H), 5.56 (s, 2H), 3.32 (s, 3H). ^{13}C-NMR (DMSO-d_6, 150 MHz, δ, TMS = 0): 164.37, 163.89, 161.57, 155.71, 152.81, 142.58, 138.38, 134.99, 132.92, 132.01, 129.41, 124.28, 123.52, 123.08, 122.10, 119.84, 116.51, 115.07, 113.81, 91.54, 62.81, 55.22. ESI-HRMS [M − H]$^-$: (*m*/*z*) Calcd. for $C_{26}H_{19}N_4O_5$: 467.1355. Found: 467.1366.

N-(2,6-Dimethylphenyl)-4-(4-(((2-oxo-2H-chromen-4-yl)oxy)methyl)-1H-1,2,3-triazol-1-yl)benzamide (**5j**): Yield 68%; m.p. 281–284 °C. ^{1}H-NMR (DMSO-d_6, 600 MHz, δ, TMS = 0): 9.95 (s, 1H), 9.21 (s, 1H), 8.24 (d, *J* = 8.7 Hz, 2H), 8.15 (d, *J* = 8.7 Hz, 2H), 7.86 (dd, *J* = 7.9, 1.6 Hz, 1H), 7.67 (ddd, *J* = 8.6, 7.3, 1.6 Hz, 1H), 7.43 (dd, *J* = 8.4, 1.0 Hz, 1H), 7.36 (ddd, *J* = 8.1, 7.3, 1.1 Hz, 1H), 7.14 (s, 3H), 6.23 (s, 1H), 5.56 (s, 2H), 2.21 (s, 6H). ^{13}C-NMR (DMSO-d_6, 150 MHz, δ, TMS = 0): 164.42, 163.93, 161.62, 152.88, 142.65, 138.58, 135.69, 135.16, 134.40, 132.96, 129.39, 127.87, 126.91, 124.32, 123.60, 123.14, 120.06, 116.56, 115.13, 91.61, 62.89, 18.17. ESI-HRMS [M − H]$^-$: (*m*/*z*) Calcd. for $C_{27}H_{21}N_4O_4$: 465.1563. Found: 465.1563.

N-(2,4-Dimethylphenyl)-4-(4-(((2-oxo-2H-chromen-4-yl)oxy)methyl)-1H-1,2,3-triazol-1-yl)benzamide (**5k**): Yield 71%; m.p. 270–273 °C. ^{1}H-NMR (DMSO-d_6, 600 MHz, δ, TMS = 0): 1H-NMR (600 MHz, DMSO-d_6) δ 9.98 (s, 1H), 9.21 (s, 1H), 8.21 (d, *J* = 8.3 Hz, 2H), 8.13 (d, *J* = 8.7 Hz, 2H), 7.86 (dd, *J* = 7.9, 1.6 Hz, 1H), 7.67 (ddd, *J* = 8.6, 7.3, 1.7 Hz, 1H), 7.43 (dd, *J* = 8.4, 1.0 Hz, 1H), 7.36 (ddd, *J* = 8.2, 7.3, 1.1 Hz, 1H), 7.22 (d, *J* = 7.9 Hz, 1H), 7.14–7.07 (m, 1H), 7.06–7.00 (m, 1H), 6.23 (s, 1H), 5.56 (s, 2H), 2.29 (s, 3H), 2.21 (s, 3H). ^{13}C-NMR (DMSO-d_6, 150 MHz, δ, TMS = 0): 164.47, 164.26, 161.67, 152.91, 142.68, 138.54, 135.43, 134.66, 133.75, 133.68, 133.01, 131.01, 129.53, 126.73, 126.70, 124.37, 123.60, 123.17, 119.97, 116.60, 115.16, 91.63, 62.90, 20.68, 17.96. ESI-HRMS [M − H]$^-$: (*m*/*z*) Calcd. for $C_{27}H_{21}N_4O_4$: 465.1563. Found: 465.1568.

N-(3-Chloro-4-fluorophenyl)-4-(4-(((2-oxo-2H-chromen-4-yl)oxy)methyl)-1H-1,2,3-triazol-1-yl)benzamide (**5l**): Yield 73%; m.p. 277–279 °C. ^{1}H-NMR (DMSO-d_6, 600 MHz, δ, TMS = 0): 10.58 (s, 1H), 9.22 (s, 1H), 8.20 (d, *J* = 8.8 Hz, 2H), 8.16 (d, *J* = 8.8 Hz, 2H), 8.10 (dd, *J* = 6.9, 2.6 Hz, 1H), 7.85 (dd, *J* = 7.9, 1.6 Hz, 1H), 7.75 (ddd, *J* = 9.0, 4.3, 2.6 Hz, 1H), 7.67 (ddd, *J* = 8.6, 7.3, 1.7 Hz, 1H), 7.48–7.40 (m, 2H), 7.36 (ddd, *J* = 8.2, 7.3, 1.1 Hz, 1H), 6.22 (s, 1H), 5.56 (s, 2H). ^{13}C-NMR (DMSO-d_6, 150 MHz, δ, TMS = 0): 164.50, 164.35, 161.56, 152.81, 142.61, 138.70, 136.23, 134.33, 132.92, 129.58, 124.27, 123.54, 123.06, 121.87, 120.79, 120.74, 119.92, 117.00, 116.86, 116.51, 115.06, 91.54, 62.80. ESI-HRMS [M − H]$^-$: (*m*/*z*) Calcd. for $C_{25}H_{15}N_4O_4FCl$: 489.0766. Found: 489.0780.

Methyl 3-(4-(4-(((2-oxo-2H-chromen-4-yl)oxy)methyl)-1H-1,2,3-triazol-1-yl)benzamido)benzoate (**5m**): Yield 55%; m.p. 256–258 °C. ^{1}H-NMR (DMSO-d_6, 600 MHz, δ, TMS = 0):10.62 (s, 1H), 9.23 (s, 1H), 8.48 (t, *J* = 2.0 Hz, 1H), 8.24 (d, *J* = 8.7 Hz, 2H), 8.16 (d, *J* = 8.7 Hz, 2H), 8.11 (ddd, *J* = 8.2, 2.3, 1.1 Hz, 1H), 7.85 (dd, *J* = 8.0, 1.6 Hz, 1H), 7.72 (dt, *J* = 7.7, 1.4 Hz, 1H), 7.67 (ddd, *J* = 8.6, 7.3, 1.6 Hz, 1H), 7.54 (t, *J* = 7.9 Hz, 1H), 7.43 (dd, *J* = 8.3, 1.1 Hz, 1H), 7.36 (ddd, *J* = 8.2, 7.3, 1.1 Hz, 1H), 6.23 (s, 1H), 5.56 (s, 2H), 3.88 (s, 3H). ^{13}C-NMR (DMSO-d_6, 150 MHz, δ, TMS = 0): 166.11, 164.53, 164.35, 161.56, 152.81, 142.60, 139.41, 138.65, 134.48, 132.91, 130.08, 129.61, 129.22, 124.84, 124.47, 124.27, 123.54, 123.07, 120.91, 119.89,

116.50, 115.06, 91.54, 62.81, 52.27. ESI-HRMS [M − H]$^-$: (m/z) Calcd. for $C_{27}H_{19}N_4O_6$: 495.1305. Found: 495.1327.

N-Benzyl-4-(4-(((2-oxo-2H-chromen-4-yl)oxy)methyl)-1H-1,2,3-triazol-1-yl)benzamide (**5n**): Yield 60%; m.p. 249–251 °C. ^{1}H-NMR (DMSO-d_6, 600 MHz, δ, TMS = 0): 9.22 (t, J = 6.0 Hz, 1H), 9.19 (s, 1H), 8.14 (d, J = 8.8 Hz, 2H), 8.09 (d, J = 8.7 Hz, 2H), 7.84 (dd, J = 7.9, 1.6 Hz, 1H), 7.67 (ddd, J = 8.6, 7.3, 1.6 Hz, 1H), 7.42 (dd, J = 8.3, 1.0 Hz, 1H), 7.37–7.32 (m, 5H), 7.25 (tt, J = 5.8, 3.0 Hz, 1H), 6.22 (s, 1H), 5.55 (s, 2H), 4.52 (d, J = 5.9 Hz, 2H). ^{13}C-NMR (DMSO-d_6, 150 MHz, δ, TMS = 0): 165.15, 164.42, 161.63, 152.88, 142.63, 139.56, 138.39, 134.39, 132.95, 129.14, 128.42, 127.37, 126.91, 124.31, 123.53, 123.14, 119.92, 116.55, 115.13, 91.60, 62.89, 42.83. ESI-HRMS [M − H]$^-$: (m/z) Calcd. for $C_{26}H_{19}N_4O_4$: 451.1406. Found: 451.1404.

N-(4-Bromonaphthalen-1-yl)-4-(4-(((2-oxo-2H-chromen-4-yl)oxy)methyl)-1H-1,2,3-triazol-1-yl)benzamide (**5o**): Yield 65%; m.p. 306–309 °C. ^{1}H-NMR (DMSO-d_6, 600 MHz, δ, TMS = 0):10.69 (s, 1H), 9.25 (s, 1H), 8.33 (d, J = 8.2 Hz, 2H), 8.20 (t, J = 7.9 Hz, 3H), 8.12 (d, J = 8.5 Hz, 1H), 7.96 (d, J = 7.9 Hz, 1H), 7.86 (d, J = 7.9 Hz, 1H), 7.75 (t, J = 7.6 Hz, 1H), 7.68 (q, J = 7.1 Hz, 2H), 7.60 (d, J = 7.9 Hz, 1H), 7.43 (d, J = 8.3 Hz, 1H), 7.37 (t, J = 7.6 Hz, 1H), 6.24 (s, 1H), 5.57 (s, 2H). ^{13}C-NMR (DMSO-d_6, 150 MHz, δ, TMS = 0): 165.15, 164.36, 161.57, 152.81, 142.63, 138.69, 134.19, 134.02, 132.92, 131.61, 130.42, 129.74, 129.71, 128.10, 127.13, 126.74, 124.70, 124.28, 124.26, 123.54, 123.08, 119.91, 119.51, 116.51, 115.07, 91.55, 62.82. ESI-HRMS [M − H]$^-$: (m/z) Calcd. for $C_{29}H_{18}N_4O_4Br$: 567.0491. Found: 567.0511.

3.2. In-Vitro Cytotoxicity Study (MTT Assay)

The cells suspended in the corresponding culture medium were inoculated in 96-well microtiter plates at a density of 1500–3000 cells per well, and incubated for 24 h at 37 °C in a humidified atmosphere with 95% air and 5% CO_2. Different concentrations (100, 50, 25, 12.5 and 6.25 μM) of DOX, *cis*-Pt or the test compounds were added in the line of cell (MDA-MB-231) in 96 h incubation. The cells were incubated for 4 h with 20 μL of 5 mg/mL MTT solution. Supernatant from each well was carefully removed, and the media were then replaced with 100 μL of dimethyl sulfoxide to dissolve the purple colored formazan crystals formed in the wells, and their absorbance were measured at 492 nm with a microplate reader (Synergy-HT, BioTek Instruments, Winooski, VT, USA); 100 μL DMSO was set as the blank control. The hypoxic condition was achieved by placing cells in a sealed hypoxia incubator chamber (Catalog Number 27310, Stemcell Technologies, Inc., Vancouver, BC, Canada) filled with 5% CO_2 and 95% N_2 [23]. For the hypoxia group, Different concentrations (100, 50, 25, 12.5 and 6.25 μM) of DOX, *cis*-Pt or the test compounds were added in the line of cell (MDA-MB-231) in 24 h incubation under hypoxia condition. Then the cells were moved into normoxic condition and cultured for additional 72 h.

3.3. Molecular Docking Simulations

The experimental crystallographic structures of CA IX complex were from the Protein DataBank (PDB ID: 4ZAO). Both the protein and the ligand were prepared by adding partial charges and polar hydrogen atoms with the assistance of AutoDock Tools (version 1.5.6, The Scripps Research Institute, San Diego, CA, USA). A grid box size of 62.0, 66.0, 66.0 Å was generated to cover the active pocket of the receptor. We set the energy evaluations as the maximum of 2.5×10^6 in the grid point spacing of 0.375 Å and performed 200 independent runs of genetic algorithm. All other parameters were set to default unless stated otherwise. Two-dimensional and Three-dimensional schematic representation of protein-ligand interaction was generated using Discovery Studio 2016 client (v16.1.0.15350, Dassault Systemes Biovia Company, Boston, MA. USA).

4. Conclusions

In conclusion, a series of compounds containing a 4-substituted coumarin moiety were designed, synthesized and evaluated for their antitumor activities. These compounds exhibited moderate antiproliferative activities against the MDA-MB-231 cell line under both normoxic and hypoxic conditions, and all displayed better activity than the parent compound 4-hydroxycoumarin. Furthermore, compound **5e** showed the most prominent anticancer activities of this series of compounds with about 5000-fold more activity than 4-hydroxycoumarin and 20-fold more than doxorubicin. Meanwhile, almost all of the compounds revealed a general promotion of proliferation inhibiting activity under hypoxia, contrasted with normoxia. Molecular docking simulations were performed to elucidate the mode(s) of binding for compound **5e**. The results showed compound that **5e** had potential to inhibit CA IX. Further studies on the mechanism of these compounds' anticancer activity are underway.

Supplementary Materials: ^{1}H and ^{13}C-NMR spectra of these compounds are available in the supplementary materials.

Author Contributions: R.A. performed experiments, analyzed data and drafted the manuscript. Z.H. performed experiments, analyzed data and revised the manuscript. J.-T.L. performed experiments. H.-N.Y. used software for molecular docking. Y.-H.M. and C.G. conceived the work, gave critical comments and revised the manuscript.

Funding: This research was funded by the National Natural Science Foundation of China grant number [81573292].

Conflicts of Interest: The authors declare no conflict of interest.

References

1. Shi, Y.; Zhou, C.H. Synthesis and evaluation of a class of new coumarin triazole derivatives as potential antimicrobial agents. *Bioorg. Med. Chem. Lett.* **2011**, *21*, 956–960. [CrossRef] [PubMed]

2. Beillerot, A.; Domínguez, J.C.R.; Kirsch, G.; Bagrel, D. Synthesis and protective effects of coumarin derivatives against oxidative stress induced by doxorubicin. *Bioorg. Med. Chem. Lett.* **2008**, *18*, 1102–1105. [CrossRef] [PubMed]

3. Wu, L.; Wang, X.; Xu, W.; Farzaneh, F.; Xu, R. The Structure and Pharmacological Functions of Coumarins and Their Derivatives. *Curr. Med. Chem.* **2009**, *16*, 4236–4260. [CrossRef] [PubMed]

4. Piazzi, L.; Cavalli, A.; Colizzi, F.; Belluti, F.; Bartolini, M.; Mancini, F.; Recanatini, M.; Andrisano, V.; Rampa, A. Multi-target-directed coumarin derivatives: HAChE and BACE1 inhibitors as potential anti-Alzheimer compounds. *Bioorg. Med. Chem. Lett.* **2008**, *18*, 423–426. [CrossRef] [PubMed]

5. Yamaguchi, T.; Fukuda, T.; Ishibashi, F.; Iwao, M. The first total synthesis of lamellarin α 20-sulfate, a selective inhibitor of HIV-1 integrase. *Tetrahedron Lett.* **2006**, *47*, 3755–3757. [CrossRef]

6. Creaven, B.S.; Egan, D.A.; Karcz, D.; Kavanagh, K.; McCann, M.; Mahon, M.; Noble, A.; Thati, B.; Walsh, M. Synthesis, characterisation and antimicrobial activity of copper(II) and manganese(II) complexes of coumarin-6,7-dioxyacetic acid (cdoaH$_2$) and 4-methylcoumarin-6,7-dioxyacetic acid (4-MecdoaH$_2$): X-ray crystal structures of [Cu(cdoa)(phen)$_2$]·8.8H$_2$O and [Cu(4-Mecdoa)(phen)$_2$]·13H$_2$O (phen=1,10-phenanthroline). *J. Inorg. Biochem.* **2007**, *101*, 1108–1119. [PubMed]

7. Creaven, B.S.; Devereux, M.; Karcz, D.; Kellett, A.; McCann, M.; Noble, A.; Walsh, M. Copper(II) complexes of coumarin-derived Schiff bases and their anti-Candida activity. *J. Inorg. Biochem.* **2009**, *103*, 1196–1203. [CrossRef] [PubMed]

8. Basanagouda, M.; Jambagi, V.B.; Barigidad, N.N.; Laxmeshwar, S.S.; Devaru, V. Narayanachar Synthesis, structure-activity relationship of iodinated-4-aryloxymethyl-coumarins as potential anti-cancer and anti-mycobacterial agents. *Eur. J. Med. Chem.* **2014**, *74*, 225–233. [CrossRef] [PubMed]

9. Le Lain, R.; Barrell, K.J.; Saeed, G.S.; Nicholls, P.J.; Simons, C.; Kirby, A.; Smith, H.J. Some coumarins and triphenylethene derivatives as inhibitors of human testes microsomal 17β-hydroxysteroid dehydrogenase (17β-HSD type 3): Further studies with tamoxifen on the rat testes microsomal enzyme. *J. Enzyme Inhib. Med. Chem.* **2002**, *17*, 93–100. [CrossRef] [PubMed]

10. Zhang, W.; Li, Z.; Zhou, M.; Wu, F.; Hou, X.; Luo, H.; Liu, H.; Han, X.; Yan, G.; Ding, Z.; et al. Synthesis and biological evaluation of 4-(1,2,3-triazol-1-yl)coumarin derivatives as potential antitumor agents. *Bioorg. Med. Chem. Lett.* **2014**, *24*, 799–807. [CrossRef] [PubMed]

11. Swietach, P.; Patiar, S.; Supuran, C.T.; Harris, A.L.; Vaughan-Jones, R.D. The role of carbonic anhydrase 9 in regulating extracellular and intracellular pH in three-dimensional tumor cell growths. *J. Biol. Chem.* **2009**, *284*, 20299–20310. [CrossRef] [PubMed]

12. Maresca, A.; Supuran, C. Coumarins incorporating hydroxy-and chloro-moieties selectively inhibit the transmembrane, tumor-associated carbonic anhydrase isoforms IX and XII over the cytosolic ones I and II. *Bioorg. Med. Chem. Lett.* **2010**, *20*, 4511–4514. [CrossRef] [PubMed]

13. Zhong, B.; Cai, X.; Chennamaneni, S.; Yi, X.; Liu, L.; Pink, J.J.; Dowlati, A.; Xu, Y.; Zhou, A.; Su, B. From COX-2 inhibitor nimesulide to potent anti-cancer agent: Synthesis, in vitro, in vivo and pharmacokinetic evaluation. *Eur. J. Med. Chem.* **2012**, *47*, 432–444. [CrossRef] [PubMed]

14. Yang, S.M.; Huang, Z.N.; Zhou, Z.S.; Hou, J.; Zheng, M.Y.; Wang, L.J.; Jiang, Y.; Zhou, X.Y.; Chen, Q.Y.; Li, S.-H.; et al. Structure-based design, structure–activity relationship analysis, and antitumor activity of diaryl ether derivatives. *Arch. Pharm. Res.* **2015**, *38*, 1761–1773. [CrossRef] [PubMed]

15. Pingaew, R.; Saekee, A.; Mandi, P.; Nantasenamat, C.; Prachayasittikul, S.; Ruchirawat, S.; Prachayasittikul, V. Synthesis, biological evaluation and molecular docking of novel chalcone-coumarin hybrids as anticancer and antimalarial agents. *Eur. J. Med. Chem.* **2014**, *85*, 65–76. [CrossRef] [PubMed]

16. MatwiJczuk, A.; Janik, E.; Luchowski, R.; Niewiadomy, A.; Gruszecki, W.I.; Gagoś, M. Spectroscopic studies of the molecular organization of 4-([1,2,4] triazolo [4,3-a] pyridin-3-yl)-6-methylbenzene-1,3-diol in selected solvents. *J. Lumin.* **2018**, *194*, 208–218. [CrossRef]

17. Singh, H.; Singh, J.V.; Gupta, M.K.; Saxena, A.K.; Sharma, S.; Nepali, K.; Bedi, P.M.S. Triazole tethered isatin-coumarin based molecular hybrids as novel antitubulin agents: Design, synthesis, biological investigation and docking studies. *Bioorg. Med. Chem. Lett.* **2017**, *27*, 3974–3979. [CrossRef] [PubMed]

18. Gieling, R.G.; Babur, M.; Mamnani, L.; Burrows, N.; Telfer, B.A.; Carta, F.; Winum, J.Y.; Scozzafava, A.; Supuran, C.T.; Williams, K.J. Antimetastatic effect of sulfamate carbonic anhydrase IX inhibitors in breast carcinoma xenografts. *J. Med. Chem.* **2012**, *55*, 5591–5600. [CrossRef] [PubMed]

19. Maresca, A.; Temperini, C.; Pochet, L.; Masereel, B.; Scozzafava, A.; Supuran, C.T. Deciphering the mechanism of carbonic anhydrase inhibition with coumarins and thiocoumarins. *J. Med. Chem.* **2010**, *53*, 335–344. [CrossRef] [PubMed]

20. Maresca, A.; Scozzafava, A.; Supuran, C.T. 7,8-Disubstituted- but not 6,7-disubstituted coumarins selectively inhibit the transmembrane, tumor-associated carbonic anhydrase isoforms IX and XII over the cytosolic ones i and II in the low nanomolar/subnanomolar range. *Bioorg. Med. Chem. Lett.* **2010**, *20*, 7255–7258. [CrossRef] [PubMed]

21. Tiwari, S.V.; Siddiqui, S.; SeiJas, J.A.; Vazquez-Tato, M.P.; Sarkate, A.P.; Lokwani, D.K.; NikalJe, A.P.G. Microwave-assisted facile synthesis, anticancer evaluation and docking study of *N*-((5-(substituted methylene amino)-1,3,4-thiadiazol-2-yl)methyl) benzamide derivatives. *Molecules* **2017**, *22*, 995. [CrossRef] [PubMed]

22. Maresca, A.; Temperini, C.; Vu, H.; Pham, N.B.; Poulsen, S.A.; Scozzafava, A.; Quinn, R.J.; Supuran, C.T. Non-zinc mediated inhibition of carbonic anhydrases: Coumarins are a new class of suicide inhibitors. *J. Am. Chem. Soc.* **2009**, *131*, 3057–3062. [CrossRef] [PubMed]

23. Yu, H.; Hou, Z.; Tian, Y.; Mou, Y.; Guo, C. Design, synthesis, cytotoxicity and mechanism of novel dihydroartemisinin-coumarin hybrids as potential anti-cancer agents. *Eur. J. Med. Chem.* **2018**, *151*, 434–449. [CrossRef] [PubMed]

Sample Availability: Samples of the compounds **5a–5e** are available from the authors.

Article

Synthesis of Novel Pyrazole Derivatives and Their Tumor Cell Growth Inhibitory Activity

Ying-Jie Cui, Long-Qian Tang, Cheng-Mei Zhang and Zhao-Peng Liu *

Institute of Medicinal Chemistry, Key Laboratory of Chemical Biology (Ministry of Education),
School of Pharmaceutical Sciences, Shandong University, Jinan 250012, China; yingjiecui@sina.cn (Y.-J.C.);
lqtang.student@sina.com (L.-Q.T.); zhangcm@sdu.edu.cn (C.-M.Z.)
* Correspondence: liuzhaop@sdu.edu.cn; Tel.: +86-531-8838-2006; Fax: +86-531-8838-2548

Academic Editors: Carla Boga and Gabriele Micheletti
Received: 31 October 2018; Accepted: 10 January 2019; Published: 13 January 2019

Abstract: To find novel antitumor agents, a series of 1H-benzofuro[3,2-c]pyrazole derivatives **4a-e** were designed and synthesized. The treatment of 6-methoxybenzofuran-3(2H)-one **3** with LiHMDS in anhydrous tetrahydrofuran (THF) followed by reaction with 3-substitued phenyl isothiocyanate gave the thioamide intermediates, which underwent condensation with hydrazine monohydrate in dioxane/EtOH (1:1) to provide the benzofuropyrazole derivatives **4a–e** as well as the unexpected pyrazole derivatives **5a–e**. In tumor cell growth inhibitory assay, all the benzofuropyrazole derivatives were not active against the breast tumor MCF-7 cell, only **4a** was highly active and more potent than ABT-751 against the leukemia K562 (GI_{50} = 0.26 µM) and lung tumor A549 cells (GI_{50} = 0.19 µM), while other benzofuropyrazoles showed very weak inhibitory activity. In contrast, the pyrazoles **5a-e** were in general more potent than the benzofuropyrazoles **4a–e**. Compound **5a** exhibited a similar tendency to that of **4a** with high potency against K562 and A549 cells but weak effects on MCF-7 cell. Both pyrazoles **5b** and **5e** exhibited high inhibitory activities against K562, MCF-7 and A549 cells. The most active compound **5b** was much more potent than ABT-751 against K562 and A549 cells with GI_{50} values of 0.021 and 0.69 µM, respectively. Moreover, **5b** was identified as a novel tubulin polymerization inhibitor with an IC_{50} of 7.30 µM.

Keywords: privileged structure; pyrazoles; tubulin inhibitors; antitumor; antitumor agents

---

1. Introduction

Privileged structures are defined as molecular frameworks that are able to provide useful ligands for multiple types of receptors or enzymes through proper structural modifications. In combination with their favorable drug-like properties, privileged structures or scaffolds are widely used in rational drug design to find new lead compounds or drug candidates [1–3]. Pyrazole derivatives represent one of the most active classes of compounds that possess a wide spectrum of biological activities, including antibacterial and antifungal [4,5], antitumor [6,7], anti-inflammatory and analgesic [8,9], antitubercular [10], antiviral [11,12], anti-Alzheimer's [13,14], α-glucosidase inhibitory [15], anti-diabetic [16], antileishmanial [17,18], anti-malarial [19], radioimaging [20], acaricidal and insecticidal [21,22] activities. As a privileged scaffold, pyrazole has been recently widely used in the design of anticancer agents for a multiple of tumor targets [23].

N-(2-((4-Hydroxyphenyl)amino)pyridin-3-yl)-4-methoxybenzenesulfonamide (ABT-751, **1**, Figure 1) is an orally available sulfonamide tubulin inhibitor under clinical investigations for the treatment of cancers [24,25]. On its X-ray crystal structures with tubulin, ABT-751 interacted with all the three pockets of tubulin at the colchicine binding site [26]. Based on the binding mode of **1** with tubulin and the pyrazole pharmacophore, our group designed and synthesized a series of indenopyrazoles as potential tubulin polymerization inhibitors targeting the colchicine binding site [27]. The indenopyrazole analogue **2** (Figure 1) was found to compete with colchicine in binding to the tubulin colchicine site and inhibit the polymerization of tubulin. In vitro, **2** displayed nanomolar potency against a variety of tumor cell lines, arrested tumor cells in G2/M phase through the regulation of cell cycle-related proteins, and induced tumor cell apoptosis through the activation of caspase pathways. Furthermore, **2** was effective for multidrug resistance tumor cells and inhibited phosphatase and tensin homolog (PTEN) phosphorylation and PTEN/Akt/NF-κB signaling [28]. In vivo, **2** demonstrated its potency in non-small cell lung cancer (NSCLC) and vincristine-resistance human oral epidermoid carcinoma cell (KB/V) xenograft models without obvious side effects [27,28]. In this study, we replaced the indenopyrazole core with the 1*H*-benzofuro[3,2-*c*]pyrazole framework and designed a series of the 1*H*-benzofuro[3,2-*c*]pyrazole derivatives **4a–e**. In the preparation of **4a–e**, the partial cleavage of the furan ring was observed and a series of 5-methoxy-2-(3-(phenylamino)-1H-pyrazol-5-yl)phenol derivatives **5a–e** were isolated. We reported here the synthesis and preliminary results of their tumor cell growth inhibitory activity as well as the identification of pyrazole **5b** as a novel tubulin polymerization inhibitor.

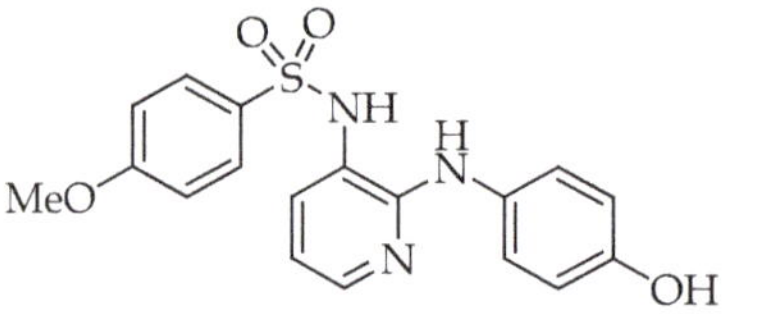

Figure 1. Tubulin inhibitors **1** and **2**.

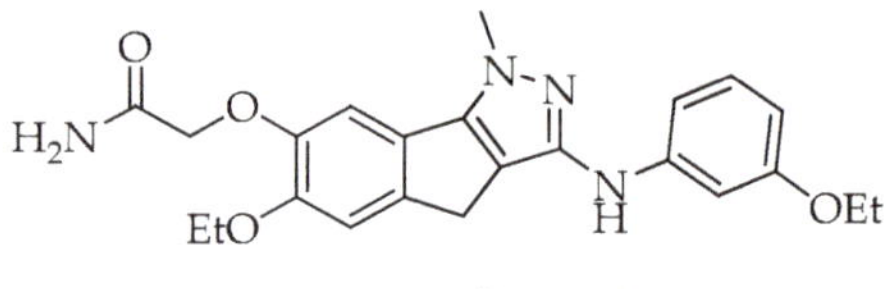

2. Results and Discussion

2.1. Chemistry

The synthetic route towards the benzofuropyrazole derivatives was shown in Scheme 1. The 6-methoxybenzofuran-3-(2*H*)-one **3** was prepared according to the reported method in three steps [29–31] (please refer to the Supplementary Materials). The Hoesch reaction of resorcinol with chloroacetonitrile in the presence of anhydrous $ZnCl_2$ and HCl gas generated an imine intermediate, which upon hydrolysis, provided 2-chloro-1-(2,4-dihydroxyphenyl)ethanone in 94% yield. Treatment of the chloromethyl ketone with a mild base, CH_3COONa, gave the cyclized 6-hydroxybenzofuran-3-(2*H*)-one (54%). Methylation of 6-hydroxybenzofuran-3-(2*H*)-one with Me_2SO_4 produced the 6-methoxybenzofuran-3-(2*H*)-one **3** in 86% yield. After the deprotonation of the α-proton of carbonyls in **3** with lithium hexamethyldisilazide (LiHMDS), the resulting enolates were reacted with 3-substitued phenyl isothiocyanates to give the thioamide intermediates, which underwent condensation with hydrazine monohydrate in dioxane/EtOH (1:1) to form the benzofuropyrazole derivatives **4a–e** in 11% to 30% yield. In this process, the partial cleavage of the furan ring occurred, a series of 5-methoxy-2-(3-(phenylamino)-1*H*-pyrazol-5-yl)phenol derivatives **5a–e** were also isolated in 13% to 31% yield. The structures of benzofuropyrazoles **4a–e** and pyrazoles **5a–e** were determined by ^{1}H nuclear magnetic resonance (NMR), ^{13}C-NMR and electrospray ionization mass spectrometry (ESI-MS). In the ^{1}H NMR spectrum of benzofuropyrazole **4c**, the pyrazole 1-NH appeared at 11.91 ppm, the aniline NH at 8.05 as a singlet, the amide NH at 8.28 (q, *J* = 4.6 Hz) in corresponding with the N-methyl at 2.77 (d, *J* = 4.6 Hz). The seven protons at the two phenyl rings appeared at 6.42 to 7.82 ppm. In comparison with **4c**, in the ^{1}H NMR spectrum of pyrazole **5c**, there were two additional peaks at 10.23 ppm for the phenol OH, 6.24 ppm for the pyrazole 4-H, supporting its estimated structure.

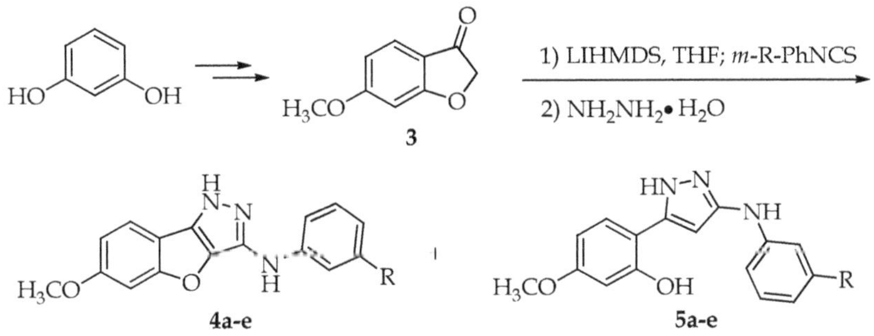

Scheme 1. Synthesis of benzofuropyrazole and pyrazole derivatives.

2.2. Tumor Cell Growth Inhibitory Activity

All the synthesized compounds were evaluated for their tumor cell growth inhibitory activity against human breast cancer MCF-7 cell, human erythroleukemia K562 cell and human lung cancer A549 cell by the conventional MTT (3-(4,5-dimethyl-2-thiazolyl)-2,5-diphenyl-2H-tetrazolium bromide) assay. ABT-751 was used as positive control.

As shown in Table 1, the breast tumor MCF-7 cell was not sensitive to the benzofuropyrazole derivatives **4a–e**. For the K562 and A549 cells, only **4a** exhibited high potency and was more potent than ABT-71 with the GI_{50} of 0.26 and 0.19 μM, respectively, while other benzofuropyrazoles showed moderate or weak activity. It seems the substitution of the ethoxy at the aniline ring with the electron-withdrawing ester, amide, and cyano groups was not tolerated among the benzofuropyrazole series. In contrast, all the three tumor cell lines were sensitive to the pyrazole analogues **5a–e**. The methyl ester **5b** was the most active that inhibited the K562, MCF-7, and A549 cell growth with GI_{50} values of 0.021, 1.7 and 0.69 μM, respectively. Both compounds **5a** and **5b** were highly active against the K562 and A549 cells, and were 5- to 35-fold more potent than ABT-751. The cyano derivative **5e** was also highly potent against the three tumor cell lines, although it showed slightly less potency than ABT-751. Unlike the benzofuropyrazole derivatives, the substitution of the ethoxy at the aniline ring with an electron-withdrawing ester, amide, and cyano group in the pyrazole series was well tolerated, and even preferred, indicating that the benzofuropyrazole derivatives **4a–e** and the pyrazoles **5a–e** might involve different mechanisms of action.

Table 1. Tumor cell growth inhibitory activity of **4a–e** and **5a–e**.

Compound	R	GI_{50} (μM)		
		K562	MCF-7	A549
4a	OCH_2CH_3	0.26 ± 0.04	>20	0.19 ± 0.08
4b	$COOCH_3$	5.46 ± 1.04	>20	>20
4c	$CONHCH_3$	5.11 ± 0.31	>20	15.11 ± 2.18
4d	$CONH_2$	9.01 ± 1.81	>20	10.08 ± 2.21
4e	CN	13.53 ± 0.41	>20	17.01 ± 2.76
5a	OCH_2CH_3	0.046 ± 0.007	16.72 ± 2.6	0.92 ± 0.17
5b	$COOCH_3$	0.021 ± 0.004	1.7 ± 0.43	0.69 ± 0.18
5c	$CONHCH_3$	7.33 ± 1.004	7.78 ± 0.87	9.46 ± 2.03
5d	$CONH_2$	14.77 ± 2.62	5.8 ± 0.202	10.9 ± 0.99
5e	CN	1.45 ± 0.047	2.27 ± 0.34	3.24 ± 0.99
ABT-751		0.74 ± 0.078	0.88 ± 0.24	4.58 ± 0.04

2.3. In Vitro Tubulin Polymerization Inhibitory Activity

The pyrazole derivative **5b** showed the best tumor cell growth inhibitory activity among all the tested compounds. To investigate whether **5b** was a tubulin inhibitor, the tubulin polymerization inhibition assay was carried out. At 37 °C, tubulin will polymerize into microtubules, which is followed by the observed fluorescence enhancement due to the incorporation of a fluorescent reporter into microtubules as polymerization occurs [32]. As shown in Figure 2, **5b** inhibited the tubulin polymerization in a concentration-dependent way with a calculated IC_{50} of 7.30 μM. Therefore, **5b** may be a good lead for further structural modification to find more potent tubulin inhibitors based on the privileged pyrazole structure.

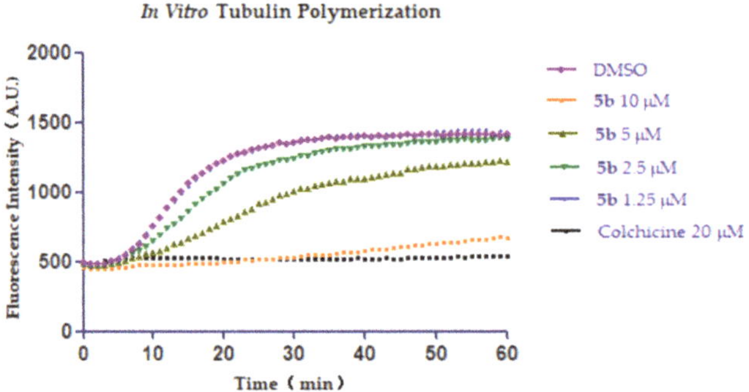

Figure 2. Effect of **5b** on tubulin polymerization in vitro. Purified tubulin protein at 2 mg/mL in a reaction buffer was incubated at 37 °C in the presence of 1% dimethyl sulfoxide (DMSO), test compound **5b** at 1.25, 2.5, 5, or 10 μM or colchicine at 20 μM. The fluorescence intensity was measured every 60 s for 60 min and is presented as increases in the polymerized microtubule.

3. Materials and Methods

3.1. General Chemical Experimental Procedures

Melting points were determined on an X-6 micromelting point apparatus (Beijing Tech. Co., Ltd.). ^{1}H- and ^{13}C-NMR spectra were recorded on Bruker-400 NMR or Bruker-600 NMR spectrometers. All spectra were recorded at room temperature for DMSO or $CDCl_3$ solutions. ESI-MS was performed on an API 4000 instrument. Thin-layer chromatography (TLC) was performed on silica gel GF254 plates. Silica gel GF254 and silica gel (200−300 mesh) from Qingdao Haiyang Chemical Company were used for TLC and column chromatography, respectively. All reagents were commercially available and were used as purchased without further purification. All reactions involving oxygen- or moisture sensitive compounds were carried out under a dry N_2 atmosphere. Unless otherwise noted, reagents were added by syringe. Tetrahydrofuran (THF) was distilled from sodium/benzophenone immediately prior to use.

3-((6-Methoxy-1H-benzofuro[3,2-c]pyrazol-3-yl)amino)ethoxybenzene (**4a**) *and* *3-((5-(2-hydroxy-4-methoxyphenyl)-1H-pyrazol-3-yl)amino)ethoxybenzene* (**5a**) A solution of 6-methoxybenzofuran-3-(2H)-one **3** (150 mg, 0.9 mmol) in anhydrous THF (5 mL) was cooled to −78 °C under nitrogen atmosphere. LiHMDS (1.09 mL, 1.1 mmol, 1.0 M THF solution) was added dropwise. The mixture was stirred at −78 °C for 2 h and then warmed to −45 °C in 45 min. After a solution of the 1-ethoxy-3-isothiocyanatobenzene (164 mg, 1.0 mmol) in anhydrous THF (3 mL) was added, the resulting mixture was stirred at room temperature overnight. Water (30 mL) was added, and the mixture was extracted with EtOAc (3 × 30 mL). The organic layer was washed with brine, dried over anhydrous Na_2SO_4. After filtration and evaporation, flash column chromatography on silica gel (hexane/EtOAc = 15:1) gave the resulting thioamide intermediate, which was dissolved in 1,4-dioxane

(3 mL) and ethanol (3 mL). Hydrazine hydrate (0.46 mL, 7.3 mmol) was added dropwise. The mixture was heated to 50 °C and stirred for 24 h. Water (40 mL) was added, and the mixture was extracted with EtOAc (3 × 40 mL). The organic layer was washed with brine, and dried over anhydrous Na_2SO_4. After filtration and evaporation, the residue was purified by column chromatography on silica gel (hexane/EtOAc = 4:1) to give **4a** (32 mg, 11%) and **5a** (38 mg, 13%).

4a: Brown solid, m.p.: 62–64 °C. ^{1}H-NMR (CDCl$_3$) δ 7.32 (d, J = 8.6 Hz, 1H, Ar-H), 7.08 (t, J = 8.1 Hz, 1H, Ar-H), 6.56 (d, J = 2.4 Hz, 1H, Ar-H), 6.46 (dd, J = 2.4, 8.6 Hz, 1H, Ar-H), 6.36 (d, J = 8.2 Hz, 1H, Ar-H), 6.25 (d, J = 8.1 Hz, 1H, Ar-H), 6.21 (s, 1H, Ar-H), 5.09 (s, 1H, 3-NH), 3.96 (q, J = 6.9 Hz, 2H, OCH$_2$), 3.81 (s, 3H, 6'-OCH$_3$), 1.37 (t, J = 6.9 Hz, 3H, OCH$_2$CH$_3$). ^{13}C-NMR (CDCl$_3$) δ 161.6, 160.3, 156.6, 146.0, 137.8, 130.2, 129.6, 127.9, 125.8, 111.5, 109.5, 107.0, 105.1, 103.7, 101.0, 63.3, 55.3, 14.9. MS (ESI) calcd. for $C_{18}H_{19}N_3O_3$ [M + NH$_4$]$^+$: 341.1, found: 341.4.

5a: Brown solid, m.p.: 82–84 °C. ^{1}H-NMR (CDCl$_3$) δ 7.39 (d, J = 8.6 Hz, 1H, Ar-H), 7.17 (t, J = 8.0 Hz, 1H, Ar-H), 6.57–6.49 (m, 5H, Ar-H), 6.22 (s, 1H, 4'-H), 5.79 (s, 1H, 3-NH), 4.01 (q, J = 6.9 Hz, 2H, 1-OCH$_2$), 3.79 (s, 3H, 4″-OCH$_3$), 1.41 (t, J = 6.9 Hz, 3H, 1-CH$_3$). ^{13}C-NMR (CDCl$_3$) δ 160.7, 160.2, 157.0, 143.9, 130.4, 127.6, 109.8, 108.7, 107.1, 106.5, 102.8, 101.7, 90.4, 63.3, 55.3, 15.4. MS (ESI) calcd. for $C_{18}H_{19}N_3O_3$ [M + H]$^+$: 326.1, found: 326.4.

Methyl 3-((6-methoxy-1H-benzofuro[3,2-c]pyrazol-3-yl)amino)benzoate (**4b**) *and methyl 3-((5-(2-hydroxy-4-methoxyphenyl)-1H-pyrazol-3-yl)amino)benzoate* (**5b**) According to the procedures described for the synthesis of **4a** and **5a**, compounds **4b** and **5b** were prepared from 3 (100 mg, 0.6 mmol), LiHMDS (0.7 mL, 0.7 mmol), methyl 3-isothiocyanatobenzoate (135 mg, 0.7 mmol) and hydrazine hydrate (0.3 mL, 4.8 mmol). The crude residue was purified by column chromatography on silica gel (hexane/EtOAc = 3:1) to give **4b** (36 mg, 18%) and **5b** (33 mg, 16%).

4b: Brown solid, m.p.: 103–105 °C. ^{1}H-NMR (DMSO-d_6) δ 11.96 (s, 1H, 1'-NH), 8.19 (s, 1H, 3-NH), 8.13 (s, 1H, Ar-H), 7.61 (s, 1H, Ar-H), 7.43 (d, J = 8.3 Hz, 1H, Ar-H), 7.33 (d, J = 4.3 Hz, 2H, Ar-H), 6.50 (dd, J = 1.9, 8.5 Hz, Ar-H), 6.42 (d, J = 2.0 Hz, 1H, Ar-H), 3.84 (s, 3H, COOCH$_3$), 3.74 (s, 3H, OCH$_3$). ^{13}C-NMR (DMSO-d_6) δ 167.2, 161.1, 156.9, 145.9, 144.6, 134.1, 130.6, 129.6, 128.3, 119.8, 119.2, 115.6, 110.4, 109.1, 106.2, 103.6, 55.5, 52.5. MS (ESI) calcd. for $C_{18}H_{15}N_3O_4$ [M + NH$_4$]$^+$: 355.1, found: 355.5.

5b: Yellow solid, m.p.: 185–186 °C. ^{1}H-NMR (DMSO-d_6) δ 12.01 (s, 1H, 1'-NH), 10.24 (s, 1H, 2″-OH), 8.67 (s, 1H, 3-NH), 8.13 (s, 1H, Ar-H), 7.57 (s, 1H, Ar-H), 7.53 (d, J = 6.8 Hz, 1H, Ar-H), 7.31 (s, 2H, Ar-H), 6.51 (s, 1H, Ar-H), 6.50 (dd, J = 1.9, 6.9 Hz, 1H, Ar-H), 6.23 (s, 1H, 4'-H), 3.84 (s, 3H, COOCH$_3$), 3.74 (s, 3H, OCH$_3$). ^{13}C-NMR (DMSO-d_6) δ 167.2, 160.3, 155.8, 151.6, 144.8, 139.8, 130.7, 129.4, 128.4, 119.7, 118.8, 115.5, 110.0, 105.8, 102.1, 93.0, 55.5, 52.2. MS (ESI) calcd. for $C_{18}H_{17}N_3O_4$ [M + H]$^+$: 340.1, found: 340.4.

3-((6-Methoxy-1H-benzofuro[3,2-c]pyrazol-3-yl)amino)-N-methylbenzamide (**4c**) *and 3-((5-(2-hydroxy-4-methoxyphenyl)-1H-pyrazol-3-yl)amino)-N-methylbenzamide* (**5c**) According to the procedures described for the synthesis of **4a** and **5a**, compounds **4c** and **5c** were prepared from 3 (100 mg, 0.6 mmol), LiHMDS (0.7 mL, 0.7 mmol), 3-isothiocyanato-N-methylbenzamide (134 mg, 0.7 mmol) and hydrazine hydrate (0.3 mL, 4.8 mmol). The crude residue was purified by column chromatography on silica gel (hexane/EtOAc = 1:1) to give **4c** (30 mg, 15%) and **5c** (33 mg, 16%).

4c: Brown solid, m.p.: 180–182 °C. ^{1}H-NMR (DMSO-d_6) δ 11.91 (s, 1H, 1'-NH), 8.28 (q, J = 4.6 Hz, 1H, CONH), 8.05 (s, 1H, 3-NH), 7.82 (s, 1H, Ar-H), 7.55 (s, 1H, Ar-H), 7.43 (d, J = 5.7 Hz, 1H, Ar-H), 7.27 (t, J = 7.8 Hz, 1H, Ar-H), 7.15 (d, J = 7.6 Hz, 1H, Ar-H), 6.49 (dd, J = 2.6, 8.6 Hz, 1H, Ar-H), 6.42 (d, J = 2.6 Hz, 1H, Ar-H), 3.74 (s, 3H, OCH$_3$), 2.77 (d, J = 4.6 Hz, 3H, N-CH$_3$). ^{13}C-NMR (DMSO-d_6) δ 167.8, 161.1, 156.9, 146.1, 144.5, 135.9, 134.2, 129.1, 128.3, 117.4, 116.7, 114.5, 110.5, 109.1, 106.1, 103.6, 55.5, 26.7. MS (ESI) calcd. for $C_{18}H_{16}N_4O_3$ [M + NH$_4$]$^+$: 354.1, found: 354.4.

5c: Brown solid, m.p.: 184–186 °C. ^{1}H-NMR (DMSO-d_6) δ 11.94 (s, 1H, 1'-NH), 10.23 (s, 1H, OH), 8.52 (q, J = 3.6 Hz, 1H, CONH), 8.25 (s, 1H, 3-NH), 7.81 (s, 1H, Ar-H), 7.52 (s, 2H, Ar-H), 7.25 (s, 1H, Ar-H), 7.11 (s, 1H, Ar-H), 6.51 (s, 1H, Ar-H), 6.49 (dd, J = 1.6, 6.9 Hz, 1H, Ar-H), 6.24 (s, 1H, 4'-H), 3.74 (s, 3H, OCH$_3$), 2.76 (d, J = 3.6 Hz, 3H, N-CH$_3$). ^{13}C-NMR (DMSO-d_6) δ 170.0, 160.3, 155.8, 151.8, 144.6, 139.7, 136.1, 129.0, 128.4, 117.6, 116.4, 114.3, 110.2, 105.8, 102.1, 93.0, 55.5, 26.7. MS (ESI) calcd. for C$_{18}$H$_{18}$N$_4$O$_3$ [M + H]$^+$: 339.1, found: 339.1.

3-((6-Methoxy-1H-benzofuro[3,2-c]pyrazol-3-yl)amino)benzamide (**4d**) *and* *3-((5-(2-hydroxy-4-methoxyphenyl)-1H-pyrazol-3-yl)amino)benzamide* (**5d**) According to the procedures described for the synthesis of **4a** and **5b**, compounds **4d** and **5d** were prepared from **3** (100 mg, 0.6 mmol), LiHMDS (0.7 mL, 0.7 mmol), 3-isothiocyanatobenzamide (124 mg, 0.7 mmol) and hydrazine hydrate (0.3 mL, 4.8 mmol). The crude residue was purified by column chromatography on silica gel (hexane/EtOAc = 1:3) to give **4d** (59 mg, 30%) and **5d** (61 mg, 31%).

4d: Brown solid, m.p.: 172–174 °C. ^{1}H-NMR (DMSO-d_6) δ 12.02 (s, 1H, 1'-NH), 8.12 (s, 1H, 3-NH), 7.82 (s, 2H, Ar-H), 7.52 (s, 1H, Ar-H), 7.43 (d, J = 2.9 Hz, 1H, Ar-H), 7.26–7.21 (m, 3H, 1-CONH$_2$, Ar-H), 6.51 (d, J = 5.7 Hz, 1H, Ar-H), 6.44 (s, 1H, Ar-H), 3.74 (s, 3H, OCH$_3$). ^{13}C-NMR (DMSO-d_6) δ 169.0, 161.0, 156.9, 146.1, 144.5, 135.6, 134.2, 129.0, 128.3, 117.7, 117.2, 114.8, 110.5, 109.2, 106.1, 103.6, 55.5. MS (ESI) calcd. for C$_{17}$H$_{14}$N$_4$O$_3$ [M + NH$_4$]$^+$: 340.1, found: 340.1.

5d: Brown solid, m.p.: 128–130 °C. ^{1}H-NMR (DMSO-d_6) δ 11.84 (s, 1H, 1'-NH), 8.56 (s, 1H, 3-NH), 7.83 (s, 1H, Ar-H), 7.75 (s, 1H, Ar-H), 7.54 (d, J = 6.8 Hz, 1H, Ar-H), 7.43 (s, 1H, Ar-H), 7.26-7.21 (m, 3H, 1-CONH$_2$, Ar-H), 6.51 (s, 1H, Ar-H), 6.49 (d, J = 6.8 Hz, 1H, Ar-H), 6.29 (s, 1H, 4'-H), 3.74 (s, 3H, OCH$_3$). ^{13}C-NMR (DMSO-d_6) δ 169.1, 160.3, 144.5, 135.8, 129.1, 128.3, 117.8, 114.6, 110.3, 105.8, 102.1, 55.5. MS (ESI) calcd. for C$_{17}$H$_{16}$N$_4$O$_3$ [M + H]$^+$: 325.1, found: 325.1.

3-((6-Methoxy-1H-benzofuro[3,2-c]pyrazol-3-yl)amino)benzonitrile (**4e**) *and* *3-((5-(2-hydroxy-4-methoxyphenyl)-1H-pyrazol-3-yl)amino)benzonitrile* (**5e**) According to the procedures described for the synthesis of **4a** and **5b**, compounds **4e** and **5e** were prepared from **3** (100 mg, 0.6 mmol), LiHMDS (0.7 mL, 0.7 mmol), 3-isothiocyanatobenzonitrile (112 mg, 0.7 mmol) and hydrazine hydrate (0.3 mL, 4.8 mmol). The crude residue was purified by column chromatography on silica gel (hexane/EtOAc = 1:4) to give **4e** (31 mg, 17%) and **5e** (33 mg, 18%).

4e: White solid, m.p.: 200–202 °C. ^{1}H-NMR (DMSO-d_6) δ 12.01 (s, 1H, 1'-NH), 8.39 (s, 1H, 3-NH), 7.91(s, 1H, Ar-H), 7.53(d, J = 6.9 Hz, 1H, Ar-H), 7.43-7.38 (m, 2H, Ar-H), 7.15 (d, J = 7.4 Hz, 1H, Ar-H), 6.51 (dd, J = 2.5, 8.6 Hz, 1H, Ar-H), 6.44 (d, J = 2.5 Hz, 1H, Ar-H), 3.75 (s, 3H, OCH$_3$). ^{13}C-NMR (DMSO-d_6) δ 161.1, 156.9, 145.2, 145.0, 134.0, 130.5, 128.5, 121.7, 120.0, 119.9, 117.2, 112.0, 110.3, 109.7, 106.2, 103.6, 55.5. MS (ESI) calcd. for C$_{17}$H$_{12}$N$_4$O$_2$ [M + NH$_4$]$^+$: 322.1, found: 322.1.

5e: White solid, m.p.: 166–168 °C. ^{1}H-NMR (DMSO-d_6) δ 12.05 (s, 1H, 1'-NH), 10.28 (s, 1H, OH), 8.93 (s, 1H, 3-NH), 7.99 (s, 1H, Ar-H), 7.55 (s, 1H, Ar-H), 7.53 (s, 1H, Ar-H), 7.38 (s, 1H, Ar-H), 7.12 (d, J = 4.4 Hz, 1H, Ar-H), 6.53 (s, 1H, Ar-H), 6.51 (dd, J = 1.3, 5.7 Hz, 1H, Ar-H), 6.25 (s, 1H, 4'-H), 3.78 (s, 3H, OCH$_3$). ^{13}C-NMR (DMSO-d_6) δ 160.4, 155.8, 151.2, 145.0, 140.0, 130.4, 128.5, 121.3, 120.0, 117.3, 112.0, 109.8, 105.8, 102.1, 92.8, 55.5. MS (ESI) calcd. for C$_{17}$H$_{14}$N$_4$O$_2$ [M + H]$^+$: 307.1, found: 307.3.

3.2. MTT Assay

The human tumor cell lines, were grown in Roswell Park Memorial Institute (RPMI) 1640 medium and supplemented with 10% foetal bovine serum in the 37 °C in an atmosphere containing 5% CO$_2$. All the synthesized compounds were assayed by conventional 3-(4,5-dimethyl-2-thiazolyl)-2,5-diphenyl-2H-tetrazolium bromide (MTT) method. In brief, the exponentially growing cells were seeded into 96-well cell plates at a density of 4−4.5 × 10^3 cells per well and allowed to adhere overnight. Cells were incubated with various concentrations of the test compounds for 72 h. Then 20 µL of MTT (2.5 mg/mL) was added, the cells were incubated at 37 °C for another 4 h. The reduced MTT crystals were dissolved in DMSO, and the absorbance was measured at

570 nm by a microplate spectrophotometer. The growth in inhibitory effects of each compound were expressed as GI_{50} values, which represent the molar drug concentrations required to cause 50% tumor cell growth inhibition.

3.3. In Vitro Tubulin Polymerization Inhibition Assay

The fluorescence-based in vitro tubulin polymerization assay was performed using the Tubulin Polymerization Assay Kit (BK011P, Cytoskeleton, USA) according to the manual. The tubulin reaction mix contained 2 mg/mL porcine brain tubulin (>99% pure), 2 mM $MgCl_2$, 0.5 mM ethylene glycol-bis(2-aminoethylether)-*N,N,N',N'*-tetraacetic acid (EGTA), 1 mM guanosine triphosphate (GTP), and 15% glycerol. First, a 96-well plate was incubated with 5 µL of inhibitors in different concentrations at 37 °C for 1 min. Then 50 µL of the tubulin reaction mix was added. Immediately, the increase in fluorescence was monitored by excitation at 355 nm and emission at 460 nm in a multimode reader.

4. Conclusions

In the synthesis of 1*H*-benzofuro[3,2-*c*]pyrazole derivatives **4a–e**, the furan ring-opening was observed, and a series of pyrazole derivatives **5a–e** were identified. In the tumor cell growth inhibitory assay, only **4a** was highly active towards the K562 and A549 cells, while other benzofuropyrazole derivatives were not active or showed weak activity. In general, the pyrazoles **5a-e** were more potent than the corresponding benzofuropyrazole derivatives. Compound **5a** exhibited a similar tendency to that of **4a** with high potency against K562 and A549 cells but weak effects on MCF-7 cells. Both pyrazoles **5b** and **5e** exhibited high inhibitory activities against K562, MCF-7 and A549 cells. The most active compound **5b** was 5- to 35-fold more potent than ABT-751 in the inhibition of A549 and K562 cells. In addition, **5b** inhibited tubulin polymerization inhibition with an IC_{50} of 7.30 µM. These results indicated that **5b** was a novel tubulin polymerization inhibitor and it may be a good lead for the discovery of novel pyrazoles as potent anticancer agents.

Supplementary Materials: Supplementary materials are available online.

Author Contributions: Conceptualization, Y.-J.C. and Z.-P.L.; methodology, Y.-J.C.; validation, Y.-J.C., L.-Q.T., C.-M.Z. and Z.-P.L; formal analysis, L.-Q.T.; investigation, Y.-J.C., L.-Q.T. and C.-M.Z.; resources, C.-M.Z.; data curation, Y.-J.C.; writing—original draft preparation, Y.-J.C.; writing—review and editing, Z.-P.L.; visualization, Y.-J.C.; supervision, Z.-P.L.; project administration, Z.-P.L.; funding acquisition, Z.-P.L.

Funding: This work was partially supported by the National Natural Science Foundation of China (NSFC, Grant No. 81573275) and the key research and development program of Shandong province (2017CXGC1401).

Conflicts of Interest: The authors declare no conflict of interest.

Abbreviations

THF	Tetrahydrofuran
MCF-7	Breast cancer cell
K562	Human erythroleukemia cell
A549	Lung cancer cell
NSCLC	Non-small cell lung cancer
KB/V	Vincristine-resistance human oral epidermoid carcinoma cell
LiHMDS	Lithium bis(trimethylsilyl)amide
PTEN	Phosphatase and tensin homolog
DMSO	Dimethyl sulfoxide
RPMI	Roswell park memorial institute
EGTA	Ethylene glycol-bis(2-aminoethylether)-*N,N,N',N'*-tetraacetic acid
GTP	Guanosine triphosphate

References

1. Duarte, C.D.; Barreiro, E.J.; Fraga, C.A. Privileged structures: A useful concept for the rational design of new lead drug candidates. *Mini Rev. Med. Chem.* **2007**, *7*, 1108–1119. [CrossRef] [PubMed]
2. Costantino, L.; Barlocco, D. Privileged structures as leads in medicinal chemistry. *Curr. Med. Chem.* **2006**, *13*, 65–85. [CrossRef] [PubMed]
3. Newman, D.J.; Cragg, G.M. Making sense of structures by utilizing mother nature's chemical libraries as leads to potential drugs. *Nat. Prod.* **2014**, 397–411. [CrossRef]
4. Akbas, E.; Berber, I.; Sener, A.; Hasanov, B. Synthesis and antibacterial activity of 4-benzoyl-1-methyl-5-phenyl-1*H*-pyrazole-3-carboxylic acid and derivatives. *Farmaco* **2005**, *60*, 23–26. [CrossRef] [PubMed]
5. Prasath, R.; Bhavana, P.; Sarveswari, S.; Ng, S.W.; Tiekink, E.R.T. Efficient ultrasound-assisted synthesis, spectroscopic, crystallographic and biological investigations of pyrazole-appended quinolinyl chalcones. *J. Mol. Struct.* **2015**, *1081*, 201–210. [CrossRef]
6. Kamal, A.; Shaik, A.B.; Jain, N.; Kishor, C.; Nagabhushana, A.; Supriya, B.; Kumar, G.B.; Chourasiya, S.S.; Suresh, Y.; Mishra, R.K.; et al. Design and synthesis of pyrazole–oxindole conjugates targeting tubulin polymerization as new anticancer agents. *Eur. J. Med. Chem.* **2015**, *92*, 501–513. [CrossRef] [PubMed]
7. Xu, Y.; Liu, X.-H.; Saunders, M.; Pearce, S.; Foulks, J.M.; Parnell, K.M.; Clifford, A.; Nix, R.N.; Bullough, J.; Hendrickson, T.F.; et al. Discovery of 3-(trifluoromethyl)-1*H*-pyrazole-5-carboxamide activators of the M2 isoform of pyruvate kinase (PKM2). *Bioorg. Med. Chem. Lett.* **2014**, *24*, 515–519. [CrossRef]
8. El-Moghazy, S.; Barsoum, F.; Abdel-Rahman, H.; Marzouk, A. Synthesis and anti-inflammatory activity of some pyrazole derivatives. *Med. Chem. Res.* **2012**, *21*, 1722–1733. [CrossRef]
9. Selvam, T.P.; Kumar, P.V.; Saravanan, G.; Prakash, C.R. Microwave-assisted synthesis, characterization and biological activity of novel pyrazole derivatives. *J. Saudi. Chem. Soc.* **2014**, *18*, 1015–1021. [CrossRef]
10. Pathak, V.; Maurya, H.K.; Sharma, S.; Srivastava, K.K.; Gupta, A. Synthesis and biological evaluation of substituted 4,6-diarylpyrimidines and 3,5-diphenyl-4,5-dihydro-1*H*-pyrazoles as anti-tubercular agents. *Bioorg. Med. Chem. Lett.* **2014**, *24*, 2892–2896. [CrossRef]
11. Jia, H.; Bai, F.; Liu, N.; Liang, X.; Zhan, P.; Ma, C.; Jiang, X.; Liu, X. Design, synthesis and evaluation of pyrazole derivatives as non-nucleoside hepatitis B virus inhibitors. *Eur. J. Med. Chem.* **2016**, *123*, 202–210. [CrossRef] [PubMed]
12. Liu, G.-N.; Luo, R.-H.; Zhou, Y.; Zhang, X.-J.; Li, J.; Yang, L.-M.; Zheng, Y.-T.; Liu, H. Synthesis and anti-HIV-1 activity evaluation for novel 3a,6a-dihydro-1*H*-pyrrolo[3,4-c]pyrazole-4,6-dione derivatives. *Molecules* **2016**, *21*, 1198. [CrossRef] [PubMed]
13. Khoobi, M.; Ghanoni, F.; Nadri, H.; Moradi, A.; Hamedani, M.P.; Moghadam, F.H.; Emami, S.; Vosooghi, M.; Zadmard, R.; Foroumadi, A. New tetracyclic tacrine analogs containing pyrano[2,3-c]pyrazole: Efficient synthesis, biological assessment and docking simulation study. *Eur. J. Med. Chem.* **2015**, *89*, 296–303. [CrossRef] [PubMed]
14. Nencini, A.; Castaldo, C.; Comery, T.A.; Dunlop, J.; Genesio, E.; Ghiron, C.; Haydar, S.; Maccari, L.; Micco, I.; Turlizzi, E.; et al. Design and synthesis of a hybrid series of potent and selective agonists of α7 nicotinic acetylcholine receptor. *Eur. J. Med. Chem.* **2014**, *78*, 401–418. [CrossRef]
15. Chaudhry, F.; Naureen, S.; Huma, R.; Shaukat, A.; Al-Rashida, M.; Asif, N.; Ashraf, M.; Munawar, M.A.; Khan, M.A. In search of new α-glucosidase inhibitors: Imidazolylpyrazole derivatives. *Bioorg. Chem.* **2017**, *71*, 102–109. [CrossRef] [PubMed]
16. Hernández-Vázquez, E.; Ocampo-Montalban, H.; Cerón-Romero, L.; Cruz, M.; Gómez-Zamudio, J.; Hiriart-Valencia, G.; Villalobos-Molina, R.; Flores-Flores, A.; Estrada-Soto, S. Antidiabetic, antidyslipidemic and toxicity profile of ENV-2: A potent pyrazole derivative against diabetes and related diseases. *Eur. J. Pharmacol.* **2017**, *803*, 159–166. [CrossRef]
17. Tuha, A.; Bekhit, A.A.; Seid, Y. Screening of some pyrazole derivatives as promising antileishmanial agent. *Afr. J. Pharm. Pharmacol.* **2017**, *11*, 32–37.
18. Reviriego, F.; Olmo, F.; Navarro, P.; Marín, C.; Ramírez-Macías, I.; García-España, E.; Albelda, M.T.; Gutiérrez-Sánchez, R.; Sánchez-Moreno, M.; Arán, V.J. Simple dialkyl pyrazole-3,5-dicarboxylates show in vitro and in vivo activity against disease-causing trypanosomatids. *Parasitology* **2017**, *144*, 1133–1143. [CrossRef]

19. Balaji, S.N.; Ahsan, M.J.; Jadav, S.S.; Trivedi, V. Molecular modelling, synthesis, and antimalarial potentials of curcumin analogues containing heterocyclic ring. *Arab. J. Chem.* **2015**. [CrossRef]

20. Fujinaga, M.; Yamasaki, T.; Nengaki, N.; Ogawa, M.; Kumata, K.; Shimoda, Y.; Yui, J.; Xie, L.; Zhang, Y.; Kawamura, K.; et al. Radiosynthesis and evaluation of 5-methyl-N-(4-[^{11}C]methylpyrimidin-2-yl)-4-(1*H*-pyrazol-4-yl)thiazol-2-amine ([^{11}C]ADX88178) as a novel radioligand for imaging of metabotropic glutamate receptor subtype 4 (mGluR4). *Bioorg. Med. Chem. Lett.* **2016**, *26*, 370–374. [CrossRef]

21. Dai, H.; Xiao, Y.-S.; Li, Z.; Xu, X.-Y.; Qian, X.-H. The thiazoylmethoxy modification on pyrazole oximes: Synthesis and insecticidal biological evaluation beyond acaricidal activity. *Chin. Chem. Lett.* **2014**, *25*, 1014–1016. [CrossRef]

22. Dai, H.; Chen, J.; Li, H.; Dai, B.; He, H.; Fang, Y. Synthesis and bioactivities of novel pyrazole oxime derivatives containing a 5-trifluoromethylpyridyl moiety. *Molecules* **2016**, *21*, 276. [CrossRef] [PubMed]

23. Karrouchi, K.; Radi, S.; Ramli, Y.; Taoufik, J.; Mabkhot, Y.N.; Al-aizari, F.A.; Ansar, M. Synthesis and pharmacological activities of pyrazole derivatives: A review. *Molecules* **2018**, *23*, 134. [CrossRef]

24. Lee, H.-Y.; Pan, S.-L.; Su, M.-C.; Liu, Y.-M.; Kuo, C.-C.; Chang, Y.-T.; Wu, J.-S.; Nien, C.-Y.; Mehndiratta, S.; Chang, C.-Y.; et al. Furanylazaindoles: Potent anticancer agents in vitro and in vivo. *J. Med. Chem.* **2013**, *56*, 8008–8018. [CrossRef] [PubMed]

25. Chen, N.E.; Maldonado, N.V.; Khankaldyyan, V.; Shimada, H.; Song, M.M.; Maurer, B.J.; Reynolds, C.P. Reactive oxygen species mediates the synergistic activity of fenretinide combined with the microtubule inhibitor ABT-751 against multidrug-resistant recurrent neuroblastoma xenografts. *Mol. Cancer Ther.* **2016**, *15*, 2653–2664. [CrossRef]

26. Dorleans, A.; Gigant, B.; Ravelli, R.B.; Mailliet, P.; Mikol, V.; Knossow, M. Variations in the colchicine-binding domain provide insight into the structural switch of tubulin. *Proc. Natl. Acad. Sci. USA* **2009**, *106*, 13775–13779. [CrossRef]

27. Liu, Y.-N.; Wang, J.-J.; Ji, Y.-T.; Zhao, G.-D.; Tang, L.-Q.; Zhang, C.-M.; Guo, X.-L.; Liu, Z.-P. Design, synthesis, and biological evaluation of 1- methyl-1,4-dihydroindeno[1,2-c]pyrazole analogues as potential anticancer agents targeting tubulin colchicine binding site. *J. Med. Chem.* **2016**, *59*, 5341–5535. [CrossRef]

28. Zhang, Y.; Gong, F.-L.; Lu, Z.-N.; Wang, H.-Y.; Cheng, Y.-N.; Liu, Z.-P.; Yu, L.-G.; Zhang, H.-H.; Guo, X.-L. DHPAC, a novel synthetic microtubule destabilizing agent, possess high anti-tumor activity in vincristine-resistant oral epidermoid carcinoma in vitro and in vivo. *Int. J. Biochem. Cell, B.* **2017**, *93*, 1–11. [CrossRef]

29. Luo, W.; Su, Y.B.; Hong, C.; Tian, R.-G.; Su, L.-P.; Yue-Qiao Wang, Y.-Q.; Li, Y.; Yue, J.-J.; Wang, C.-J. Design, synthesis and evaluation of novel 4-dimethylamine flavonoid derivatives as potential multi-functional anti-Alzheimer agents. *Eur. J. Med. Chem.* **2013**, *21*, 7275–7282. [CrossRef]

30. Ferreira, J.A.; Nel, J.W.; Brandt, E.V.; Bezuidenhoudt, B.; Ferreira, D. Oligomeric isoflavonoids. Part 3. Daljanelins A-D, the first pterocarpan- and isoflavanoid-neoflavonoid analogs. *J. Chem. Soc. Perkin Trans.* **1995**, *1*, 1049–1056. [CrossRef]

31. Muzychka, O.V.; Kobzar, O.L.; Popova, A.V.; Frasinyuk, M.S.; Vovk, A.I. Carboxylated aurone derivatives as potent inhibitors of xanthine oxidase. *Bioorg. Med. Chem.* **2017**, *25*, 3606. [PubMed]

32. Bonne, D.; Heusele, C.; Simon, C.; Pantaloni, D. 4′,6-Diamidino-2-phenylindole, a fluorescent probe for tubulin and mictrotubules. *J. Biol. Chem.* **1985**, *260*, 2819–2825. [PubMed]

Sample Availability: Samples of the compounds **4a–e** and **5a–e** are available from the authors.

Article

Synthesis, Antitumor Activity, and Docking Analysis of New Pyrido[3′,2′:4,5]furo(thieno)[3,2-*d*]pyrimidin-8-amines

Samvel N. Sirakanyan [1,*], Domenico Spinelli [2,*], Athina Geronikaki [3], Elmira K. Hakobyan [1], Harutyun Sahakyan [4], Erik Arabyan [5], Hovakim Zakaryan [5], Lusine E. Nersesyan [1], Anahit S. Aharonyan [1], Irina S. Danielyan [1], Rafayel E. Muradyan [1] and Anush A. Hovakimyan [1]

[1] Scientific Technological Center of Organic and Pharmaceutical Chemistry of National Academy of Science of Republic of Armenia, Institute of Fine Organic Chemistry of A.L. Mnjoyan, 0014 Yerevan, Armenia; hakobyan.elmira@mail.ru (E.K.H.); lusinenersesyan@gmail.com (L.E.N.); AAaharonyan@gmail.com (A.S.A.); irina_danielyan@mail.ru (I.S.D.); muradyanraf@inbox.ru (R.E.M.); aaa.h.87@mail.ru (A.A.H.)

[2] Dipartimento di Chimica G. Ciamician, Alma Mater Studiorum-Università di Bologna, Via F. Selmi 2, 40126 Bologna, Italy

[3] School of Pharmacy, Aristotle University of Thessaloniki, 54124 Thessaloniki, Greece; geronik@pharm.auth.gr

[4] Department of Bioengineering, Bioinformatics and Molecular Biology, Russian-Armenian University, 0051 Yerevan, Armenia; sahakyanhk@gmail.com

[5] Institute of Molecular Biology of NAS, Hasratyan 7, 0014 Yerevan, Armenia; ear20@mail.ru (E.A.); h_zakaryan@mb.sci.am (H.Z.)

* Correspondence: shnnr@mail.ru (S.N.S.); domenico.spinelli@unibo.it (D.S.); Tel.: +374-91-32-15-99 (S.N.S.); +39-051-209-9478 (D.S.)

Academic Editors: Carla Boga and Gabriele Micheletti
Received: 2 September 2019; Accepted: 29 October 2019; Published: 31 October 2019

Abstract: Continuing our research in the field of new heterocyclic compounds, herein we report on the synthesis and antitumor activity of new amino derivatives of pyrido[3′,2′:4,5](furo)thieno[3,2-*d*]pyrimidines as well as of two new heterocyclic systems: furo[2–*e*]imidazo[1,2-*c*]pyrimidine and furo[2,3-*e*]pyrimido[1,2-*c*]pyrimidine. Thus, by refluxing the 8-chloro derivatives of pyrido[3′,2′:4,5]thieno(furo)[3,2-*d*]pyrimidines with various amines, the relevant pyrido[3′,2′:4,5]thieno(furo)[3,2-*d*]pyrimidin-8-amines were obtained. Further, the cyclization of some amines under the action of phosphorus oxychloride led to the formation of new heterorings: imidazo[1,2-*c*]pyrimidine and pyrimido[1,2-*c*]pyrimidine. The possible antitumor activity of the newly synthesized compounds was evaluated in vitro. The biological tests evidenced that some of them showed pronounced antitumor activity. A study of the structure–activity relationships revealed that the compound activity depended mostly on the nature of the amine fragments. A docking analysis was also performed for the most active compounds.

Keywords: pyrido[3′,2′:4,5]furo(thieno)[3,2-*d*]pyrimidin-8-amines; amination; furo[2,3-*e*]imidazo[1,2-*c*]pyrimidine; furo[2,3-*e*]pyrimido[1,2-*c*]pyrimidine; DNA methylation; sarcoma 180; antitumor activity

1. Introduction

Enzymatic DNA methylation by DNA methyltransferases is an important constituent of the cell epigenetic regulatory system that modulates gene expression without altering the DNA base sequence. It is tissue- (cell) and age-specific and is involved in the regulation of all genetic functions, including transcription, DNA replication and repair, gene transposition, and cell and sex differentiation.

The methylation pattern of DNA is inherited, and significant distortions result in defects in growth and development. There is no doubt that some changes in DNA methylation induce cancer, premature aging, apoptosis, and death [1]. Malignant cells have a different DNA methylation pattern and a different set of expressed DNA methyltransferase activities compared to normal cells [2]. De novo methylation of CpG islands in the promoter regions of tumor-suppressor genes may lead to transcriptional silencing through a complex process involving histone deacetylation and chromatin condensation, and thus it represents a tumorigenic event that is functionally equivalent to genetic changes such as mutation and deletion [3]. The aberrant methylation of promoters has already been described in hundreds of genes associated with tumorigenesis and could serve as an important biomarker if new methods applicable to clinical practice are sufficiently advanced [4]. The DNA methylation profile of genomes or separate genes already serves as a true marker of different cancer varieties even at the early stages of carcinogenesis [1]. DNA methylation biomarkers with diagnostic, prognostic, and predictive power are already in clinical trials [5,6]. DNA methylation is interesting from a diagnostic viewpoint because it may be easily detected in DNA released from neoplastic and preneoplastic lesions into sera, urine, or sputum; and from a therapeutic viewpoint because epigenetically silenced genes may be reactivated by inhibitors of DNA methylation and/or histone deacetylase. A better understanding of epigenetic mechanisms leading to tumor formation and chemoresistance may eventually improve current cancer treatment regimens and be instructive for a more rational use of anticancer agents [3,7].

Data in the literature have indicated that amino derivatives of fused furo(thieno)[3,2-*d*]pyrimidines can show a broad spectrum of biological activity. In particular, these compounds are PI3 kinase p110a [8] and phosphodiesterase type 4 [9] inhibitors, and they possess antimicrobial [10] and antifungal [11] activities.

We have previously reported on the synthesis and biological activity of some amino derivatives of condensed furo- and thieno [3,2-*d*]pyrimidines [12–16]. Particularly, the studies showed that these compounds are distinguished by high anticonvulsant [12–14] and antibacterial activity [15,16]. Continuing our studies in the field of searching for new heterocyclic compounds of biological interest, herein we report on the synthesis and antitumor activity of new amino derivatives of pyrido[3′,2′:4,5](furo)thieno[3,2-*d*]pyrimidines as well as new heterocyclic systems: furo[2,3-*e*]imidazo[1,2-*c*]pyrimidine and furo[2,3-*e*]pyrimido[1,2-*c*]pyrimidine.

It should be mentioned that our first investigations into the amino derivatives of condensed pyrido[3′,2′:4,5]furo(thieno)[3,2-*d*]pyrimidines were reported some years ago during an international conference [17,18].

2. Results and Discussion

2.1. Chemistry

Here, 3-chloro-1-(2-furyl)-5,6,7,8-tetrahydroisoquinoline-4-carbonitrile (**1**) [19] was used as a starting compound: it was treated with ethyl mercaptoacetate in the presence of sodium ethoxide, thus obtaining ethyl 1-amino-5-(2-furyl)-6,7,8,9-tetrahydrothieno[2,3-*c*]isoquinoline-2-carboxylate (**2**). The presence of two functional groups in the thiophene ring of compound **2** allowed for the cyclocondensation of the latter through the action of formamide. As a result, the corresponding 5-(2-furyl)-1,2,3,4-tetrahydropyrimido[4′,5′:4,5]thieno[2,3-*c*]isoquinolin-8(9*H*)-one (**3**) was obtained. In turn, compound **3**, through a reaction with phosphorus oxychloride, led to 8-chloro-5-(2-furyl)-1,2,3,4tetrahydropyrimido[4′,5′:4,5]thieno[2,3-*c*]isoquinoline (**4**), which contained an "activated" chlorine atom that could be easily displaced by nucleophiles. Thus, further reaction of 8-chlorothieno[3,2-*d*]pyrimidine (**4**) with various amines in ethanol led to the formation of a series of 5-(2-furyl)-1,2,3,4-tetrahydropyrimido[4′,5′:4,5]thieno[2,3-*c*]isoquinolin-8-amines (**5a–f**) in very high yields (Scheme 1, from 79% to 88%).

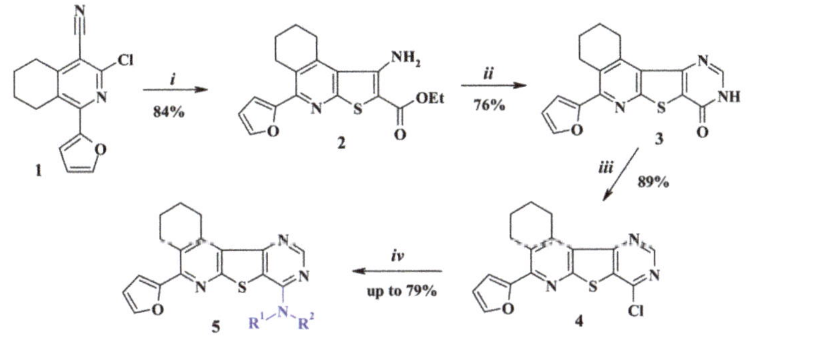

Scheme 1. Reagents and conditions: (i) $HSCH_2COOEt$, NaOEt/EtOH, reflux 5 h; (ii) $HCONH_2$, reflux 4 h; (iii) $POCl_3$, reflux 5 h; (iv) NHR^1R^2, EtOH, reflux 8 h.

In turn, the nucleophilic substitution of 8-chloro-2,2-dimethyl-1,4-dihydro-2*H*-pyrano[4″,3″:4′,5′]pyrido[3′,2′:4,5]furo(thieno)[3,2-*d*]pyrimidines (**6a–c**), the second starting compounds previously synthesized by us [13], with a series of amines gave (in very high yields, from 74% to 89%) the aimed-for pyrano[4″,3″:4′,5′]pyrido[3′,2′:4,5]furo(thieno)[3,2-*d*]pyrimidin-8-amines (**7a–m**) [13,20] (Scheme 2).

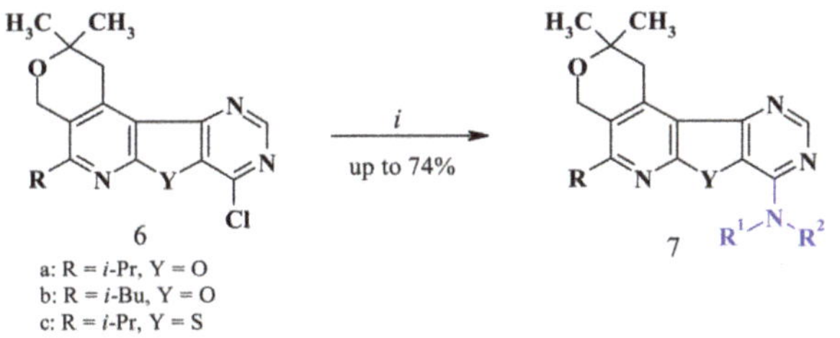

Scheme 2. Reagents and conditions: (i) NHR^1R^2, EtOH, reflux 8 h.

In the following step, the new heterocyclic systems, i.e., 11-isobutyl-8,8-dimethyl2,3,7,10-tetrahydro-8H-imidazo[1,2-c]pyrano[4″,3″:4′,5′]pyrido[3′,2′:4,5]furo[2,3-e]pyrimidine (**8**) and 12-isobutyl-9,9-dimethyl3,4,8,11-tetrahydro-2H,9H-pyrano[4″,3″:4′,5′]pyrido[3′,2′:4,5]furo[2,3-e]pyrimido[1,2-c]pyrimidine (**9**), were synthesized from compounds **7j,k** by simple refluxing with phosphorus oxychloride followed by treatment with a base (Scheme 3). The disappearance of NH and OH groups in the ^{1}H-NMR spectrum of compounds **8** and **9** supported the cyclized structures.

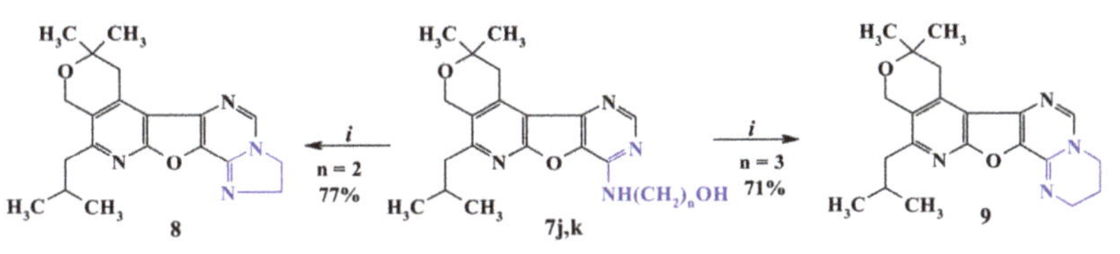

Scheme 3. Reagents and conditions: (i) $POCl_3$, reflux 10 h, then KOH.

The copies of ^{1}H-NMR and ^{13}C-NMR spectra for all new synthesized compounds were submitted along with the manuscript (Supplementary Materials).

2.2. Biology

It is known that DNA methylation in cancer plays a variety of roles, helping to change the healthy regulation of gene expression to a disease pattern. DNA hypermethylation in neoplasia is a specific epigenetic sign that can guide the diagnosis and treatment of this disease. We studied the effect of synthesized substances on tumor DNA in vitro. After the incubation of tumor cells with the studied compounds, the isolated DNA was hydrolyzed to nitrogen bases. In the studied DNA, the nucleotide composition (guanine (G), cytosine (C), 5-methylcitosine (5-MC), adenine (A), thymine (T)) complied with the rules of Chargaff. Isolated DNA belonged to the AT type, and the amount of basic bases (G + C + 5-MC) within them was 42.24–44.64 mol %, which is typical for eukaryotic DNA. Almost the same content of basic bases (G, C, A, and T) was observed, while a clear difference between DNA samples was found only in relation to the content of 5-MC, which indicated the demethylating ability of the compounds.

A distinct difference between tumor DNA samples after treatment with compounds **5** and **7** was observed only for the 5-MC content, and the data are given in Table 1. Most of the compounds inhibited the methylation level of tumor DNA. The highest activity was found for three compounds (**5f**, **7b**, and **5b**), which reduced the 5-MC content of tumor DNA by 89.1%, 62.5%, and 60.9%, respectively. Under identical experimental conditions, doxorubicin suppressed the level of methylation of tumor DNA by 67.2%. Meanwhile, two compounds (**7i,j**) did not affect DNA methylation. One compound (**7a**) showed the opposite effect (the degree of methylation of tumor DNA increased). The obtained results are presented in Table 1.

DNA demethylation in the tumor tissue under the influence of compounds **5a–f** and **7b–h,k–m** can be explained by the enzymatic demethylation of 5-MC. It is known that demethylating agents may reactivate tumor suppressor genes aberrantly methylated in tumor cells, since DNA methylation is reversible, leading to the inhibition of tumor growth. For example, decitabine and 2′-deoxy-5-azacytidine can inhibit DNA methyltransferases and reverse the epigenetic silencing of aberrantly methylated genes [21,22].

The increase in methylation level under the influence of compound **7a** could be explained by its damaging effect on tumor DNA. It is assumed that single-stranded DNA formed during replication or repair might serve as a nucleation site for de novo methylation by DNA methyltransferase, which often occurs in tumors [23].

Table 1. Content and inhibition of 5-MC of DNA samples of tumors under the influence of the studied compounds **5** and **7** in vitro.

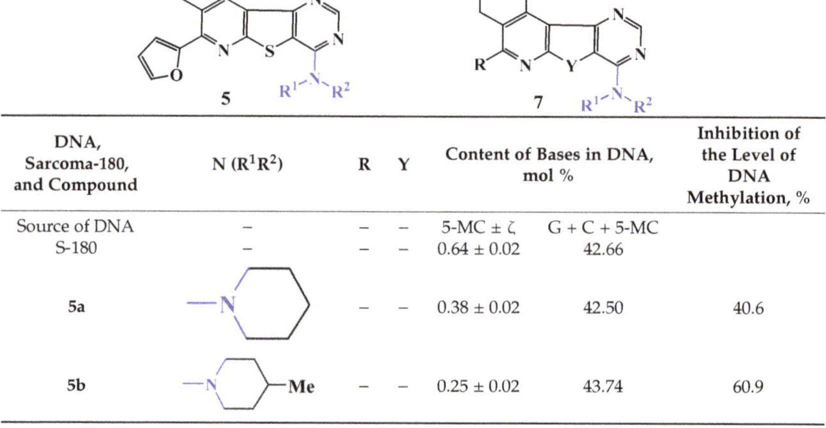

DNA, Sarcoma-180, and Compound	N (R¹R²)	R	Y	Content of Bases in DNA, mol %		Inhibition of the Level of DNA Methylation, %
				5-MC ± ζ	G + C + 5-MC	
Source of DNA S-180	– –	–	–	0.64 ± 0.02	42.66	
5a		–	–	0.38 ± 0.02	42.50	40.6
5b	–Me	–	–	0.25 ± 0.02	43.74	60.9

Table 1. *Cont.*

DNA, Sarcoma-180, and Compound	N (R¹R²)	R	Y	Content of Bases in DNA, mol %		Inhibition of the Level of DNA Methylation, %
5c	—N(morpholine ring, N–O)	–	–	0.49 ± 0.02	42.44	23.0
5d	—NH–CH₂–CH(Me)Me	–	–	0.42 ± 0.03	42.58	34.4
5e	—NH–CH₂CH₂–N(Et)Et	–	–	0.46 ± 0.02	42.44	28.1
5f	—NH–CH₂–(tetrahydrofuran-2-yl)	–	–	0.07 ± 0.02	44.02	89.1
7a	—NH–CH₂CH₂–OH	*i*-Pr	O	0.85 ± 0.03	43.50	–
7b	—NH–CH₂CH₂–N(Me)Me	*i*-Pr	O	0.24 ± 0.01	44.16	62.5
7c	—NH–CH₂CH₂–N(Et)Et	*i*-Pr	O	0.37 ± 0.01	42.72	42.2
7d	—NH–CH₂CH₂–(morpholine)	*i*-Pr	O	0.33 ± 0.01	43.44	48.4
7e	—NH–CH₂–(3,4-dimethoxyphenyl, OMe, OMe)	*i*-Pr	O	0.28 ± 0.01	44.00	56.2
7f	—NH–CH₂–(tetrahydrofuran-2-yl)	*i*-Pr	O	0.50 ± 0.02	42.82	21.9
7g	—NH–CH₂–(furan-2-yl)	*i*-Pr	O	0.38 ± 0.02	43.32	40.6
7h	—NH–CH₂–(pyridin-3-yl)	*i*-Pr	O	0.42 ± 0.02	43.20	34.4
7i	—NH–CH₂CH₂–N(Me)Me	*i*-Pr	O	0.52 ± 0.02	44.64	–
7j	—NH–CH₂CH₂–OH	*i*-Bu	O	0.54 ± 0.01	43.20	–
7k	—NH–CH₂CH₂CH₂–OH	*i*-Bu	O	0.52 ± 0.01	42.70	18.8
7l	—N(piperazine)–C(=O)–OEt	*i*-Pr	S	0.45 ± 0.03	43.76	29.7
7m	—NH–CH₂–(tetrahydrofuran-2-yl)	*i*-Pr	S	0.40 ± 0.02	43.06	37.5
Doxorubicin				0.21 ± 0.01	43.41	67.2

Number of definitions: seven. These changes were reliable ($p < 0.05$) compared to the control.

A study of the structure–activity relationships revealed that the activity of the tested compounds depended mostly on the nature of the amine fragment. Thus, the presence of the 2-tetrahydrofurylmethyl (**5f** and **7f,m**) group seemed to have a beneficial effect on antitumor activity. In the series of 5-(2-furyl)-1,2,3,4-tetrahydropyrimido[4′,5′:4,5]thieno[2,3-*c*]isoquinolin-8-amines (**5**), the replacement of the piperidine group (**5a**) with 4-methylpiperidine (**5b**) led to a very active compound. Among the pyrano[4″,3″:4′,5′]pyrido[3′,2′:4,5]furo[3,2-*d*]pyrimidin-8-amines (**7a–k**), the highest activity was displayed by compound **7b**, which had a $CH_2CH_2N(Me)_2$ group. The lengthening of the alkyl chain with one more CH_2 group (**7i**) led to a loss of activity. Compounds containing a CH_2CH_2OH group (**7a,j**) had no activity. Replacing the furan ring (**7f**) with a thiophene one (**7m**) also increased antitumor activity. The introduction of 2-morpholinoethyl (**7d**) as well as $CH_2CH_2C_6H_3$-*m,p*-OMe (**7e**) seemed to play a positive role in activity.

To test if compounds **5b**, **5f**, and **7b** had an inhibitory effect on cell proliferation, the antiproliferative activity was assessed in both a cancer cell line (HeLa) and in normal cells (Vero). As shown in Figure 1A,B, all three compounds were active against cancer and normal cells. Compound **5f** was more potent against HeLa (IC_{50} = 4.4 ± 0.3 µg/mL) and Vero (IC_{50} = 6.4 ± 0.4 µg/mL) cells than compounds **7b** (HeLa: IC_{50} = 5.4 ± 0.2 µg/mL; Vero: 13.7 ± 0.4 µg/mL) and **5b** (HeLa: 8.4 ± 0.2 ug/mL; Vero: 16.8 ± 0.6 µg/mL) were. In general, all tested compounds were less cytotoxic against Vero, indicating low cytotoxicity toward normal cells.

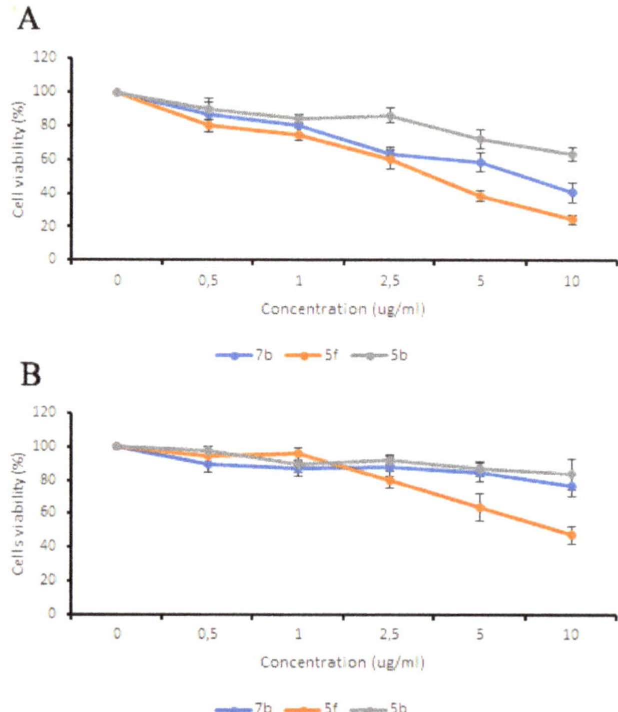

Figure 1. Effects of compounds **5b**, **5f**, and **7b** on the proliferation of HeLa (**A**) and Vero (**B**) cells. Cell viability was calculated through a comparison to the control, which contained <1% of DMSO (dimethyl sulfoxide). Representative data with means ± SD from three independent experiments.

2.3. Docking Analysis

In order to identify possible binding sites and the mechanisms of action of the active compounds, molecular docking was performed. DNA (cytosine-5)-methyltransferase 1 (DNMT1) plays a crucial role in DNA methylation [24]. With this in mind, we predicted its binding pockets using DNMT1 structures and performed a docking analysis on each of those pockets, including the sinefungin binding site, which is a well-known DNMT1 inhibitor [25]. According to our results, compounds **5b,f** and **7b** had the highest affinity to BAH2 (bromo-adjacent homology domain 2) of DNMT1. BAH2 consists of about 110 amino acids and folds with a tilted β-barrel, forming a cavity [26]. In DNMT1, this domain contributes to DNA binding and interacts with the main catalytic MTase domain. Moreover, BAH2 is present in a number of proteins involved in the DNA methylation process [27]. Thus, the docking showed that compounds **5b,f** and **7b** could bind in the cavity of the BAH2 domain with −14.73, −15.59, and −10.64 kcal/mol, respectively (Table 2). Hydrophobic interactions also contributed to the complex stabilization of binding energy.

Table 2. Binding energies, Internal Coordinate Mechanics (ICM) binding scores, and Protein Data Bank (PDB) structure IDs for which docking showed the highest results.

Compound	Binding Energy (kcal/mol)	ICM Score	PDB ID
5b	−14.73	−34.61	4WXX
5f	−15.59	−36.41	4DA4
7b	−10.64	−28.01	4YOC

Compounds **5b** and **5f** had very similar binding positions (Figure 2A,D and Figure 2B,E), and they both could bind with amino acids Y933, V935, V939, L941, F946, I996, I999, V1018, A1049, and V1051 via hydrophobic interactions. However, the binding mode of **7b** was slightly different, as it was located in the side of the pocket (Figure 2C,F) and interacted mainly with amino acids I999, Y923, I996, I1014, F1053, V910, A926, V939, Y933, and K928. The estimated binding energies of the investigated compounds showed approximately the same results as the DNA methylation tests. Compound **5f** had the highest binding energy, although the other two compounds did not fit well in the in vitro experiments.

Docking with the sinefungin binding site did not show a high binding energy or Internal Coordinate Mechanics (ICM) score: this allows us to suggest that these compounds do not interact with this site and have another mechanism of action that is different from sinefungin [25]. It is known that DNMT1 can adopt several conformational states under different DNA binding conditions [26]. Probably, the interaction of the active compounds with the BAH2 domain changes its flexibility and affects its function.

Methods

A docking analysis was performed using ICM-Pro 3.8–7 [28]. To imitate protein flexibility, we used a 4D docking method with several crystallographic structures of DNMT1 (PDB: 3SWR, 4WXX, 4YOC, 3AV6, 3PT6, 4DA4, and 5GUV) [29]. Protein structures were minimized, and hydrogens were added: 3D structures for **5b,f** and **7b** were built, adding hydrogens, assigned charges, and relaxed geometry. The docking was performed by simulating ligand flexibility and with docking effort 10. Binding pockets were identified using icmPocketFinder [30]. We performed the docking analysis 10 times for each identified binding site and evaluated the results using a standard ICM score and the estimated binding energy [31].

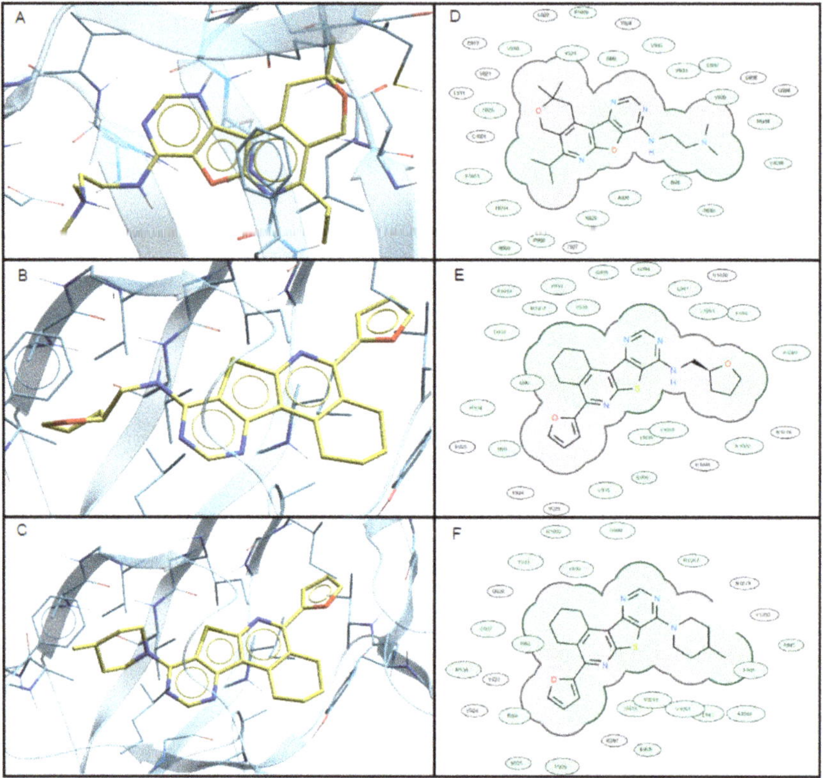

Figure 2. Docking positions of **7b** (**A**), **5f** (**B**), and **5b** (**C**) in the cavity of the bromo-adjacent homology domain 2 (BAH2) domain of DNA (cytosine-5)-methyltransferase 1 (DNMT1) and their 2D interaction diagrams (**D–F**). In the 2D diagrams, the broken thick lines around the ligand shape indicate an accessible surface, and gray and green colors represent hydrophobic and hydrophilic areas, respectively.

3. Materials and Methods

3.1. General Information

^{1}H- and ^{13}C-NMR spectra were recorded on a Varian Mercury 300VX spectrometer in DMSO (dimethyl sulfoxide)/CCl$_4$ (1/3) solution (300 MHz for ^{1}H and 75 MHz for ^{13}C, respectively). Chemical shifts were reported as δ (parts per million) relative to TMS (tetramethylsilane) as an internal standard. IR spectra were recorded on a Nicolet Avatar 330-FT-IR spectrophotometer, and the reported wave numbers are given in cm^{-1}. Elemental analyses were performed on an Elemental Analyzer Euro EA 3000. Melting points were determined on a Boetius microheating stage. Compounds **1** [19], **6** [13], and **7a,b,d,h,i,k** [13,20] have already been described.

Ethyl 1-amino-5-(2-furyl)-6,7,8,9-tetrahydrothieno[2,3-c]isoquinoline-2-carboxylate (**2**): To a solution of sodium ethoxide (0.28 g (12 mmol) of sodium in absolute ethanol (30 mL)), ethyl 2-mercaptoacetate (1.44 g, 12 mmol) and compound **1** (2.59 g, 10 mmol) were added. The mixture was refluxed for 5 h, cooled, and poured onto ice. The formed crystals were filtered off, washed with water, dried, and recrystallized from ethanol. Light-yellow solid, yield 84%, m.p. 192–194 °C; IR ν/cm^{-1}: 3485, 3346 (NH$_2$), 1664 (C=O). ^{1}H-NMR (300 MHz, DMSO/CCl$_4$, 1/3) δ 7.64 (dd, J = 1.8, 0.8 Hz, 1H, 5-CH$_{furyl}$), 7.03 (dd, J = 3.5, 0.8 Hz, 1H, 3-CH$_{furyl}$), 6.63 (br, 2H, NH$_2$), 6.55 (dd, J =3.5, 1.8 Hz, 1H, 4-CH$_{furyl}$), 4.30 (q, J = 7.1 Hz, 2H, C$\underline{H}_2$CH$_3$), 3.38–3.32 (m, 2H, 9-CH$_2$), 3.07–3.01 (m, 2H, 6-CH$_2$), 1.94–1.78 (m,

4H, 7,8-CH$_2$), 1.39 (t, *J* = 7.1 Hz, 3H, CH$_2$C<u>H</u>$_3$). ^{13}C NMR (75 MHz, DMSO/CCl$_4$, 1/3) δ 164.5, 157.5, 153.3, 149.7, 147.9, 144.5, 142.8, 124.7, 122.4, 112.2, 110.9, 59.2, 26.7, 26.6, 21.6, 21.1, 14.1. Anal. calcd (Analytically calculated) for C$_{18}$H$_{18}$N$_2$O$_3$S: C 63.14; H 5.30; N 8.18%. Found: C 63.47; H 5.49; N 8.42%.

5-(2-Furyl)-1,2,3,4-tetrahydropyrimido[4′,5′:4,5]thieno[2,3-c]isoquinolin-8(9H)-one (**3**): A mixture of compound **2** (3.42 g, 10 mmol) and formamide (30 mL) was refluxed for 5 h. After cooling, the separated crystals were filtered off, washed with water, dried, and recrystallized from dimethylformamide. Gray solid, yield 76%, m.p. > 360 °C; IR *v*/cm^{-1}: 3108 (NH), 1642 (C=O). ^{1}H-NMR (300 MHz, DMSO/CCl$_4$, 1/3) δ 12.88 (br s, 1H, NH), 8.33 (s, 1H, 10-CH), 7.94 (dd, *J* = 1.8, 0.8 Hz, 1H, 5-CH$_{furyl}$), 7.17 (dd, *J* = 3.5, 0.8 Hz, 1H, 3-CH$_{furyl}$), 6.72 (dd, *J* = 3.5, 1.8 Hz, 1H, 4-CH$_{furyl}$), 3.59–3.53 (m, 2H, 1-CH$_2$), 3.06–2.99 (m, 2H, 4-CH$_2$), 1.88–1.80 (m, 4H, 2,3-CH$_2$). Anal. calcd for C$_{17}$H$_{13}$N$_3$O$_2$S: C 63.14; H 4.05; N 12.99%. Found: C 63.52; H 4.28; N 13.25%.

8-Chloro-5-(2-furyl)-1,2,3,4-tetrahydropyrimido[4′,5′:4,5]thieno[2,3-c]isoquinoline (**4**). A mixture of compound **3** (3.23 g, 10 mmol) and phosphorus oxychloride (30 mL) was refluxed for 4 h. The excess phosphorus oxychloride was distilled off to give a dry residue. Ice water and then ammonia solution were added. The separated crystals were filtered off, washed with water, dried, and recrystallized from a mixture of ethanol–chloroform (1:2). Light-yellow solid, yield 89%, m.p. 206–208 °C. ^{1}H-NMR (300 MHz, DMSO/CCl$_4$, 1/3) δ 8.98 (s, 1H, 10-CH), 7.73 (dd, *J* = 1.8, 0.8 Hz, 1H, 5-CH$_{furyl}$), 7.20 (dd, *J* = 3.5, 0.8 Hz, 1H, 3-CH$_{furyl}$), 6.62 (dd, *J* = 3.5, 1.8 Hz, 1H, 4-CH$_{furyl}$), 3.64–3.59 (m, 2H, 1-CH$_2$), 3.17–3.11 (m, 2H, 4-CH$_2$), 2.00–1.86 (m, 4H, 2,3-CH$_2$). ^{13}C NMR (75 MHz, DMSO/CCl$_4$, 1/3) δ 158.9, 157.5, 153.3, 153.2, 152.8, 149.4, 147.9, 143.9, 129.6, 126.7, 122.3, 113.9, 111.4, 27.5, 26.6, 21.8, 20.7. Anal. calcd for C$_{17}$H$_{12}$ClN$_3$OS: C 59.73; H 3.54; N 12.29%. Found: C 60.09; H 3.76; N 12.54%.

3.2. General Method for the Preparation of Compounds 5a–f and 7c,e–g,j,l,m

A mixture of compound **4** or **6** (5 mmol) and the corresponding amine (11 mmol) in absolute ethanol (30 mL) was refluxed for 8 h. The ethanol was distilled off to dryness, and water (50 mL) was added to the residue. The separated crystals were filtered off, washed with water, dried, and recrystallized from ethanol.

5-(2-Furyl)-8-piperidin-1-yl-1,2,3,4-tetrahydropyrimido[4′,5′:4,5]thieno[2,3-c]isoquinoline (**5a**): Cream solid, yield 84%, m.p. 167-169 °C. ^{1}H-NMR (300 MHz, DMSO/CCl$_4$, 1/3) δ 8.52 (s, 1H, 10-CH), 7.68 (dd, *J* = 1.7, 0.7 Hz, 1H, 5-CH$_{furyl}$), 7.07 (dd, *J* = 3.5, 0.7 Hz, 1H, 3-CH$_{furyl}$), 6.58 (dd, *J* = 3.5, 1.7 Hz, 1H, 4-CH$_{furyl}$), 3.95–3.89 (m, 4H, N(CH$_2$)$_2$), 3.71–3.65 (m, 2H, 1-CH$_2$), 3.16–3.10 (m, 2H, 4-CH$_2$), 1.97–1.84 (m, 4H, 2,3-CH$_2$), 1.82–1.69 (m, 6H, 3CH$_2$, C$_5$H$_{10}$N). ^{13}C NMR (75 MHz, DMSO/CCl$_4$, 1/3) δ 157.4, 157.1, 156.5, 153.4, 153.0, 148.2, 146.6, 143.0, 125.9, 122.9, 112.8, 112.4, 111.1, 46.8, 27.4, 26.7, 25.5, 24.2, 21.9, 21.0. Anal. calcd for C$_{22}$H$_{22}$N$_4$OS: C 67.67; H 5.68; N 14.35%. Found: C 68.01; H 5.88; N 14.58%.

5-(2-Furyl)-8-(4-methylpiperidin-1-yl)-1,2,3,4-tetrahydropyrimido[4′,5′:4,5]thieno[2,3-c]isoquinoline (**5b**): Light-yellow solid, yield 79%, m.p. 110–112 °C. ^{1}H-NMR (300 MHz, DMSO/CCl$_4$, 1/3) δ 8.51 (s, 1H, 10-CH), 7.68 (dd, *J* = 1.8, 0.7 Hz, 1H, 5-CH$_{furyl}$), 7.07 (dd, *J* = 3.4, 0.7 Hz, 1H, 3-CH$_{furyl}$), 6.59 (dd, *J* = 3.4, 1.8 Hz, 1H, 4-CH$_{furyl}$), 4.77–4.68 (m, 2H, NCH$_2$), 3.71–3.64 (m, 2H, 1-CH$_2$), 3.16–3.05 (m, 4H, NCH$_2$, 4-CH$_2$), 1.99–1.69 and 1.38–1.22 (both m, 7H and 2H, C$_5$H$_9$N, 2,3-CH$_2$), 1.01 (d, *J* = 6.2 Hz, 3H, CH$_3$). ^{13}C NMR (75 MHz, DMSO/CCl$_4$, 1/3) δ 157.4, 157.1, 156.5, 153.4, 153.1, 148.2, 146.6, 143.0, 125.9, 122.9, 112.8, 112.4, 111.1, 46.0, 33.7, 30.7, 27.5, 26.7, 21.9, 21.4, 21.0. Anal. calcd for C$_{23}$H$_{24}$N$_4$OS: C 68.29; H 5.98; N 13.85%. Found: C 68.61; H 6.15; N 14.07%.

5-(2-Furyl)-8-morpholin-4-yl-1,2,3,4-tetrahydropyrimido[4′,5′:4,5]thieno[2,3-c]isoquinoline (**5c**): Cream solid, yield 88%, m.p. 222-224 °C. ^{1}H-NMR (300 MHz, DMSO/CCl$_4$, 1/3) δ 8.58 (s, 1H, 10-CH), 7.70 (dd, *J* = 1.7, 0.8 Hz, 1H, 5-CH$_{furyl}$), 7.10 (dd, *J* = 3.4, 0.8 Hz, 1H, 3-CH$_{furyl}$), 6.59 (dd, *J* = 3.4, 1.7 Hz, 1H, 4-CH$_{furyl}$), 3.96–3.91 (m, 4H, O(CH$_2$)$_2$), 3.83–3.78 (m, 4H, N(CH$_2$)$_2$), 3.73–3.66 (m, 2H, 1-CH$_2$), 3.18–3.11 (m, 2H, 4-CH$_2$), 1.99–1.86 (m, 4H, 2,3-CH$_2$). ^{13}C NMR (75 MHz, DMSO/CCl$_4$, 1/3) δ 157.6, 157.5, 156.8,

153.3, 152.9, 148.4, 146.8, 143.2, 126.2, 122.7, 113.2, 112.6, 111.1, 65.8, 46.0, 27.5, 26.7, 21.9, 20.9. Anal. calcd for $C_{21}H_{20}N_4O_2S$: C 64.27; H 5.14; N 14.28%. Found: C 64.64; H 5.37; N 14.54%.

5-(2-Furyl)-N-isobutyl-1,2,3,4-tetrahydropyrimido[4′,5′:4,5]thieno[2,3-c]isoquinolin-8-amine (**5d**): Colorless solid, yield 83%, m.p. 96–98 °C; IR v/cm^{-1}: 3400, 3241 (NH). ^{1}H-NMR (300 MHz, DMSO/CCl$_4$, 1/3) δ 8.45 (s, 1H, 10-CH), 7.68 (dd, J = 1.7, 0.7 Hz, 1H, 5-CH$_{furyl}$), 7.41 (t, J = 5.7 Hz, 1H, NH), 7.06 (dd, J = 3.4, 0.7 Hz, 1H, 3-CH$_{furyl}$), 6.58 (dd, J = 3.4, 1.7 Hz, 1H, 4-CH$_{furyl}$), 3.72–3.66 (m, 2H, 1-CH$_2$), 3.39–3.33 (m, 2H, NHC$\underline{H}_2$), 3.14–3.08 (m, 2H, 4-CH$_2$), 2.06 (sp, J = 6.7 Hz, 1H, C$\underline{H}$(CH$_3$)$_2$), 1.98–1.85 (m, 4H, 2,3-CH$_2$), 0.98 (sp, J = 6.7 Hz, 6H, CH(C$\underline{H}_3$)$_2$). ^{13}C NMR (75 MHz, DMSO/CCl$_4$, 1/3) δ 157.9, 156.7, 154.1, 153.8, 153.2, 147.7, 146.3, 142.8, 125.8, 123.9, 113.5, 112.2, 111.0, 47.7, 27.4, 27.2, 26.7, 22.1, 21.0, 20.0, 19.9. Anal. calcd for $C_{21}H_{22}N_4OS$: C 66.64; H 5.86; N 14.80%. Found: C 66.99; H 6.07; N 15.04%.

N,N-Diethyl-N′-[5-(2-furyl)-1,2,3,4-tetrahydropyrimido[4′,5′:4,5]thieno[2,3-c]isoquinolin-8-yl]ethane-1, 2-diamine (**5e**): Colorless solid, yield 81%, m.p. 137-139 °C; IR v/cm^{-1}: 3365 (NH). ^{1}H-NMR (300 MHz, DMSO/CCl$_4$, 1/3) δ 8.49 (s, 1H, 10-CH), 7.68 (dd, J = 1.8, 0.8 Hz, 1H, 5-CH$_{furyl}$), 7.09 (br, 1H, NH), 7.07 (dd, J = 3.4, 0.8 Hz, 1H, 3-CH$_{furyl}$), 6.59 (dd, J = 3.4, 1.8 Hz, 1H, 4-CH$_{furyl}$), 3.73–3.67 (m, 2H, 1-CH$_2$), 3.65–3.56 (m, 2H, NHC$\underline{H}_2$), 3.16–3.10 (m, 2H, 4-CH$_2$), 2.70 (br t, J = 7.2 Hz, 2H, NHCH$_2$C$\underline{H}_2$), 2.59 (q, J = 7.1 Hz, 4H, N(C$\underline{H}_2$CH$_3$)$_2$), 1.99–1.86 (m, 4H, 2,3-CH$_2$), 1.06 (t, J = 7.1 Hz, 6H, N(CH$_2$C$\underline{H}_3$)$_2$). ^{13}C NMR (75 MHz, DMSO/CCl$_4$, 1/3) δ 157.8, 156.4, 154.1, 153.9, 153.2, 147.7, 146.3, 142.9, 125.8, 123.9, 113.5, 112.2, 111.0, 51.1, 46.7, 38.2, 27.2, 26.7, 22.0, 21.0, 11.8. Anal. calcd for $C_{23}H_{27}N_5OS$: C 65.53; H 6.46; N 16.61%. Found: C 65.91; H 6.66; N 16.88%.

5-(2-Furyl)-N-(tetrahydrofuran-2-ylmethyl)-1,2,3,4-tetrahydropyrimido[4′,5′:4,5]thieno[2,3-c]isoquinolin-8-amine (**5f**): Colorless solid, yield 87%, m.p. 146–148 °C; IR v/cm^{-1}: 3318 (NH). ^{1}H-NMR (300 MHz, DMSO/CCl$_4$, 1/3) δ 8.48 (s, 1H, 10-CH), 7.69 (dd, J = 1.7, 0.7 Hz, 1H, 5-CH$_{furyl}$), 7.43 (t, J = 5.6 Hz, 1H, NH), 7.08 (dd, J = 3.4, 0.7 Hz, 1H, 3-CH$_{furyl}$), 6.59 (dd, J = 3.4, 1.7 Hz, 1H, 4-CH$_{furyl}$), 4.20–4.10 (m, 1H, OCH$_{furyl}$), 3.91–3.83 (m, 1H, OC$\underline{H}_2$), 3.74–3.52 (m, 5H, NHC$\underline{H}_2$, OC$\underline{H}_2$, 1-CH$_2$), 3.16–3.09 (m, 2H, 4-CH$_2$), 2.06–1.62 (m, 8H, 2,3-CH$_2$, 3,4-CH$_2$-furyl). ^{13}C NMR (75 MHz, DMSO/CCl$_4$, 1/3) δ 157.9, 156.6, 154.2, 153.7, 153.2, 147.8, 146.3, 142.9, 125.8, 123.8, 113.6, 112.3, 111.1, 76.7, 66.8, 44.1, 28.5, 27.2, 26.7, 25.0, 22.1, 21.0. Anal. calcd for $C_{22}H_{22}N_4O_2S$: C 65.00; H 5.46; N 13.78%. Found: C 65.34; H 5.64; N 14.01%.

N,N-Diethyl-N′-(5-isopropyl-2,2-dimethyl-1,4-dihydro-2H-pyrano[4″,3″:4′,5′]pyrido[3′,2′:4,5]furo[3,2-d] pyrimidin-8-yl)ethane-1,2-diamine (**7c**): Colorless solid, yield 85%, m.p. 146-148 °C; IR v/cm^{-1}: 3230 (NH). ^{1}H-NMR (300 MHz, DMSO/CCl$_4$, 1/3) δ 8.34 (s, 1H, 10-CH), 7.41 (br, 1H, NH), 4.86 (s, 2H, OCH$_2$), 3.66–3.57 (m, 2H, NHCH$_2$), 3.32 (s, 2H, 1-CH$_2$), 3.05 (sp, J = 6.6 Hz, 1H, CH(CH$_3$)$_2$), 2.75–2.65 (m, 2H, CH$_2$N(CH$_2$)$_2$), 2.64–2.54 (m, 4H, N(CH$_2$)$_2$), 1.35 (s, 6H, C(CH$_3$)$_2$), 1.29 (d, J = 6.6 Hz, 6H, CH(CH$_3$)$_2$), 1.05 (d, J = 7.1 Hz, 6H, N(CH$_2$CH$_3$)$_2$). ^{13}C NMR (75 MHz, DMSO/CCl$_4$, 1/3) δ 161.4, 160.6, 152.9, 148.5, 140.5, 132.9, 121.8, 110.5, 68.6, 59.6, 51.2, 46.6, 37.8, 36.8, 30.3, 25.9, 21.3, 11.7. Anal. calcd for $C_{23}H_{33}N_5O_2$: C 67.12; H 8.08; N 17.02%. Found: C 67.48; H 8.29; N 17.27%.

N-[2-(3,4-Dimethoxyphenyl)ethyl]-5-isopropyl-2,2-dimethyl-1,4-dihydro-2H-pyrano[4″,3″:4′,5′]pyrido [3′,2′:4,5]furo[3,2-d]pyrimidin-8-amine (**7e**): Colorless solid, yield 82%, m.p. 207-209 °C; IR v/cm^{-1}: 3227 (NH). ^{1}H-NMR (300 MHz, DMSO/CCl$_4$, 1/3) δ 8.36 (s, 1H, 10-CH), 7.85 (br, 1H, NH), 6.82–6.75 (m, 3H, C$_6$H$_3$), 4.87 (s, 2H, OCH$_2$), 3.83–3.72 (m, 2H, NHCH$_2$), 3.79 (s, 3H, OCH$_3$), 3.75 (s, 3H, OCH$_3$), 3.33 (s, 2H, 1-CH$_2$), 3.05 (sp, J = 6.6 Hz, 1H, CH(CH$_3$)$_2$), 2.93–2.87 (m, 2H, CH$_2$C$_6$H$_3$), 1.35 (s, 6H, C(CH$_3$)$_2$), 1.29 (d, J = 6.6 Hz, 6H, CH(CH$_3$)$_2$). ^{13}C NMR (75 MHz, DMSO/CCl$_4$, 1/3) δ 161.4, 160.6, 152.9, 148.7, 148.5, 147.2, 140.5, 132.9, 131.7, 121.8, 120.3, 112.6, 111.8, 110.5, 68.6, 59.6, 55.2, 55.1, 41.4, 36.8, 34.6, 30.3, 25.9, 21.3. Anal. calcd for $C_{27}H_{32}N_4O_4$: C 68.05; H 6.77; N 11.76%. Found: C 68.38; H 6.96; N 11.98%.

5-Isopropyl-2,2-dimethyl-N-(tetrahydrofuran-2-ylmethyl)-1,4-dihydro-2H-pyrano[4″,3″:4′,5′]pyrido[3′,2′:4,5] furo[3,2-d]pyrimidin-8-amine (**7f**): Colorless solid, yield 76%, m.p. 164-166 °C; IR v/cm^{-1}: 3334, 3277 (NH). ^{1}H-NMR (300 MHz, DMSO/CCl$_4$, 1/3) δ 8.33 (s, 1H, 10-CH), 7.62 (br t, J = 5.8 Hz, 1H, NH), 4.86 (s, 2H, OCH$_2$), 4.16–4.07 (m, 1H, OCH$_{furyl}$), 3.90–3.82 (m, 1H, OCH$_{2furyl}$), 3.72–3.59 (m, 3H, NHCH$_2$, OCH$_{2furyl}$), 3.32 (s, 2H, 1-CH$_2$), 3.05 (sp, J = 6.6 Hz, 1H, CH(CH$_3$)$_2$), 2.04–1.64 (m, 4H, 3,4-CH$_{2furyl}$),

1.34 (s, 6H, C(CH$_3$)$_2$), 1.29 (d, J = 6.6 Hz, 6H, CH(CH$_3$)$_2$). ^{13}C NMR (75 MHz, DMSO/CCl$_4$, 1/3) δ 161.5, 160.6, 152.8, 148.6, 140.5, 132.8, 121.8, 110.5, 76.6, 68.6, 66.9, 59.6, 43.6, 36.8, 30.3, 28.4, 25.9, 25.0, 21.3. Anal. calcd for C$_{22}$H$_{28}$N$_4$O$_3$: C 66.64; H 7.12; N 14.13%. Found: C 67.01; H 7.34; N 14.39%.

N-(2-Furylmethyl)-5-isopropyl-2,2-dimethyl-1,4-dihydro-2H-pyrano[4″,3″:4′,5′]pyrido[3′,2′:4,5]furo[3,2-d] pyrimidin-8-amine (**7g**): Colorless solid, yield 88%, m.p. 180-182 °C; IR ν/cm^{-1}: 3286, 3217 (NH). ^{1}H-NMR (300 MHz, DMSO/CCl$_4$, 1/3) δ 8.38 (s, 1H, 10-CH), 8.26 (t, J = 6.0 Hz, 1H, NH), 7.37 (dd, J = 1.9, 1.0 Hz, 1H, 5-CH$_{furyl}$), 6.29 (dd, J = 3.3, 1.9 Hz, 1H, 3-CH$_{furyl}$), 6.25 (dd, J = 3.3, 1.0 Hz, 1H, 4 CH$_{furyl}$), 4.88 (s, 2H, OCH$_2$), 4.76 (d, J = 6.0, 2H, NHCH$_2$), 3.32 (s, 2H, 1-CH$_2$), 3.05 (sp, J = 6.6 Hz, 1H, CH(CH$_3$)$_2$), 1.35 (s, 6H, C(CH$_3$)$_2$), 1.29 (d, J = 6.6 Hz, 6H, CH(CH$_3$)$_2$). ^{13}C NMR (75 MHz, DMSO/CCl$_4$, 1/3) δ 161.6, 160.7, 152.8, 152.2, 148.1, 144.2, 140.8, 140.6, 132.9, 121.8, 110.4, 109.8, 106.4, 68.6, 59.6, 36.8, 36.4, 30.3, 25.8, 21.3. Anal. calcd for C$_{22}$H$_{24}$N$_4$O$_3$: C 67.33; H 6.16; N 14.28%. Found: C 67.64; H 6.36; N 14.52%.

2-[(5-Isobutyl-2,2-dimethyl-1,4-dihydro-2H-pyrano[4″,3″:4′,5′]pyrido[3′,2′:4,5]furo[3,2-d]pyrimidin-8-yl) amino]ethanol (**7j**): Colorless solid, yield 74%, m.p. 219-220 °C; IR ν/cm^{-1}: 3226 (NH). ^{1}H-NMR (300 MHz, DMSO/CCl$_4$, 1/3) δ 8.33 (s, 1H, 10-CH), 7.54 (br, 1H, NH), 4.79 (s, 2H, OCH$_2$), 4.45 (br t, J = 5.4 Hz, 1H, OH), 3.67–3.62 (m, 4H, NHCH$_2$CH$_2$), 3.31 (s, 2H, 1-CH$_2$), 2.56 (d, J = 7.1 Hz, 2H, CHCH$_2$), 2.33–2.19 (m, 1H, CH(CH$_3$)$_2$), 1.33 (s, 6H, C(CH$_3$)$_2$), 0.98 (d, J = 6.8 Hz, 6H, CH(CH$_3$)$_2$). ^{13}C NMR (75 MHz, DMSO/CCl$_4$, 1/3) δ 160.3, 156.1, 152.9, 148.7, 143.8, 140.3, 132.9, 123.4, 110.4, 68.7, 60.0, 59.8, 42.8, 42.0, 36.7, 27.3, 25.9, 22.2. Anal. calcd for C$_{20}$H$_{26}$N$_4$O$_3$: C 64.84; H 7.07; N 15.12%. Found: C 65.19; H 7.25; N 15.35%.

Ethyl 4-(5-isopropyl-2,2-dimethyl-1,4-dihydro-2H-pyrano[4″,3″:4′,5′]pyrido[3′,2′:4,5]thieno[3,2-d]pyrimidin-8 -yl)piperazine-1-carboxylate (**7l**): Colorless solid, yield 84%, m.p. 217-219 °C; IR ν/cm^{-1}: 1711 (C=O). ^{1}H-NMR (300 MHz, DMSO/CCl$_4$, 1/3) δ 8.60 (s, 1H, 10-CH), 4.88 (s, 2H, OCH$_2$), 4.12 (q, J = 7.1 Hz, 2H, CH$_2$CH$_3$), 4.00–3.94 (m, 4H, N(CH$_2$)$_2$), 3.64–3.59 (m, 4H, N(CH$_2$)$_2$), 3.56 (s, 2H, 1-CH$_2$), 3.08 (sp, J = 6.6 Hz, 1H, CH(CH$_3$)$_2$), 1.35 (s, 6H, C(CH$_3$)$_2$), 1.31 (d, J = 6.6 Hz, 6H, CH(CH$_3$)$_2$), 1.29 (t, J = 7.1 Hz, 3H, CH$_2$CH$_3$). ^{13}C NMR (75 MHz, DMSO/CCl$_4$, 1/3) δ 163.6, 158.7, 157.4, 156.7, 154.0, 153.1, 141.6, 122.9, 122.1, 112.7, 68.6, 60.5, 59.6, 45.3, 43.0, 42.9, 36.9, 30.4, 26.0, 21.2, 14.2. Anal. calcd for C$_{24}$H$_{31}$N$_5$O$_3$S: C 61.38; H 6.65; N 14.91%. Found: C 61.71; H 6.82; N 15.16%.

5-Isopropyl-2,2-dimethyl-N-(tetrahydrofuran-2-ylmethyl)-1,4-dihydro-2H-pyrano[4″,3″:4′,5′]pyrido[3′,2′:4,5] thieno[3,2-d]pyrimidin-8-amine (**7m**): Colorless solid, yield 89%, m.p. 179-181 °C; IR ν/cm^{-1}: 3299 (NH). ^{1}H-NMR (300 MHz, DMSO/CCl$_4$, 1/3) δ ^{1}H NMR (300 MHz, DMSO/CCl$_4$, 1/3) δ 8.83 (s, 1H, 10-CH), 7.80 (t, J = 5.6 Hz, 1H, NH), 5.23 (s, 2H, OCH$_2$), 4.56–4.47 (m, 1H, OCH$_{furyl}$), 4.27–4.19 (m, 1H, OCH$_2$), 4.10–3.92 (m, 3H, NHCH$_2$, OCH$_2$), 3.91 (s, 2H, 1-CH$_2$), 3.43 (sp, J = 6.6 Hz, 1H, CH(CH$_3$)$_2$), 2.42–1.98 (m, 4H, 3,4-CH$_2$-furyl), 1.70 (s, 6H, C(CH$_3$)$_2$), 1.69 (d, J = 6.6 Hz, 6H, CH(CH$_3$)$_2$). ^{13}C NMR (75 MHz, DMSO/CCl$_4$, 1/3) δ 162.5, 159.1, 156.6, 154.1, 153.7, 141.1, 122.9, 122.3, 112.9, 76.7, 68.6, 66.8, 59.6, 44.1, 36.7, 30.3, 28.5, 26.1, 25.0, 21.3. Anal. calcd for C$_{22}$H$_{28}$N$_4$O$_2$S: C 64.05; H 6.84; N 13.58%. Found: C 64.41; H 7.04; N 13.81%.

3.3. General Method for the Preparation of Compounds 8 and 9

A mixture of compound **7j** or **7k** (5 mmol) and phosphorus oxychloride (30 mL) was refluxed for 4 h. The excess of phosphorus oxychloride was distilled off to give a dry residue. Ice water was added, and then the mixture was treated with potassium hydroxide solution (pH = 8–9). The separated crystals were filtered off, washed with water, dried, and recrystallized from ethanol.

11-Isobutyl-8,8-dimethyl-2,3,7,10-tetrahydro-8H-imidazo[1,2-c]pyrano[4″,3″:4′,5′]pyrido[3′,2′:4,5]furo[2,3-e] pyrimidine (**8**): Light-yellow solid, yield 77%, m.p. 270-272 °C. ^{1}H-NMR (300 MHz, DMSO/CCl$_4$, 1/3) δ 8.00 (s, 1H, 5-CH), 4.77 (s, 2H, OCH$_2$), 4.27–4.18 (m, 2H, 2-CH$_2$), 4.09–4.00 (m, 2H, 3-CH$_2$), 3.20 (s, 2H, CH$_2$), 2.54 (d, J = 7.1 Hz, 2H, CHCH$_2$), 2.34–2.19 (m, 1H, CH(CH$_3$)$_2$), 1.31 (s, 6H, C(CH$_3$)$_2$), 0.98 (d, 6H, J = 6.6 Hz, CH(CH$_3$)$_2$). ^{13}C NMR (75 MHz, DMSO/CCl$_4$, 1/3) δ 160.1, 155.1, 145.5, 144.7, 139.1, 137.9,

134.0, 123.6, 110.5, 68.6, 60.0, 53.2, 46.3, 41.9, 36.5, 27.4, 25.8, 22.2. Anal. calcd for $C_{20}H_{24}N_4O_2$: C 68.16; H 6.86; N 15.90%. Found: C 68.53; H 7.05; N 16.14%.

12-Isobutyl-9,9-dimethyl-3,4,8,11-tetrahydro-2H,9H-pyrano[4″,3″:4′,5′]pyrido[3′,2′:4,5]furo[2,3-e]pyrimido [1,2-c]pyrimidine (**9**): Colorless solid, yield 71%, m.p. 282–284 °C. ^{1}H-NMR (300 MHz, DMSO/CCl$_4$, 1/3) δ 7.60 (s, 1H, 6-CH), 4.76 (s, 2H, OCH$_2$), 4.07–4.01 (m, 2H, 2-CH$_2$), 3.56 (t, *J* = 5.6 Hz, 2H, 4-CH$_2$), 3.18 (s, 2H, CH$_2$), 2.53 (d, *J* = 7.1 Hz, 2H, CHC<u>H</u>$_2$), 2.34–2.19 (m, 1H, C<u>H</u>(CH$_3$)$_2$), 2.06–1.96 (m, 2H, 3-CH$_2$), 1.31 (s, 6H, C(CH$_3$)$_2$), 0.98 (d, 6H, *J* = 6.6 Hz, CH(C<u>H</u>$_3$)$_2$). ^{13}C NMR (75 MHz, DMSO/CCl$_4$, 1/3) δ 149.6, 144.1, 137.1, 128.4, 128.1, 127.8, 123.9, 113.2, 100.9, 58.6, 50.0, 35.9, 33.2, 31.9, 26.4, 17.4, 15.8, 12.1, 10.0. Anal. calcd for $C_{21}H_{26}N_4O_2$: C 68.83; H 7.15; N 15.29%. Found: C 69.18; H 7.36; N 15.56%.

3.4. Cell Viability Assay

Vero and HeLa cells were seeded in a 96-well culture plate at a density of 20,000 cells/well. The plates were preincubated in a 5% CO$_2$/95% air-humidified atmosphere at 37 °C for 24 h to allow for adaptation of the cells prior to the addition of the test compounds. All compounds were dissolved in dimethyl sulfoxide (DMSO) prior to dilution (1 mg/mL stock solution). The 50% inhibitory concentration (IC$_{50}$) was determined over a range of concentrations (0.5 μg/mL to 10 μg/mL). All cells were incubated for 72 h, and then cell viability was estimated using an MTT (3-(4,5-dimethylthiazol-2-yl)-2,5-diphenyltetrazolium bromide) assay, as previously described [32]. Colorimetric measurements were performed on a microplate reader at 570 nm (Tecan Spectra II, Switzerland). The percentage of viable cells was calculated for each concentration as $((OD_T/OD_C) \times 100)$, where OD_T and OD_C corresponded to the absorbance of the treated and control cells, respectively. The IC$_{50}$ was calculated by using linear regression analysis.

3.5. Antitumor Assay

The influence of synthesized compounds on the methylation of DNA of tumor sarcoma 180 (S-180) was studied in vitro. Transplanted S-180 was isolated from animals after slaughter by decapitation under ether narcosis. A 3×10^{-6} M solution of compounds (solvent: carboxymethylcellulose) was added to the tumor homogenate (12.5 mL of a solution per 10 g of tumor), and the mixture was incubated for 24 h at 37 °C. After the incubation, tumor DNA was extracted using the phenol–chloroform method [33]. Hydrolysis to the nitrogen bases was carried out in sealed glass ampoules in 85% formic acid at 176 °C for 1 h (0.1 mL of acid per 1 mg of DNA). The separation of nitrogen bases (guanine (G), cytosine (C), 5-methylcitosine (5-MC), adenine (A), and thymine (T)) was produced by thin-layer chromatography in the solvent *n*-butanol/water/ammonia. Spectrophotometry of the eluates of all bases was made against eluates from the respective control areas of the chromatograms. In our studies, the well-known antitumor drug doxorubicin (manufactured by EBEWE Pharma Ges.m.b.H. Nfg. KG, 4866 Unterach, Austria) was used as a standard preparation. The data were statistically processed using Student's and Fischer's tests ($p < 0.05$).

4. Conclusions

Through a multistage organic synthesis, new tetraheterocyclic compounds were obtained, and their antitumor activity was investigated. The results of the biological investigations revealed that all new synthesized amino derivatives of the pyrido[3′,2′:4,5]furo(thieno)[3,2-d]pyrimidines (with the exception of compounds **7a,i,j**) showed DNA demethylation activity, and three of them (**5b,f** and **7b**) showed pronounced antitumor activity. Moreover, several amino derivatives of two new five-membered heterocyclic systems, imidazo[1,2-c]pyrimidine and pyrimido[1,2-c]pyrimidine, were synthesized.

Supplementary Materials: The copies of ^{1}H-NMR and ^{13}C-NMR spectra for all new synthesized compounds were submitted along with the manuscript.

Author Contributions: Conceptualization: S.N.S. designed the experiments and wrote the chemical part of manuscript text. A.A.H. performed the experiments on the synthesis of all compounds and analyzed the results. E.K.H. helped conduct the literature data and participated in the synthesis. E.A. performed the cell viability assay. H.Z. analyzed the data of antiproliferative activity and wrote the corresponding part of manuscript text. I.S.D. and A.S.A. performed the antitumor assay and made data curation. L.E.N. worked with the literature data and analyzed the results of antitumor activity. R.E.M. wrote the part of antitumor activity of manuscript text. H.S. performed the docking analysis. D.S. reviewed the manuscript. A.G. edited the manuscript.

Funding: This research received no external funding.

Conflicts of Interest: The authors declare no conflicts of interest.

References

1. Vanyushin, B.F. Enzymatic DNA methylation: What it is needed for in the cell. *JSM Genet. Genom.* **2016**, *3*, 1010–1014.

2. Vanyushin, B.F. Enzymatic DNA methylation is an epigenetic control for genetic functions of the cell. *Biochemistry* **2005**, *70*, 488–499. [CrossRef] [PubMed]

3. Worm, J.; Guldberg, P. DNA methylation: An epigenetic pathway to cancer and a promising target for anticancer therapy. *J. Oral Pathol. Med.* **2002**, *31*, 443–449. [CrossRef] [PubMed]

4. Bartosik, M.; Ondrouskova, E. Novel approaches in DNA methylation studies–MS-HRM analysis and electrochemistry. *Klin. Onkol.* **2016**, *29*, 64–71. [CrossRef]

5. Mikeska, T.; Craig, J.M. DNA methylation biomarkers: Cancer and beyond. *Genes* **2014**, *5*, 821–864. [CrossRef]

6. Beltran-Garcia, J.; Osca-Verdegal, R.; Mena-Molla, S.; Garcia-Gimenez, J.L. Epigenetic IVD tests for personalized precision medicine in cancer. *Front. Genet.* **2019**, *10*, 1–14. [CrossRef]

7. Howell, P.M.; Liu, Z.; Khong, H.T. Demethylating agents in the treatment of cancer. *Pharmaceuticals* **2010**, *3*, 2022–2044. [CrossRef]

8. Hayakawa, M.; Kaizawa, H.; Moritomo, H.; Koizumi, T.; Ohishi, T.; Yamano, M.; Okada, M.; Ohta, M.; Tsukamoto, S.; Raynaud, F.I.; et al. Synthesis and biological evaluation of pyrido[3′,2′:4,5]furo[3,2-*d*] pyrimidine derivatives as novel PI3 kinase p110a inhibitors. *Bioorganic Med. Chem. Lett.* **2007**, *17*, 2438–2442. [CrossRef]

9. Taltavull, J.; Serrat, J.; Gracia, J.; Gavalda, A.; Andres, M.; Cordoba, M.; Miralpeix, M.; Vilella, D.; Beleta, J.; Ryder, H.; et al. Synthesis and biological activity of pyrido[3′,2′:4,5]thieno[3,2-*d*]pyrimidines as phosphodiesterase type 4 Inhibitors. *J. Med. Chem.* **2010**, *53*, 6912–6922. [CrossRef]

10. Agarwal, A.; Louise-May, S.; Thanassi, J.A.; Podos, S.D.; Cheng, J.; Thoma, C.; Liu, C.; Wiles, J.A.; Nelson, D.M.; Phadke, A.S.; et al. Small molecule inhibitors of *E. coli* primase, a novel bacterial target. *Bioorg. Med. Chem. Lett.* **2007**, *17*, 2807–2810. [CrossRef]

11. Kjellerup, L.; Gordon, S.; Cohrt, K.O.; Brown, W.D.; Fuglsang, A.T.; Winther, A.-M.L. Identification of antifungal H^{+}-ATPase inhibitors with effect on the plasma membrane potential. *Antimicrob. Agents Chemother.* **2017**, *61*, 1–14. [CrossRef] [PubMed]

12. Sirakanyan, S.N.; Ovakimyan, A.A.; Noravyan, A.S.; Dzhagatspanyan, I.A.; Shakhatuni, A.A.; Nazaryan, I.M.; Hakopyan, A.G. Synthesis and antyconvulsive activity of 8-amino substidutet cyclopenta[4′,5′] pyrido[3′,2′:4,5]thieno[3,2-*d*]pyrimidines. *Pharm. Chem. J.* **2013**, *47*, 130–134. [CrossRef]

13. Sirakanyan, S.N.; Ovakimyan, A.A.; Noravyan, A.S.; Minasyan, N.S.; Dzhagatspanyan, I.A.; Nazaryan, I.M.; Hakopyan, A.G. Synthesis and neurotropic activity of 8-amino derivatives of condensed thieno[3,2-*d*]- and furo[3,2-*d*]pyrimidines. *Pharm. Chem. J.* **2014**, *47*, 655–659. [CrossRef]

14. Sirakanyan, S.N.; Akopyan, E.K.; Paronikyan, R.G.; Akopyan, A.G.; Ovakimyan, A.A. Synthesis and anticonvulsant activity of 7(8)-amino derivatives of condensed thieno[3,2-*d*]pyrimidines. *Pharm. Chem. J.* **2016**, *50*, 296–300. [CrossRef]

15. Sirakanyan, S.N.; Geronikaki, A.; Spinelli, D.; Hakobyan, E.K.; Kartsev, V.G.; Petrou, A.; Hovakimyan, A.A. Synthesis and antimicrobial activity of new amino derivatives of pyrano[4″,3″:4′,5′]pyrido[3′,2′:4,5] thieno[3,2-d]pyrimidine. *An. Acad. Bras. Cienc.* **2018**, *90*, 1043–1057. [CrossRef]

16. Sirakanyan, S.N.; Spinelli, D.; Geronikaki, A.; Kartsev, V.G.; Hakobyan, E.K.; Hovakimyan, A.A. Synthesis and antimicrobial activity of new derivatives of pyrano[4″,3″:4′,5′]pyrido[3′,2′:4,5]thieno[3,2-*d*]pyrimidine and new heterocyclic systems. *Synth. Commun.* **2019**, *49*, 1262–1276. [CrossRef]

17. Sirakanyan, S.N.; Paronikyan, E.G.; Noravyan, A.S. *The Chemistry and Biological Activity of Nitrogen-containing Heterocycles and Alkaloids*; Iridium Press: Moscow, Russia, 2001; pp. 519–521.

18. Sirakanyan, S.N.; Paronikyan, E.G.; Noravyan, A.S. *The Chemistry and Biological Activity of Synthetic and Natural Compounds*; IBS Press: Moscow, Russia, 2003; pp. 398–401.

19. Paronikyan, E.G.; Sirakanyan, S.N.; Noravyan, A.S.; Paronikyan, R.G.; Dzhagatspanyan, I.A. Synthesis and anticonvulsant activity of pyrazolo[3,4-*b*]pyrano(thiopyrano)[4,3-*d*]pyridine and pyrazolo[3,4-*c*]isoquinoline derivatives. *Pharm. Chem. J.* **2001**, *35*, 8–10. [CrossRef]

20. Sirakanyan, S.N.; Paronikyan, E.G.; Ghukasyan, M.S.; Noravyan, A.S. Synthesis of 8-amino derivatives of condensed furo[3,2-d]pyrimidines. *Chem. Heterocycl. Compd.* **2010**, *46*, 736–741. [CrossRef]

21. Brown, R.; Plumb, J.A. Demethylation of DNA by decitabine in cancer chemotherapy. *Expert Rev. Anticancer Ther.* **2004**, *4*, 501–510. [CrossRef]

22. Fahy, J.; Jeltsch, A.; Arimondo, P.B. DNA methyltransferase inhibitors in cancer: A chemical therapeutic patent overview and selected clinical studies. *Expert Opin. Ther. Pat.* **2012**, *22*, 1427–1442. [CrossRef]

23. Christman, J.K.; Sheikhnejad, G.; Marasco, C.J.; Sufrin, J.R. 5-Methyl-2′-deoxycytidine in single-stranded DNA can act in cis to signal de novo DNA methylation. *Proc. Natl. Acad. Sci. USA* **1995**, *92*, 7347–7351. [CrossRef] [PubMed]

24. Gowher, H.; Jeltsch, A. Mammalian DNA methyltransferases: New discoveries and open questions. *Biochem. Soc. Trans.* **2018**, *46*, 1191–1202. [CrossRef] [PubMed]

25. Xie, T.; Yu, J.; Fu, W.; Wang, Z.; Xu, L.; Chang, S.; Wang, E.; Zhu, F.; Zeng, S.; Kang, Y.; et al. Insight into the selective binding mechanism of DNMT1 and DNMT3A inhibitors: A molecular simulation study. *Phys. Chem. Chem. Phys.* **2019**, *21*, 12931–12947. [CrossRef] [PubMed]

26. Ren, W.; Gao, L.; Song, J. Structural basis of DNMT1 and DNMT3A-mediated DNA methylation. *Genes* **2018**, *9*, 620. [CrossRef] [PubMed]

27. Callebaut, I.; Courvalin, J.C.; Mornon, J.P. The BAH (bromo-adjacent homology) domain: A link between DNA methylation, replication and transcriptional regulation. *FEBS Lett.* **1999**, *446*, 189–193. [CrossRef]

28. Abagyan, R.; Totrov, M.; Kuznetsov, D. ICM–A new method for protein modeling and design: Applications to docking and structure prediction from the distorted native conformation. *J. Comput. Chem.* **1994**, *15*, 488–506. [CrossRef]

29. Bottegoni, G.; Kufareva, I.; Totrov, M.; Abagyan, R. Four-dimensional docking: A fast and accurate account of discrete receptor flexibility in ligand docking. *J. Med. Chem.* **2009**, *52*, 397–406. [CrossRef]

30. Nicola, G.; Smith, C.A.; Abagyan, R. New method for the assessment of all drug-like pockets across a structural genome. *J. Comput. Biol.* **2008**, *15*, 231–240. [CrossRef]

31. Schapira, M.; Totrov, M.; Abagyan, R. Prediction of the binding energy for small molecules, peptides and proteins. *J. Mol. Recognit.* **1999**, *12*, 177–190. [CrossRef]

32. Arabyan, E.; Hakobyan, A.; Kotsinyan, A.; Karalyan, Z.; Arakelov, V.; Arakelov, G.; Nazaryan, K.; Simonyan, A.; Aroutiounian, R.; Ferreira, F.; et al. Genistein inhibits African swine fever virus replication in vitro by disrupting viral DNA synthesis. *Antivir. Res.* **2018**, *156*, 128–137. [CrossRef]

33. Vanyushin, B.F.; Masin, F.N.; Vasiliev, V.R.; Belozersky, A.N. The content of 5-methylcytosine in animal DNA: The species and tissue specifity. *Biochim. Biophys. Acta* **1973**, *299*, 397–403. [CrossRef]

Sample Availability: Samples of the compounds **1–9** are available from the authors.

Article

4,5-Diaryl 3(2*H*)Furanones: Anti-Inflammatory Activity and Influence on Cancer Growth

Dmitrii Semenok [1,2], Jury Medvedev [3], Lefki-P. Giassafaki [4], Iason Lavdas [4], Ioannis S. Vizirianakis [4], Phaedra Eleftheriou [5], Antonis Gavalas [6], Anthi Petrou [6] and Athina Geronikaki [6,*]

[1] Skolkovo Institute of Science and Technology, Skolkovo Innovation Center, 3 Nobel Street, 143026 Moscow, Russia; dmitrii.semenok@skolkovotech.ru

[2] Moscow Institute of Physics and Technology, 9 Institutsky lane, 141700 Dolgoprudny, Russia

[3] Saint-Petersburg State University, Institute of Chemistry, Universitetskiy Prospekt, 26, 198504 Petergof, Russia; j2j3@yandex.ru

[4] Department of Pharmacology, School of Pharmacy, Aristotle University of Thessaloniki, 54124 Thessaloniki, Greece; lefkigiassafaki@gmail.com (L.-P.G.); iasonlavdas@pharm.auth.gr (I.L.); ivizir@pharm.auth.gr (I.S.V.)

[5] Department of Medical Laboratory Studies, School of Health and Medical Care, Alexander Technological Educational Institute of Thessaloniki, 57400 Thessaloniki, Greece; eleftheriouphaedra@gmail.com

[6] Department of Pharmaceutical Chemistry, School of Pharmacy, Aristotle University of Thessaloniki, 54124 Thessaloniki, Greece; agavalas@pharm.auth.gr (A.G.); anthi.petrou.thessaloniki1@gmail.com (A.P.)

* Correspondence: geronik@pharm.auth.gr; Tel.: +30-2310997616

Received: 1 April 2019; Accepted: 30 April 2019; Published: 6 May 2019

Abstract: Apart from their anti-inflammatory action, COX inhibitors have gathered the interest of many scientists due to their potential use for the treatment and prevention of cancer. It has been shown that cyclooxygenase inhibitors restrict cancer cell growth and are able to interact with known antitumor drugs, enhancing their in vitro and in vivo cytotoxicity. The permutation of hydrophilic and hydrophobic aryl groups in COX inhibitors leads to cardinal changes in the biological activity of the compounds. In the present study, thirteen heterocyclic coxib-like 4,5-diarylfuran-3(2*H*)-ones and their annelated derivatives—phenanthro[9,10-b]furan-3-ones—were synthesized and studied for anti-inflammatory and COX-1/2 inhibitory action and for their cytotoxic activity on the breast cancer (MCF-7) and squamous cell carcinoma (HSC-3) cell lines. The F-derivative of the –SOMe substituted furan-3(2*H*)-ones exhibited the best activity (COX-1 IC_{50} = 2.8 μM, anti-inflammatory activity (by carrageenan paw edema model) of 54% (dose 0.01 mmol/kg), and MCF-7 and HSC-3 cytotoxicity with IC_{50} values of 10 μM and 7.5 μM, respectively). A cytotoxic effect related to the COX-1 inhibitory action was observed and a synergistic effect with the anti-neoplastic drugs gefitinib and 5-fluorouracil was found. A phenanthrene derivative exhibited the best synergistic effect with gefitinib.

Keywords: 3(2*H*)furanones; phenanthro[9,10-b]furanones; cyclooxygenase; cytotoxicity; anticancer; MCF-7; HSC-3; gefitinib; 5-fluorouracil

---

1. Introduction

Prostaglandins produced during inflammation are responsible for most of the undesirable effects of the inflammatory process and cyclooxygenase isoenzymes—COX-1 and COX-2—involved in their biosynthesis are the main targets of most anti-inflammatory drugs. As COX-2 is the isoenzyme induced during inflammation, COX-2-specific inhibitors have been developed during the last decades. Since 1999, when the first selective cyclooxygenase-2 (COX-2) inhibitors rofecoxib [1] and celecoxib [2] were launched, numerous research studies aiming to find effective and safe COX-2-associated anti-inflammatory drugs were performed. Despite the fact that in 2004 rofecoxib was withdrawn from the market due to an increased risk of cardiovascular events [3], an intensive study of properties and possible applications of coxibs continued for more than 20 years.

Currently, COX inhibitors including even rofecoxib [4] are of renewed interest due to their potential use for the treatment and prevention of cancer.

Cancer cells present cellular and genomic heterogeneity that contribute to mechanisms leading to drug resistance development and to therapy failure in clinical practice [5,6]. This heterogeneity is also modulated through interactions between malignant and normal cells that support the tumor microenvironment. Among the various hallmarks of cancer cells development, inflammation is a critical component of tumor survival, growth, and progression [7,8]. As a result, the development of selective anti-inflammatory drugs represents a promising target in cancer therapy. Inhibition of prostaglandin production by inactivation of the cyclooxygenase isoenzymes has been related in various ways with prevention of cancer cell growth, adhesion, migration and invasion [4,9–16]. As shown in the case of BrafV600E mouse melanoma [17], prostaglandin produced from the tumor cells may suppress immune response

Among the two COX isoenzymes, COX-2 was mostly related with tumor development. The antiapoptotic role of COX-2 activity in chemoresistant cancer cells was noted in a study concerning drug resistance in patients with acute myeloid leukemia during standard treatment [18]. COX-2 is overexpressed in most cancer cells and its activity is decreased as a result of effective antitumor therapy [19,20]. In addition, COX-2 inhibition was found to reduce cancer cell development [16]. On the other hand, COX-1 is mainly expressed in certain tumor cells and cancer cell lines, such as the breast cancer MCF-7 cell line. Interestingly, transfected MCF-7 cells with more aggressive properties, overexpressed COX-2 isoenzyme [21]. Overexpression of COX-1 at the first phase of tumorigenesis, followed by overexpression of COX-2 at a later phase has been supported by certain researchers [19]. Although, most researches concerned mainly COX-2 inhibitors, COX-1 inhibition also may have application in anticancer therapy [19].

Most interestingly, COX inhibitors may act synergistically with known antitumor drugs, enhancing their in vitro and in vivo cytotoxicity [10,22–24]. It was found that celecoxib in combination with 5-fluorouracil (5-FU) increased the effectiveness of the latter against esophageal squamous cell carcinoma [25]. These synergistic effects motivated recent clinical trials of combination therapy of tumors, for the treatment of duodenal neoplasia (sulindac/erlotinib) [26] and breast cancer (etodolac) [27]. Moreover, combinations of aspirin or celecoxib with an anti-PD-1 monoclonal antibody resulted in a 1.5–2.1 fold decrease in melanoma growth rate in vivo [28].

Previous studies by S. Shin and coworkers showed that regioisomeric 5-aryl-2,2-dialkyl-4-phenylfuran-3($2H$)-one derivatives (Figure 1A) have IC_{50} values of COX-2 inhibitory action comparable to that of rofecoxib [29,30]. Taking all these into account we designed and synthesized a number of novel 4,5-diarylfuran-3($2H$)-one and phenanthro[9,10-b]furan-3-one derivatives (Figure 1B) with the aim of obtaining novel COX inhibitors. The anti-inflammatory and antitumor effects, as well as synergistic effects of the compounds, were tested and the effects of different aryl moieties in the total yield and in the activities of the compounds are discussed.

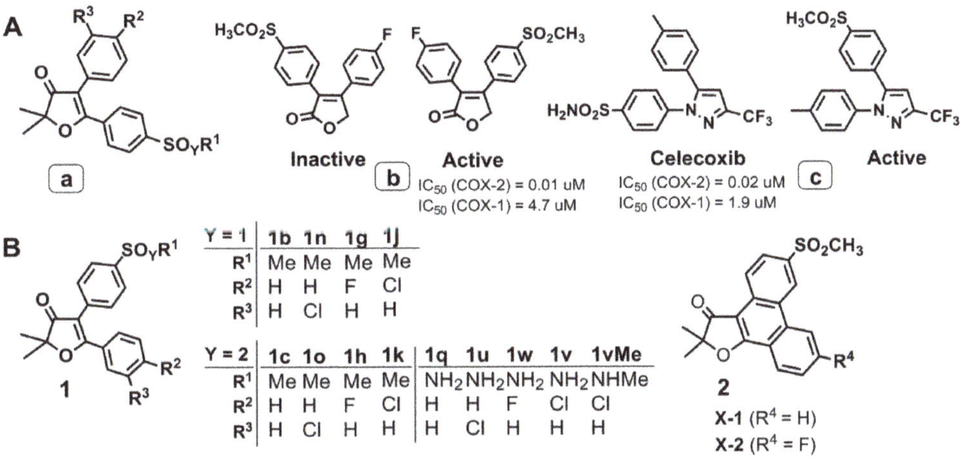

Figure 1. (**A**) Known 3(*2H*) furanones (a) [19,28], rofecoxib analogs (b), celecoxib and its regioisomer active analog (c) [29]. (**B**) Novel 3(*2H*)furanone (**1**) and phenanthro[9,10-b]furan-3-one (**2**) derivatives.

2. Results and Discussion

2.1. Design and Computer-Aided Prediction of COX-1/2 Inhibitory Activity

Since COX-2 is the main isoenzyme overexpressed in inflammation and mostly related to cancer development, COX-2 inhibition was among the desired properties of the designed compounds. The design and selection of the compounds was performed using a computer-aided approach.

Crucial structural differences between the shape of the active site of COX-1 and COX-2 are responsible for the effectiveness of COX-2 selective inhibitors. The replacement of isoleucines at position 523 and 434 of COX-1 isoenzyme by the smaller valine in COX-2 leads to the creation of a secondary pocket in the COX-2 active site [31]. The scaffold of cis-1,2-diaryl-alkene can be easily placed in the larger cavity of COX-2 isoenzyme but not in COX-1, so this scaffold should be suitable for selective inhibition of COX-2 [32,33]. Since, a large number of biological studies of different classes of human cyclooxygenase inhibitors have been carried out [34] and a considerable array of experimental data on COX-1/2 inhibition accumulated, we attempted to analyze the expected inhibitory activity of the designed compounds **1–2** by molecular docking analysis before the experimental evaluation of the compounds.

As the novel compounds were designed to have structural similarity to the coxib family, protein structures of the COX-1 and COX-2 enzymes in complex with celecoxib were selected for Docking Analysis. The *Ovis aries* COX-1 and 3KK6 [35] and *Mus musculus* COX-2 and 3LN1 [36] were obtained from the Protein Data Base (PDB). However, since both structures were not human, the 3D structures of the human isoforms were constructed using these structures as templates. The developed structures were used to predict the ability of thirteen known COX-1/2 inhibitors (E1–E14, Figure S1) [37] to form stable complexes with enzymes using docking analysis. The binding energy was predicted for each of the compounds E1–E14 and their correlation with the experimentally calculated IC_{50} value of inhibitory action was calculated.

Linear correlation was obtained with $R^2 = 0.791$ for COX-1 and $R^2 = 0.704$ for COX-2 (Figure 2).

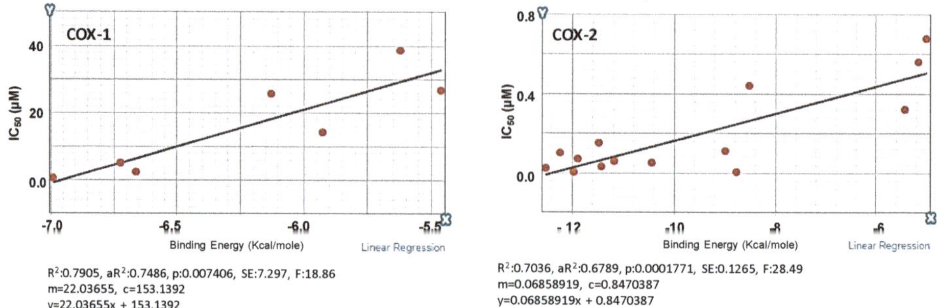

R²:0.7905, aR²:0.7486, p:0.007406, SE:7.297, F:18.86
m=22.03655, c=153.1392
y=22.03655x + 153.1392

R²:0.7036, aR²:0.6789, p:0.0001771, SE:0.1265, F:28.49
m=0.06858919, c=0.8470387
y=0.06858919x + 0.8470387

Figure 2. Correlation of the predicted binding energy in kcal/mole with the experimentally calculated IC_{50} value of inhibitory action of the reference compounds E1–E14. against COX-1 (**left**) and COX-2 (**right**).

The same enzyme structures were then used for the calculation of the binding energy for the designed compounds (Figure 1). Based on the calculated binding energies and the correlations obtained from the reference test set, an estimation of the expected inhibitory action of the designed compounds was made. The results are shown in Table 1.

Table 1. Predicted IC_{50} values of the designed compounds.

Structure	Code	R	COX-1 Binding Energy (Kcal/mole)	COX-1 IC_{50} (μM)	COX-2 Binding Energy (Kcal/mole)	COX-2 IC_{50} (μM)	S *
SO₂Me	1c	H	−5.78	25.8	−6.67	0.39	66.2
	1o	*m*–Cl	−6.18	17.0	−10.21	0.15	113.3
	1h	*p*–F	−5.27	37.0	−7.17	0.36	102.8
	1k	*p*–Cl	−4.96	43.8	−7.32	0.35	125.1
SOMe	1b	H	−5.68	27.0	−6.38	0.41	65.9
	1n	*m*–Cl	−5.97	21.6	−7.60	0.33	65.5
	1g	*p*–F	−5.94	22.2	−7.61	0.33	67.3
	1j	*p*–Cl	−6.44	11.2	−6.93	0.37	30.3
SO₂NH₂	1q	H	−6.48	10.3	−9.13	0.22	46.8
	1v	*p*–Cl	−6.19	16.7	−9.99	0.16	1–4.4
	1w	*p*–F	−5.83	24.7	−9.61	0.19	130.0
Celecoxib				14(exp)		0.04(exp)	
SO₂Me	x-1	H	−4.92	44.7	−7.28	0.28	159.6
	x-2	F	−5.23	37.9	−8.86	0.24	157.9

* selectivity.

According to the results, all compounds were expected to be COX-2 selective inhibitors with high inhibitory action. As clearly shown in Table 2, compound **1o** was predicted to be the most potent COX-2 inhibitor with the best selectivity, with compound **1j** being the best ligand towards

COX-1 isoform. Based on these estimations, it was decided to proceed with the synthesis of all the designed compounds.

2.2. Chemistry

Synthesis

Synthesis of 4,5-diarylfuran-3(2*H*)-ones and their annelated derivatives involved 5–7 stages [38,39]. The preliminary step of the synthesis was Friedel–Crafts reaction between thioanisole and benzoyl chloride in CH_2Cl_2, catalyzed by anhydrous aluminium chloride to furnish the 80–90% yield of ketone **7** (see Supporting Information, Scheme S1, and Table S1). The first step involved nucleophilic addition of the Grignard reagent derived from 2-methyl-but-3-yn-2-ol to ketone **7** to produce diol **6** in yield up to 98% (Table 2, Scheme 1).

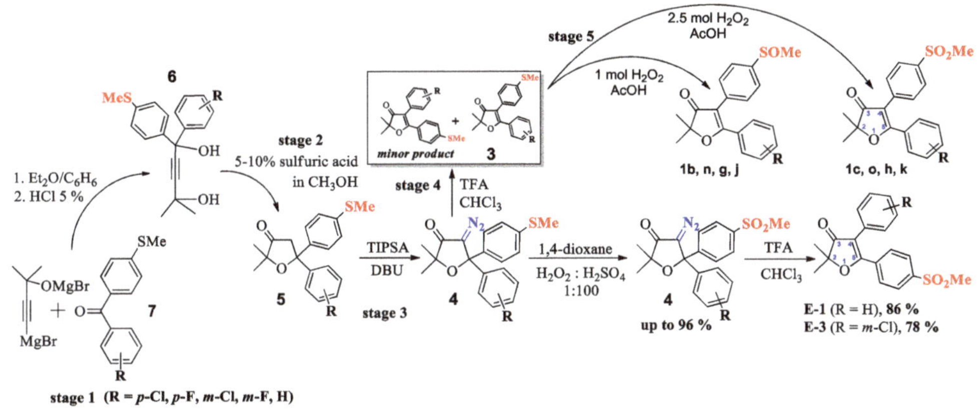

Scheme 1. Synthesis of methylsulfones and methylsulfoxides of 4,5-diarylfuran-3(2*H*)-ones (**1**) as well as reference compounds **E-1** and **E-3**. Oxidative reactions require a catalyst—sulfuric acid.

At the second stage, diol **6** was transformed into corresponding monoketone **5** by intramolecular cyclization in 5–10% sulfuric acid/methanol solution and 2,2-dimethyl-5,5-diaryldihydrofuran-3(2*H*)-one **5** formed can be used without purification on the next stage.

The third step (Scheme 1) was transformation of monoketone **5** into diazoketone **4** (racemic mixture) by diazo transfer reaction with DBU as base. Acid-catalyzed decomposition of diazoketones **4** [38,39] yielded a mixture of corresponding regioisomeric furan-3(2*H*)-ones **3** with excellent total yields of 95 to 99% (stage 4). The ratio of regioisomers strongly depends on substitution of benzene rings: predominantly migrates donor-substituted aryl [40]. Regioisomeric products of migration can be distinguished by position and intensity HNMR signals of aryl protons, e.g., by comparison of downfield doublets of *o*-protons in *p*-XPh (X = Cl, F, SO$_n$Me), which can be compared directly for both regioisomers. HNMR spectra of different regioisomers were previously published [41–43]. Complete oxidation of thiomethyl group by hydrogen peroxide required a catalyst—sulfuric acid. Compounds E-1 and E-3 (Scheme 1) were prepared by oxidation of thiomethyl group in the minor products of decomposition of corresponding diazocompounds (**4**).

The yields of desired products are presented in Tables 2 and 3 (stage-wise) and Table 4 (total).

Table 2. Yields (%) of reactions on different steps of synthesis of sulfoxides and sulfones (Scheme 1).

R	1,4-diol (6)	Dihydro furan-3-one (5)	4-diazodi- hydrofuran-3-one (4)	4,5-diaryl-3(2*H*) furanones (2)	Sulfoxides 1b,n,g,j	Sulfones 1c,o,h,k
m-Cl	57 (65) [a]	90	83-87	83% (94; 5.2:1) [c]	92%	93%
m-F	25 (33) [a]	31	52 (71) [b]	66% (75; 7.3:1) [c]	-	-
p-Cl	23 (28) [a]	91	70	65% (86; 3.2:1) [c]	89%	98%
p-F	52 (64) [a]	84	86	77% (99; 3.5:1) [c]	94%	80%
H	86–93	93	71	75% (99; 3.2:1) [c]	84%	96%

[a] The yield in brackets is given taking into account incomplete conversion of benzophenone. [b] The yield in brackets is given taking into account incomplete conversion of furanone. [c] In brackets the yield of both products (stage 4) and their ratio are given.

Another branch of Scheme 1(stage 5) led to 5-(4′-(methylsulfonyl or sulfinyl)phenyl)-4-aryl-furan-3(2*H*)-ones obtained after full or partial oxidation of key intermediate 5-(4′-(methylthio)-phenyl)-4-arylfuran-3(2*H*)-ones purified by recrystallization. It should be noted that the mixture of regioisomeric furan-3(2*H*)-ones formed on the 1,2-aryl migration stage could not be easily separated by chromatography.

Synthesis of sulfonamides (Scheme 2) differs from the synthesis of sulfones and sulfoxides starting from stage 5. The mixture of regioisomeric diazoketone **4** decomposition products (stage 4, Scheme 2) without separation was subjected to chlorosulfonation with 10-fold excess of chlorosulfonic acid for 24 to 48 h at r.t. Surprisingly, only one of the regioisomers was capable of chlorosulfonation in the *para*-position of the unsubstituted 4-Ph ring. It is interesting to note that the standard Friedel–Crafts substitution in this ring did not occur with any of typical catalysts. As a result of this highly selective reaction, a mixture of the desired sulfonyl chloride and one of the migration products was formed after hydrolysis of the excess chlorosulfonic acid with ice. However, their separation was still a difficult task. Therefore, the mixture without further purification was treated with NH_3 solution in water/THF.

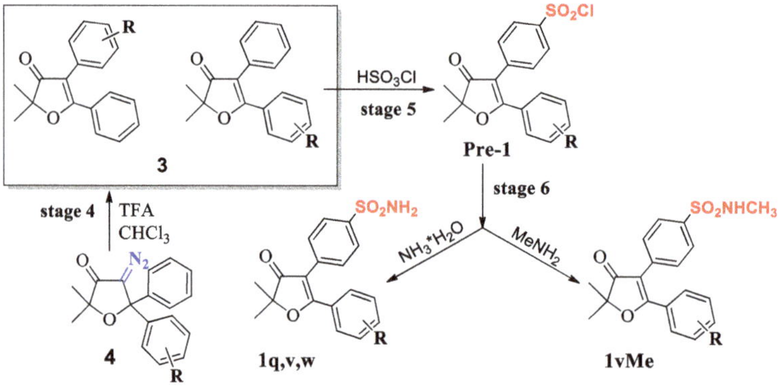

Scheme 2. Synthesis of 4,5-diarylfuran-3(2*H*)-ones sulfonamides **1**.

Table 3. Yields (%) of reactions on different steps of sulfonyl amide synthesis (Scheme 2).

R	1,4-diol (6)	dihydrofuran-3- one (5)	4-diazodihyd- rofuran-3-one (4)	4,5-diaryl-3(2*H*) furanones (3)	Sulfonyl Chloride	Sulfonyl amide (1q,1u, 1w,1v)
m-Cl	95	98	85	67 (95;2.45:1) [a]	45 [b]	79
p-Cl	96	98	87	55 (99; 1.25:1) [a]	50 [b]	71
p-F	98	96	86	47 (99; 1:1.12) [a]	62 [b]	72
H	90	95	69	99 (one product)	50	74

[a] In brackets, the yield of both 1,2-migration products and their ratio are given. [b] As a percentage of the total weight of the mixture **3**.

As a result of the conversion of the SO_2Cl group to SO_2NH_2, the polarity of the compound changed sharply and sulfonamide obtained was easily separated chromatographically from the regioisomeric diazoketone 4 decomposition product. Comparison of NMR spectra of the products and known structures of 1,2-migration products allows the unambiguously determination the structure of sulfonamides obtained.

Synthesis of phenanthro[9,10-b]furanones (Scheme 3) was carried out using key intermediate **3**, which was subjected to irradiation with hard ultraviolet light (2–3 h) in dilute solution in hexane, accompanied by intramolecular oxidative cyclization of the stilbene fragment [44]. Despite low conversion (~15%) the yield of product was increased up to 25% by respective chromatographic separation of phenanthrofuranones and irradiating the depleted reaction mixture again (see Supporting Information).

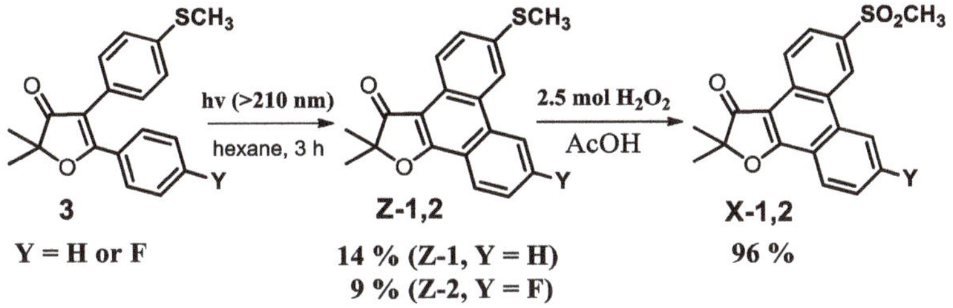

Scheme 3. Synthesis of phenanthro[9,10-b]furanones (**2**) by photochemical cyclization of the stilbene moiety of 4,5-diaryl-3(*2H*)furanones **1**.

Table 4. Total yield (%) of desired 4,5-diarylfuran-3(*2H*)-ones **1–2** starting from benzophenones.

R	Sulfones (SO₂Me) (1c,1o,1h,1k)	Sulfoxides (SOMe) (1b,1n,1g,1j)	Sulfonamides (SO₂NH₂) (1q,1u,1w,1v)	Phenanthrenes
m-Cl	33	33	27	-
p-Cl	9	8	29	-
p-F	23	27	36	3.1 (**X-2**)
H	43	37	23	6.0 (**X-1**)

The main difficulties in the synthesis arose for *p*-Cl, and, especially, for *m*-F-substituted compounds, for which only the ratio of products after 1,2-aryl migration was established.

2.3. Biological Evaluation

2.3.1. In Vitro Evaluation of COX-1/2 Inhibitory Action

The COX-1/2 inhibitory action, mainly of the most active and COX-2 selective inhibitors according to the prediction results, was evaluated in vitro using ovine COX-1 and human recombinant COX-2 isoenzymes included in the COX inhibition assay kit of Cayman as described at the experimental part. Specifically, compound **1o** with the higher predicted COX-2 activity as well as the other two most selective compounds, **1h** and **1k** (predicted selectivity: 100–125), of the –SO₂CH₃ group were tested. The less selective compound **1c** of the same group was also tested for comparison. Furthermore, the most selective, **1g**, and the less selective, **1j**, compounds of – SOCH₃ group as well as the two most selective compounds of the SO₂NH₂ group were also tested. The results are shown in Tables 5 and 6.

Table 5. Inhibition % of COX-1/2 isoenzymes *.

Compound	% Inhibition (COX-1)	% Inhibition (COX-2)
1c (H)	0	0
1h (*p*–F)	32	0
1k (*p*–Cl)	80	6
1j (*p*–Cl)	0	0
1g (*p*–F)	73	0

* Compound concentration 50 μM. Arachidonic acid concentration: 0.1 μM.

Table 6. Predicted Energy and IC_{50} values and in vitro activity of synthesized compounds against COX-1/2[1].

Comp.	Experimental IC_{50}, μM		[2] *Ovis aries* COX-1, 3KK6		[2] *Mus musculus* COX-2, 3LN1		[3] Ovine COX-1 4O1Z	[3] Human COX-2 5IKT
	COX-1	COX-2	E Kcal/mol	IC_{50} (μM)	E Kcal/mol	IC_{50} (μM)	COX-1	COX-2
1k (*p*–Cl)	29	> 50	−4.96	43.8	−7.32	0.345	−5.69	+5.91
1o (*m*–Cl)	22	71	−6.18	17.0	−10.21	0.147	−6.27	−5.49
1g (*p*–F)	**2.8**	> 50	−5.94	22.2	−7.61	0.325	−6.33	−3.50
1v (*p*–Cl)	**28**	**20**	−6.19	16.7	−9.99	0.162	−7.01	−4.36
1w (*p*–F)	70	>50	−5.83	24.7	−9.61	0.188	−4.85	−2.92
x-1							−6.09	−5.87
Naproxen	40	50						−5.66

[1] Methylsulfides ($-SCH_3$) were not tested due to their low solubility in polar solvents like water/dmso mixtures.
[2] Docking analysis using 3D structures of human isoforms, constructed based on the templates of 3KK6 *Ovis aries* COX-1 and 3LN1 and *Mus musculus* COX-2, in complex with Celecoxib. [3] Docking analysis using the human COX-2 structure 5IKT, in complex with tolfenamic acid and ovine COX-1 structure 4O1Z in complex with meloxicam.

Despite the coxib like structure and the prediction results which indicated a preference in COX-2 inhibition, most of the compounds exhibited mainly COX-1 inhibitory action with the exception of compound **1v** which showed a slightly lower IC_{50} value for COX-2 compared to COX-1 isoenzyme (20 μM vs. 28 μM). Compound **1v**, the *p*–Cl derivative of the $-SO_2NH_2$ substituted compounds was the most active COX-2 inhibitor (IC_{50} = 20 μM) followed by the *m*–Cl derivative **1o** (IC_{50} = 71 μM) of the $-SO_2 CH_3$ substituted compounds.

In contrast, compound **1g,** the *p*–F derivative of the –SOMe substituted group was the best COX-1 inhibitor with an IC_{50} value of 2.8 μM followed by two compounds of the $-SO_2 CH_3$ Me substituted group, **1o** and **1k** with IC_{50} values 22 μM and 29 μM respectively and two compounds of the $-SO_2NH_2$ substituted group **1v** and **1w** with IC_{50} values of 28 μM and 70 μM. Polarity of the substituents of both aryl moieties influences the activity as shown by the comparison between the IC_{50} values of the *p*–F derivatives **1g** and **1w** where the less polar $-SO CH_3$ Me group appears to favor inhibitory activity compared to the $-SO_2NH_2$ group. On the other hand, introduction of the less electronegative Cl atom at the *p*-position of the phenyl ring in the $-SO_2NH_2$ derivatives (**1v**) resulted in enhanced inhibition of both isoenzymes compared to the strongly electronegative F-substituted **1w**. It was found that none of the substituents favor COX-2 selective inhibition. However, compound **1v** appears to be a better COX-2 inhibitor than is the reference drug naproxen.

It should be mentioned that in general, the docking analysis failed to predict the inhibitory action of the compounds especially in case of COX-1 as well as their selectivity. This may mean that the human protein 3D structures used for the prediction are not appropriate. Docking analysis using other COX-1 protein structures such as the ovine 1EQH and 2AYL structures co-crystalized in complex with flurbiprofen and the ovine 4O1Z in complex with meloxicam did not show better correlation with the in vitro activity. COX-2 inhibition was better predicted in all cases. Docking analysis results using the human COX-2 structure 5IKT derived from its complex with tolfenamic acid are presented in

Table 6. According to the 5IKT based docking analysis **x-1** was expected to be one of the best COX-2 isoenzyme inhibitors.

Interestingly, docking analysis indicate that all studied compounds may form stable complexes with the human COX-2 or the ovine COX-1 structures which were used. However, they are not placed deep inside the active site pocket as observed with most known inhibitors but are preferably oriented near the entrance of the active site (Figure 3). Much higher binding energies were calculated in case of deeper orientation.

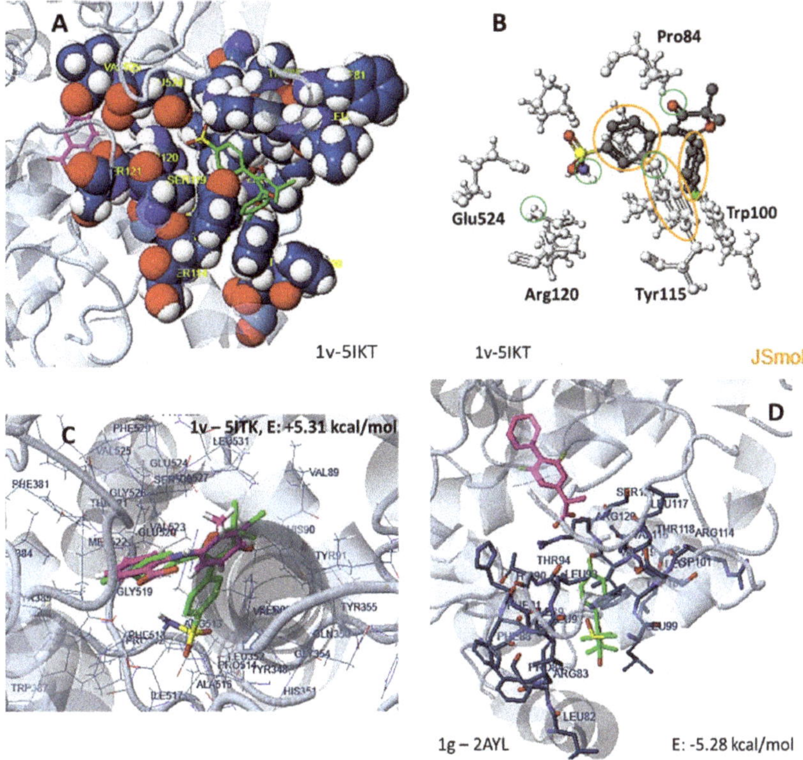

Figure 3. Docking of compound **1v** to the human COX-2 structure 5IKT, at a lower energy complex (**A**,**B**) with the compound docked at the entrance of the active site and at a high energy complex with the compound oriented in a deeper space of the active site (**C**). Docking of compound **1g** to the ovine COX-1 structure 2AYL (**D**). Studied compound in green, initial ligand in magenta. Yellow cycles: pi–pi interactions, green cycles: polar interactions. Polar interactions are also formed between Arg120 and the –SO$_2$ NH$_2$ group.

As shown in Figure 3B, pi–pi interactions between Tyr115 of the COX-2 structure 5IKT and both aryl moieties of compound **1v** are formed while Tyr115 also participates in polar interactions with the O atom of the furanone ring. Similar interactions are observed in the case of all compounds stabilizing their orientation at the entrance of the active site cleft, including the more rigid phenanthrene derivatives. On the other hand, a negative effect leading to the formation of a high energy complex (−5.31 kcal/mol) is observed when the compound is oriented deep inside the active site, mainly because of the hydrophobic interactions between Leu352 (Figure 3C, arrow) and both aryl groups of the molecule.

Polar interactions with Arg 120 and hydrophobic interactions with Ile89, Leu112 and Val116 are involved in complex stabilization between compound **1g** and the COX-1 structure 2AYL (Figure 3D).

The inability of the molecules to be placed at the inner part of the active cleft may be explained by the volume and rigidity of the molecules. However, the observed negative interactions may be enhanced by the fact that the protein structures are derived from complexes with smaller molecules. It is well known that enzymes are flexible molecules that conformationally adapt to the substrate or inhibitor, and that the volume of the active site cavity is restricted when small molecules have been used for co-crystallization.

2.3.2. In Vivo Anti-Inflammatory Activity

For the evaluation of anti-inflammatory activity of the tested compounds in vivo carrageen-induced mouse paw edema assay was used, with celecoxib (64.6%) and indomethacin (47%) as reference compounds. In general compounds in concentration of 0.1 mmol/g exhibited inhibition of inflammation varying between 38–57.7%. From the data of Table 7 it is obvious that anti-inflammatory activity of alkylsulfones and sulfoxides of furan-3(2H)-ones, including annelated derivatives are very similar.

Table 7. In vivo anti-inflammatory activity of alkylsulfones and sulfoxides of furan-3(2H)-ones, including annelated derivatives.

Class	Compound	Dose, μmol/g	CPE, % *
	1c (H)	0.1	50.2
	1o (*m*–Cl)	0.1	50.7
SO_2CH_3	**1h** (*p*–F)	0.1	48.8
	1k (*p*–Cl)	0.1	50.6
	1b (H)	0.1	47
	1n (*m*–Cl)	0.1	51
$SOCH_3$	**1g** (*p*–F)	0.1	49
	1j (*p*–Cl)	0.1	54
	1q (H)	0.1	38
SO_2NH_2	**1v** (*p*–Cl)	0.1	44.3
	1w (*p*–F)	0.1	43.7
Phen	**x-1** (H)	0.1	45
	x-2 (*p*–F)	0.1	40
Reference	**E-1** (H)	0.1	57.7
	E-3 (*m*–Cl)	0.1	52.1

* Values are the mean of three determinations, and deviation from the mean is <10% of the mean value.

Thus, the study of structure–activity relationship revealed that anti-inflammatory activity of the compounds tested is almost independent of the nature and position of substituents. It should be noted that COX-1/COX-2 inhibition does not correlate with anti-inflammatory activity, as has been observed in anti-inflammatory results of other researches. Furthermore, since other mechanisms are involved in the first steps of edema formation, linear correlation between in vivo and in vitro results cannot be expected to be found.

2.3.3. Anticancer Activity against MCF-7 Breast Cancer Cells

The anticancer treatments traditionally were based on the inhibition of DNA synthesis and function. Nowadays, many researchers turn their interest to selective inhibition of signaling pathways involved in proliferation, as another approach for anticancer therapy. Epidermal growth factor receptor (EGFR), which is overexpressed in major number of human tumors, plays a significant role in growth signaling. One representative of potent and selective EGER inhibitors is gefitinib, which is a competitive inhibitor of the ATP binding site of EGFR tyrosine kinase.

Gefitinib is an antineoplastic drug well known for its inhibitory effect in cell proliferation and induction of cell death in various cancer cell lines, including breast cancer. It inhibits the catalytic activity of numerous tyrosine kinases and may also induce cell cycle arrest and inhibit angiogenesis [45,46].

Three compounds were selected for the evaluation of their effect on cancer cell growth in the human epithelial breast adenocarcinoma cell line (MCF-7 cells) and in human tongue squamous cell carcinoma HSC-3 cell line. The p–F derivative of the SO CH_3 group (**1g**) with the best COX inhibitory action (COX-1 IC_{50}:2.8 µM), the p–F derivative of SO_2CH_3 group (**1h**) with low COX inhibitory action and one compound from the phenanthren group (**x-1**) with moderate predicted inhibitory action on both COX-1/2 isoenzymes and the best predicted selectivity over COX-2 were chosen.

The cells were grown in the presence of each of the selected compounds as well as of the two well-known anti-neoplastic drugs gefitinib and 5-fluorouracil. Celecoxib, a selective COX-2 inhibitor was used as a control. The IC_{50} values of each compound after 48 h exposure of the cells were determined.

Interestingly, all three compounds inhibited the MCF-7 cell growth (Table 8, Figure S2) with an IC_{50} value of 24 µM for **1h** and 10 µM for **1g** and **x-1**, while the IC_{50} values of the control compounds were 70 µM for gefitinib, 0.03 µM for 5-fluorouracil, and 29.2 µM for celecoxib (Figure 4). Two of the studied compounds exhibited higher cytotoxicity at MCF-7 cell line than did gefitinib (IC_{50}:70 µM), with the less active COX inhibitor, **1h**, to exhibit similar cytotoxicity under the culture conditions (IC_{50}:24 µM). As far as the HSC-3 cell line is concerned, compound **1g** exhibited remarkable growth inhibition of this cell line (64%) at the concentration of 10 µM with an IC_{50} value of 7.5 µM. The two other compounds exhibited much lower cytotoxicity at the HSC-3 cells. The inhibition potency of celecoxib and 5-fluouracil is comparable with that presented in the literature [25]. The results obtained for synergistic activity are presented in Table 8, Figure 4.

Table 8. IC_{50} values of the new synthesized compounds as well as gefitinib, 5-fluorouracil and celecoxib.

Compound	MCF-7 IC_{50} Values (µM) *	HSC-3 % Cell ** Growth	HSC-3 IC_{50} Values (µM) *
1-h	24	100 ± 10	90
1-g	10	36 ± 4	7.5
x-1	10	91 ± 10	60
Gefitinib	70		
5-Fluorouracil	0.03		
Celecoxib	29.186		

* Data are the mean average of 3 independent experiments. ** Compound concentration: 10µM.

In addition, exposure of the MCF-7 cells to each of the compounds in combination with each one of the two anti-neoplastic drugs was performed in order to study the synergistic effect of the compounds. The results are shown in Table 9 and Figures 5 and 6.

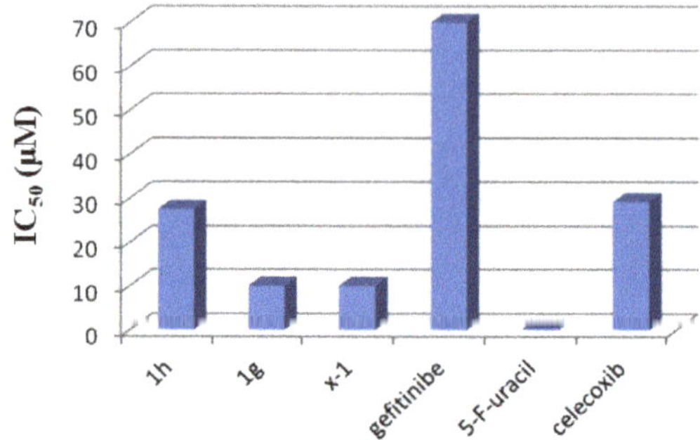

Figure 4. IC$_{50}$ values of the new synthesized compounds as well as gefitinib, 5-fluorouracil, and celecoxib.

Table 9. % MCF-7 cell growth rates after cell exposure to the combinations of each novel compound **1h**, **1g**, and **x-1** with gefitinib, 5-fluorouracil or celecoxib in their IC$_{50}$ or lower concentrations, for 48 h.

Compound * Concentrations (µM)	% Cell Growth	CI	Combination Effect
27.5 µM **1h** + 70 µM gef	45.9	>1	antagonism
10 µM **1g** + 70 µM gef	36.5	0.780 ± 0.200 < 1	synergism
10 µM **x-1** + 70 µM gef	24.3	0.260 ± 0.033 < 1	synergism
29 µM cel + 70 µM gef	56.8	>1	antagonism
21.3 µM **1h** + 0.3 µM 5-fu	27.3	0.533 ± 0.015 <1	synergism
2.9 µM **1g** + 0.3 µM 5-fu	19.1	0.058 ± 0.010 <1	synergism
3.8 µM **x-1** + 0.3 µM 5-fu	21.0	>1	antagonism
29 µM cel + 00.3 µM 5-fu	49.3	>1	antagonism
21.3 µM **1h** + 29 µM cel	38.6	>1	antagonism
2.9 µM **1g** + 29 µM cel	36.6	0.750 ± 0.025 < 1	synergism
3.8 µM **x-1** + 29 µM cel	60.6	>1	antagonism

* Abbreviations: gef: gefitinib; 5-FU: 5-fluorouracil; cel: celecoxib.

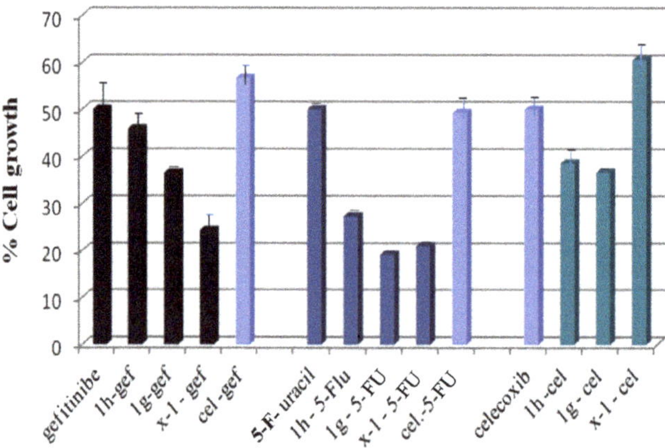

Figure 5. % MCF-7 cell growth rates after cell exposure to the combinations of each COX-2 inhibitor with gefitinib, 5-fluorouracil, or celecoxib in their IC$_{50}$ or lower concentrations, for 48 h. Cell growth was assessed using a hemocytometer (Neubauer chamber) and is expressed as a percentage (%) relative to that for the untreated, control culture (CTL). The data represent mean values of four.

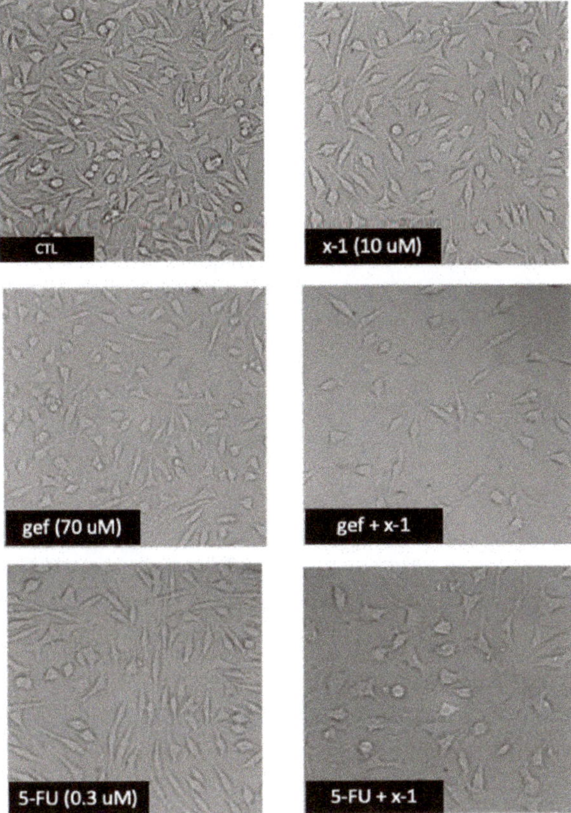

Figure 6. Phase-contrast microscopy images (32x) after MCF-7 cell treatment with the single antineoplastic drugs and synthesized compounds as well as with the combinations of drugs and compounds, for 48 h.

The two more active compounds—**1g** and **x-1**—further decreased cell growth when added in combination with gefitinib. Calculation of the Combination Index (CI) revealed a synergistic effect with CI = 0.78 and CI = 0.26 for compounds **1g** and **x-1**, respectively. The low activity compound **1h** exhibited antagonistic effect with CI = 1.95. Compound **1g** also showed a remarkable synergistic effect when added in combination with 5-FU (CI = 0.058). The low activity compound **1h** showed a lower synergistic effect (CI = 0.533), while compound **x-1** exhibited antagonistic effect. On the other hand, the known COX-2 selective inhibitor, celecoxib showed antagonistic effect with both anti-neoplastic drugs gefitinib or 5-fluorouracil with CI = 9.8 and CI = 1.57, respectively. This has also been observed elsewhere. According to previously published results, the synergistic effect of celecoxib depends on the cell culture as well as on the concentration of the anticancer drug used [25]. Decrease instead of enhancement of the anticancer effect has been mentioned in an experiment concerning cis-platin [47].

Celecoxib was also used in combination with each one of the new synthesized compounds (**1h, 1g,** and **x-1**). A synergistic effect with the most active COX-1 inhibitor **1g** was observed (CI = 0.75), while with the two other compounds **1h** and **x-1** celecoxib showed antagonistic effect with CI = 1.32 and CI = 1.90, respectively.

In conclusion, the most active and COX-1 selective inhibitor, **1g**, exhibited synergistic effect with both anti-neoplastic drugs probably due to overexpression of COX-1 in MCF-7 cells [48]. The synergistic effect of compound **1g** and celecoxib may be indicative for the involvement of both cyclooxygenase isoenzymes in cancer cell development and the value of combined COX1/2 inhibition for best inhibitory effect. Combined treatment with specific COX-1 and COX-2 inhibitors has been found to have better results in cell culture experiments, including the MCF-7 cell line, by other researchers as well [49].

The most active compounds—**1g** and **x-1**—belonged to the -SOMe derivatives of the 4,5-diaryl-furan-3(*2H*)-ones and to the phenanthrene derivatives, respectively. Usually sulfoxides are used to study cytotoxicity as ligands in complexes of rhodium and platinum [50]. The antitumor activity of sulfoxides has been studied earlier [51] against Ehrlich ascites brain cells where the sulfoxides and sulfones exhibited an approximately equal level of activity, and the IC_{50} value was comparable to that of 8-azaguanine. Phenanthrofuranones are a quite rare class of molecules in studies of anti-inflammatory and anticancer activity although some phenanthrofurane quinone derivatives have been isolated from the roots of *S. miltiorrhiza* [52] and from *Pleione bulbocodioides*—a medicinal plant with anti-inflammatory properties that is used in traditional Chinese medicine [53].

3. Materials and Methods

All the chemicals used were of analytical grade and commercially available. The kit for COX Activity Assay was purchased from Cayman (Ann Arbor, MI, USA).

3.1. Computer Simulation Methods

3.1.1. Preparation of Protein Structures and Docking Analysis Based on *Ovis aries* COX-1, 3KK6 and *Mus musculus* COX-2, 3LN1

Homology modeling and preprocessing were performed in ICM Pro Software. 3D Molecular docking was carried out using MOE v.2014.09.01 (Chemical Computing Group, www.chemcomp.com, Köln, Germany).

RSA data were obtained from the PDB for the structures of *Ovis aries* COX-1 (3KK6 [35]) and *Mus musculus* COX-2 (3LN1 [36]) in complex with Celebrex and were used for homology modeling. As a result, the 3D structures of the corresponding human isoforms were constructed using these structures as templates (*h*COX-1, FASTA code: sp_P23219_PGH1_HUMAN [54]) and *h*COX-2, FASTA code: sp_P35354_PGH2_HUMAN [55]). To simplify modeling, water molecules were excluded from the preprocessing and docking procedures. In the first step, preoptimized 3D conformations (organized second structures) of the target proteins were constructed using the selected templates, then the optimal position of amino acid residues was found by an annealing procedure using different temperature factors thereby achieving the stable state with an insignificant stress influence (Figure S3). As clearly shown in Figure S3, a relatively low stress was achieved for *h*COX-1 and *h*COX-2, except for a minor set of amino acids that are beyond the predefined binding site. The main focus during annealing was placed on amino acids that organized the binding site and neighboring area (30Å) as well as hydrogen atoms therein. As a result, minimized 3D conformations of the target proteins were obtained and subsequently used for modeling. It should be noted that the binding sites of celecoxib in the template proteins and human isoforms are very similar.

The constructed models were evaluated using a set of known selective COX-1/2 inhibitors (Figure S1). Celecoxib was also docked into the binding site (Figure 7) to obtain a reference score value (baseline point). During the evaluation, 50 different "active" conformations were generated per ligand. Docking was performed in dynamic mode with the flexible binding site. As shown in Figure 7, the docking results correlated well with the RSA data.

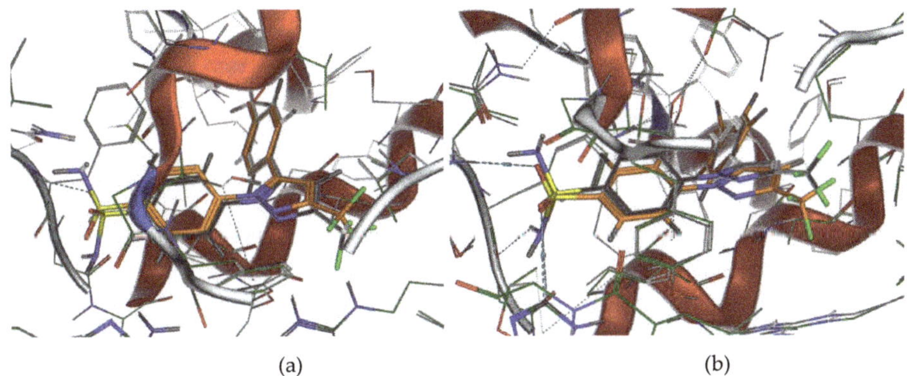

(a)   (b)

Figure 7. Internal validation of the developed model. The most adequate conformation generated for celecoxib (gray) as compared to the template location (RSA data, orange): *h*COX-2: Free binding energy: −14.27 kcal/mol (**a**); *h*COX-1: Free binding energy: −13.39 kcal/mol (**b**); RMSD = 0.13 and 0.21, respectively.

The developed model was used to predict the activity of fourteen compounds E1–E14(Figure S1) with known inhibitory action which were used as reference compounds (test set, Figure S3) Docking was carried out in dynamic fashion with the integrated 3-centered pharmacophore model (Figure 8) to restrict the pool of possible conformations. This pharmacophore hypothesis was generated based on the binding mode which was revealed for the template molecule (celecoxib) retaining all the crucial binding points.

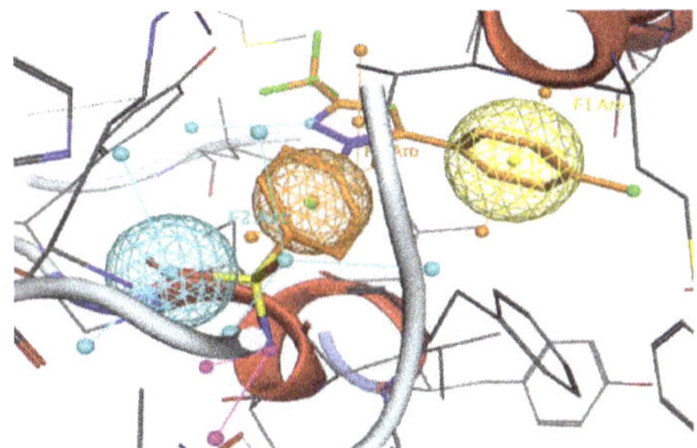

Figure 8. The common 3-centered pharmacophore model for COX-1/2 inhibitors.

3.1.2. Docking Analysis Based on Ovine COX-1 Structure 2AYL and Human COX-2 Structure 5IKT

Since human COX-2 recombinant enzyme and ovine COX-1 isoenzyme were used for the in vitro evaluation of inhibitory action, a second approach of docking analysis was performed using the human COX-2 structure 5IKT, in complex with tolfenamic acid and several ovine COX-1 structuresamong which was the structure 4O1Z in complex with meloxicam.

In all cases the target box was set around the ligand. For the human COX-2 structure, 5IKT, the target center was at x = 165.42, y = 185.73, z = 192.38, and target box dimensions were set at 30 × 30 × 30 (X × Y × Z).

Docking analysis was carried out using Molecular Docking Server [56] and AutoDock tools [57].

For the calculation of the van der Waals and the electrostatic terms AutoDock parameter set and distance-dependent dielectric functions were used. The Lamarckian genetic algorithm (LGA) and the Solis & Wets local search method [58] were used to perform docking simulation. More details are given in our previous paper [59]. Docking of the initial ligand was performed for verification of the Docking efficacy (Figures 9A and 10). Docking of known inhibitors was also performed for comparison reasons (Figure 9B).

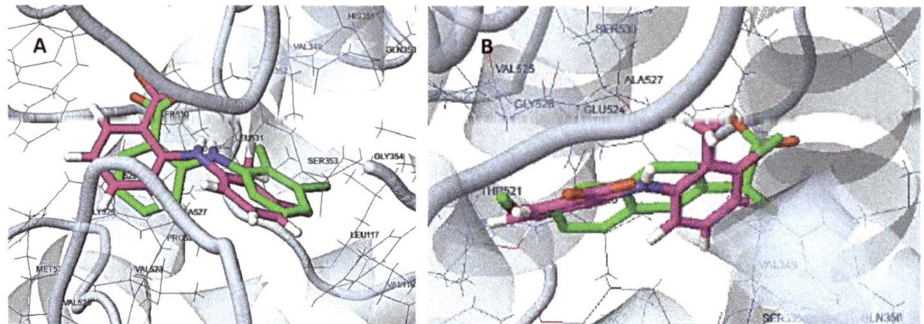

Figure 9. Docking of tolfenamic acid (**A**) and naproxen (**B**) to the structure 5IKT of human COX-2, Energy: −7.04 kcal/mole (docked molecule in green, initial molecule in magenta).

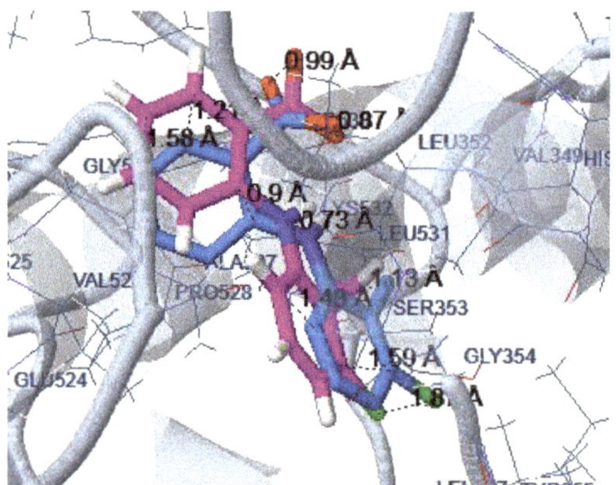

Figure 10. Docking of tolfenamic acid to the structure 5IKT of human COX-2 (docked molecule in blue, initial molecule in magenta), the distances between the same atoms of the two structures do not exceed 1.6 Å.

3.2. Chemistry

All solvent used for reactions were dried and purified before by known methods. TLC on precoated silica gel SIL G/UV$_{254}$ plates (Macherey-Nagel & Co. Düren, Germany)) were used for the reaction monitoring. All apparatus used for the compound's characterization were mentioned in our previous paper [41]

3.2.1. General Method for Preparation of Compounds **1b**, **1g**, **1j**, and **1n**

To a solution of 2,2-dimethyl-4-[4′-(methylthio)phenyl]-5-phenylfuran-3(2*H*)-one (300 mg, 0.96 mmol) in acetic acid (10 mL, 50–55 °C) a mixture of 35% hydrogen peroxide (100 mg, 0.96 mmol), 98% H$_2$SO$_4$ (0.05 mL) and acetic acid (3 mL) was added to the reaction dropwise with vigorous stirring for 1 h. The reaction mixture was stirred for an additional 3 h at 50–55 °C (control by TLC). After completion of the reaction, the mixture was poured into water, extracted with ethyl acetate (4 × 20 mL), organic phase was washed by NaHCO$_3$/water (3 × 5 mL) solution, dried over Na$_2$SO$_4$ and K$_2$CO$_3$, and, after removing the solvent, the residue was dried in vacuum for 24 h. For isolation and purification of compounds the mixtures of n-hexane/CH_2Cl_2/ethyl acetate (from 10:2:1 to 2:1:1) were used.

2,2-Dimethyl-4-[4′-(methylsulfinyl)phenyl]-5-phenylfuran-3(2H)-one (**1b** racemic mixture) Yield: 270 mg 86% colorless oily substance. ^{1}H NMR (400 MHz, 20 mg in 0.8 mL of $CDCl_3$, reference: $CHCl_3$ = 7.260 ppm), δ, ppm: 1.58 (s, 6H, 2CH_3), 2.74 (s, 3H, $SOCH_3$), 7.38 (t, 2H, $^3J_{HH}$ = 7.8 Hz, Ph), 7.49–7.51 (m, 3H, Ar), 7.63 (t, 4H, $^3J_{HH}$ = 7.3 Hz, Ar). ^{13}CNMR (100 MHz, 20 mg in 0.8 mL of $CDCl_3$, reference: $CHCl_3$ = 77.00 ppm), δ, ppm: 23.3 (CH_3), 23.4 (CH_3), 43.9 ($SOCH_3$), 87.6 ($(CH_3)_2$), 112.4 (C_4), 123.9, 127.9 (2C), 128.5 (2C), 128.7(2C), 129.5, 130.3(2C), 132.2, 144.6 (all $C_{Ar Ar'}$), 179.2 (C_5), 204.9 (C–O). IR (in CCl_4), cm^{-1}: 955 w., 1051 m., 1090 w., 1383 s., 1616 m., 1701 s. (C=O), 2931 w., 2982 w. HRMS, *m/z*, calculated for $C_{19}H_{18}NaO_3S$ 349.0869, found: 349.0876 [M + Na]$^+$.

5-(4′-Fluorophenyl) 2,2-dimethyl-4-[4-(methylsulfinyl)phenyl]-furan-3(2H)-one (**1g** racemic mixture) Yield: 27%. Slightly yellow oily substance. ^{1}H NMR (400 MHz, 25 mg in 0.8 mL of $CDCl_3$, reference: $CHCl_3$ = 7.26 ppm), δ, ppm: 1.54 (s, 6H, 2CH_3), 2.72 (s, 3H, $SOCH_3$), 7.04 (t, 2H, $^3J_{HH}$ = 8.7 Hz, Ar), 7.45–7.47 (m, 2H, Ar), 7.60–7.65 (dd, 4H, $^3J_{HH}$ = 8.7 Hz, $^4J_{HF}$ = 5.2 Hz). ^{13}C NMR (100 MHz, 25 mg in 0.7 mL of acetone-d_6, reference: $CHCl_3$ = 29.84 ppm), δ, ppm: 23.4 (2CH_3), 43.9 ($SOCH_3$), 87.9 ($H_3)_2$), 116.7 (d, 2C, $^2J_{CF}$ = 22.3 Hz), 124.8 (2C), 127.1 (d, 1C, $^4J_{CF}$ = 3.2 Hz), 129.0, 131.0 (2C), 131.8 (d, 2C, $^3J_{CF}$ = 9.1 Hz), 134.0, 145.9, 165.5 (d, 1C, $^1J_{CF}$ = 252.0 Hz), 178.3 (C_5), 204.5 (C=O). ^{19}F NMR (376 MHz, $CDCl_3$) δ, ppm: −105.4. IR (in CCl_4), cm^{-1}: 842 m, 902 w, 1049 m (S=O), 1161 m, 1242 s., 1381 s, 1515 m, 1616 s, 1703 s (C=O), 2980 w. (C-H). LCMS, *m/z*, calculated for $C_{19}H_{18}FO_3S^+$: 345.0956, found 345.1015 [M + H]$^+$.

5-(4′-Chlorophenyl)-2,2-dimethyl-4-[4-methylsulfinyl)phenyl]-furan-3(2H)-one (**1j** racemic mixture) Yiled: 8%. Slightly yellow oily substance. ^{1}H NMR (400 MHz, 25 mg in 0.8 mL of $CDCl_3$, reference: $CHCl_3$ = 7.26 ppm), δ, ppm: 1.54 (s, 6H, 2CH_3), 2.74 (s,3H, $SOCH_3$), 7.32(d, 2H, $^3J_{HH}$ = 7.4 Hz, PhCl), 7.45 (d, 2H, $^3J_{HH}$ = 7.5 Hz, PhCl), 7.58 (d, 2H, $^3J_{HH}$ = 7.4 Hz, PhSOMe), 7.73 (d, 2H, $^3J_{HH}$ = 7.5 Hz, PhSOMe). ^{13}C NMR (100 MHz, $CDCl_3$, reference77.00 ppm), δ, ppm: 23.2 (2CH_3), 43.6 ($SOCH_3$), 88.0 ($(CH_3)_2$), 112.7 (C_4), 125.2 (2C), 127.3 (2C), 128.4, 128.7 (2C), 129.0 (2C), 133.2, 134.5, 144.2, 177.5 (C_5), 204.6 (C=O). LCMS, *m/z* calculated for $C_{19}H_{18}ClO_3S^+$: 361.0660, found 361.0740 [M + H]$^+$.

5-(3′-Chlorophenyl)-2,2-dimethyl-4-[4′-(methylsulfinyl)phenyl]-furan-3(2H)-one (**1n** racemic mixture) Yield: 33%. Slightly yellow oily substance. ^{1}H NMR (400 MHz, 25 mg in 0.8 mL of $CDCl_3$, reference: $CHCl_3$ = 7.26 ppm), δ, ppm: 1.53 (s, 6H, 2CH_3), 2.79 (s, 3H, $SOCH_3$), 7.23–7.27 (m, 1H, PhCl), 7.37–7.50 (m, 4H, Ar), 7.60–7.70 (m, 3H, Ar). ^{13}C NMR (100 MHz, 20 mg in 0.7 mL of acetone-d_6, reference: 29.84 ppm), δ, ppm: 23.3 (2CH_3), 44.2 ($SOCH_3$), 88.2 ($(CH_3)_2$), 114.1 (C_4), 124.6 (2C), 127.8, 128.7, 131.0 (2C), 131.3, 132.6, 132.8, 133.5, 135.0, 147.1, 177.6 (C_5), 204.7 (C=O). IR (in CCl_4), cm^{-1}: 1051 m. (S=O), 1145 m, 1238 w, 1384 s, 1618 m, 1706 s, (C=O), 2983 w (C-H). LCMS, *m/z*, calculated for $C_{19}H_{18}ClO_3S^+$: 361.0660, found 361.0740 [M + H]$^+$.

3.2.2. General Method for Preparation of Compounds **1c**,**1h**, **1k**, and **1o**

To a solution of 2,2-dimethyl-4-[4′-(methylthio)phenyl]-5-phenylfuran-3(2H)-one (300 mg, 0.96 mmol) in acetic acid (10 mL, 50–55 °C) a mixture of 35% hydrogen peroxide (210 mg, 1.96 mmol), 98% H_2SO_4 (0.05 mL), and acetic acid (3 mL) was added to the reaction dropwise with vigorous stirring for 1 h. The reaction mixture was stirred for 30 minutes at 75–85 °C (control by TLC). After completion of the reaction, the mixture was poured into water, extracted with ethyl acetate (4 × 20 mL), organic phase was washed by $NaHCO_3$/water (3 × 5 mL solution, dried over Na_2SO_4 and K_2CO_3, and, after removing the solvent, the residue was dried under vacuum for 24 h. For isolation and purification of compounds the mixtures of n-hexane/ ethyl acetate (from 20:1 to 1:1) were used.

2,2-Dimethyl-4-[4′-(methylsulfonyl)phenyl]-5-phenylfuran-3(2H)-one (**1c**) Yield: 94%. ^{1}H NMR (400 MHz, 25 mg in 0.8 mL of $CDCl_3$, reference: $CHCl_3$ = 7.260 ppm), δ, ppm: 1.58 (s, 6H, 2CH_3), 3.04 (s, 3H, SO_2CH_3), 7.38 (t, 2H, $^3J_{HH}$ = 7.7 Hz, Ph), 7.50–7.54 (m, 3H, Ph), 7.61 (d, 2H, $^3J_{HH}$ = 8.5 Hz, PhSO$_2$Me),

7.91 (d, 2H, $^3J_{HH}$ = 8.4 Hz, PhSO$_2$Me). ^{13}C NMR (100 MHz, 25 mg in 0.8 mL of CDCl$_3$, reference: CHCl$_3$ = 77.00 ppm), δ, ppm: 23.3 (2CH$_3$), 43.5 (SO$_2$CH$_3$), 87.9 (CH$_3$)$_2$), 111.6 (C$_4$), 127.5, 128.4, 128.6, 129.2, 130.1, 132.4, 136.1, 139.0 (all C$_{ArAr'}$), 179.8 (C$_5$), 204.4 (C=O). IR (in CCl$_4$), cm^{-1}: 955 m, 1051 m., 1157 s, 1327 m., 1383 m., 1614 m., 1701 s. (C=O), 2932 w., 2982 w. HRMS, *m/z*, calculated for C$_{19}$H$_{18}$NaO$_4$S$^+$: 365.0818, found 365.0823 [M + Na]$^+$.

5-(3'-Chlorophenyl)-2,2-dimethyl-4-[4'-(methylsulfonyl)phenyl]-furan-3(2H)-one (**1o**) Yield: 33%. Colorless crystals, m.p. 219–221 °C. ^{1}H NMR (400 MHz, 20 mg in 0.8 mL of CDCl$_3$, reference: CHCl$_3$ = 7.26 ppm), δ, ppm: 1.56 (s, 6H, 2CH$_3$), 3.04 (s, 3H, SO$_2$CH$_3$), 7.29 (d, 1H, $^3J_{HH}$ = 7.8 Hz, PhCl), 7.38 (d, 1H, $^3J_{HH}$ = 7.9 Hz, PhCl), 7.47 (d, 1H, $^3J_{HH}$ = 8.0 Hz, PhCl), 7.51 (d, 2H, $^3J_{HH}$ = 8.4 Hz, PhSO$_2$Me), 7.67 (s,1H, PhCl), 7.91 (d, 2H, $^3J_{HH}$ = 8.4 Hz, PhSO$_2$Me). ^{13}C NMR (100 MHz, 25 mg in 0.8 mL of CDCl$_3$, reference: CHCl$_3$ = 77.00 ppm), δ, ppm: 23.2 (2CH$_3$), 44.4 (SO$_2$CH$_3$), 88.1 (CH$_3$)$_2$), 112.4, 126.7, 127.6 (2C), 128.0, 130.0, 130.1 (2C), 131.0, 132.3, 135.0, 135.6, 139.3, 177.8 (C$_5$), 204.2 (C=O). IR (in CCl$_4$), cm^{-1}: 957 m., 1157 s. (SO$_2$), 1330 s. (SO$_2$), 1384 m., 1616 m., 1705 s. (C=O), 2931 w., 2982 w. (C-H). LCMS, *m/z*, calculated for C$_{19}$H$_{18}$ClO$_4$S$^+$: 377.0609, found 377.0667 [M + H]$^+$.

5-(4'-Chlorophenyl)-2,2-dimethyl-4-[4'-(methylsulfonyl)phenyl]-furan-3(2H)-one (**1k**) Yield: 9%. Colorless crystals, m.p. 223–225 °C. ^{1}H NMR (400 MHz, 20 mg in 0.8 mL of CDCl$_3$, reference: CHCl$_3$ = 7.26 ppm), δ, ppm: 1.57 (s, 6H, 2CH$_3$), 3.06 (s, 3H, SO$_2$CH$_3$), 7.37 (d, 2H, $^3J_{HH}$ = 8.8 Hz, PhCl), 7.51-7.56 (m, 4H, Ar), 7.53 (dd, 4H, $^3J_{HH}$ = 8.8 Hz, $^4J_{HH}$ = 5.4 Hz), 7.92 (d, 2H, *J* = 8.8 Hz, PhSO$_2$Me). ^{13}C NMR (100 MHz, 25 mg in 0.8 mL CDCl$_3$, reference: CHCl$_3$ = 77.00 ppm), δ, ppm: 23.3 (2CH$_3$), 44.5 (SO$_2$CH$_3$), 88.1 (CH$_3$)$_2$), 112.0, 127.7 (2C), 129.2 (2C), 129.7 (2C), 130.2 (2C), 130.6, 135.9, 138.8, 139.3, 178.3 (C$_5$), 204.2 (C=O). IR (in CCl$_4$), cm^{-1}: 956 m., 1095 m., 1158 s. (SO$_2$), 1329 s. (SO$_2$), 1384 s., 1616 m., 1703 s. (C=O), 2931 w., 2982 w.(C-H). LCMS, *m/z* calculated for C$_{19}$H$_{18}$ClO$_4$S$^+$: 377.0609, found 377.0679 [M + H]$^+$.

5-(4'-Fluorophenyl)-2,2-dimethyl-4-[4'-(methylsulfonyl)phenyl]-furan-3(2H)-one (**1h**) Yield: 23%. Colorless crystals, m.p. 191–192 °C. ^{1}H NMR (400 MHz, 25 mg in 0.8 mL of CDCl$_3$, reference: CHCl$_3$ = 7.26 ppm), δ, ppm: 1.57 (s, 6H, 2CH$_3$), 3.06 (s, 3H, SO$_2$CH$_3$), 7.09 (t, 2H, $^3J_{HH}$ = 8.5 Hz, PhF), 7.52 (d, 2H, $^3J_{HH}$ = 8.0 Hz, PhSO$_2$Me), 7.63 (dd, 2H, $^3J_{HH}$ = 8.8 Hz, $^4J_{HF}$ = 5.4 Hz, PhF), 7.92 (d, 2H, $^3J_{HH}$ = 8.0 Hz, PhSO$_2$Me). ^{13}C NMR (100 MHz, 25 mg in 0.8 mL of CDCl$_3$, reference: CHCl$_3$ = 77.00 ppm), δ, ppm: 23.3 (2CH$_3$), 44.5 (SO$_2$CH$_3$), 88.0 (CH$_3$)$_2$), 116.2 d (2C, $^2J_{CF}$ = 22.0 Hz), 126.3 (1C), 127.8 (2C), 130.2 (2C), 130.9 d (2C, $^3J_{CF}$ = 9.0 Hz), 136.1, 139.3, 165.1 d (1C, $^1J_{CF}$ = 255.4 Hz), 178.4 (C$_5$), 204.2 (C=O). ^{19}F NMR (376 MHz, CDCl$_3$) δ, ppm: -104.8. IR (in CCl$_4$), cm^{-1}: 955 m., 1157 s. (SO$_2$) 1243 m., 1327 s. (SO$_2$), 1383 m., 1613 m., 1704 s. (C=O), 2933 w. (C-H). LCMS, *m/z*, calculated for C$_{19}$H$_{18}$FO$_4$S$^+$: 361.0905, found 361.0953 [M + H]$^+$.

3.2.3. General Method for Preparation of Compounds **1q**, **1v**, and **1w**

To an aqueous solution of 25% NH$_3$ (5mL) 2,2-dimethyl-4-[4'-(chlorosulfonyl)phenyl]-5-(3'-chlorophenyl)furan-3(2H)-one (200 mg) in THF (5 mL) was added and mixture was stirred at room temperature for 24 h. Then, water, ammonia and tetrahydrofuran were removed in vacuo on a rotary evaporator, the solid residue was dissolved in ethanol, then silica gel (5 g) was added, and ethanol was removed in vacuo. The silica gel with the product was washed on a Schott filter 3 times with dichloromethane (5 mL each), after which the compound was washed off the filter by ethanol after removing the alcohol and vacuum drying (0.1 torr, 24 h, 50 °C). For isolation and purification of compounds the mixtures of n-hexane/ CH$_2$Cl$_2$/ethyl acetate (from 10:2:1 to pure ethyl acetate) were applied.

5-(3'-Chlorophenyl)-2,2-dimethyl-4-[4'-(aminosulfonyl)phenyl]-furan-3(2H)-one (**1u**) Yield: 150 mg, 79%. Colorless crystals, m.p. 198–200 °C. ^{1}H NMR (300 MHz, 15 mg in 0.6 mL of (CD$_3$)$_2$CO, reference: 2.05 ppm), δ, ppm: 1.54 (s, 6H, 2CH$_3$), 6.63 (s, 2H, NH$_2$), 7.45–7.62 (m, 3H, PhCl), 7.49 (d, 2H, $^3J_{HH}$ = 8.4 Hz, PhSO$_2$Me), 7.68–7.71 (m, 1H, PhCl), 7.89 (d, 2H, $^3J_{HH}$ = 8.4 Hz, PhSO$_2$Me). ^{13}CNMR (75 MHz, 15 mg in 0.6 mL of (CD$_3$)$_2$CO, reference: 206.26 ppm), δ, ppm: 204.54, 177.90, 144.07, 135.09,

134.77, 132.77, 132.70, 131.37, 130.61, 128.67, 127.86, 127.07, 113.86, 88.33, 23.30. HRMS, *m/z*, calculated for $C_{18}H_{17}ClNO_4S^+$: 378.0562, found: 378.0563 [M + H]$^+$.

5-(4′-Phenyl)-2,2-dimethyl-4-[4′-(aminosulfonyl)phenyl]-furan-3(2H)-one (**1q**) Yield: 23%.Colorless crystals, m.p. 158–160 °C. ^{1}H NMR (400 MHz, 5 mg in 0.7 mL of acetone-d_6, reference 2.05 ppm), δ, ppm: 1.53 (s, 6H, 2CH$_3$), 6.60 (s, 2H, -NH$_2$), 7.44-7.49 (m, 4H, Ph), 7.58 (t,1H, $^3J_{HH}$ = 7.4 Hz, Ph), 7.65 (d, 2H, $^3J_{HH}$ = 7.2 Hz, PhSO$_2$Me), 7.87 (d, 2H, $^3J_{HH}$ = 8.5 Hz, PhSO$_2$Me). ^{13}CNMR (100 MHz, 15 mg in 0.7 mL of acetone-d_6, reference: 29.84 ppm), δ, ppm: 23.4 (2C), 88.0 (C$_2$), 113.0 (C$_4$), 127.0 (2C), 129.2 (2C), 129.6 (2C), 130.6 (2C), 130.7, 133.0, 135.3, 143.7, 179.7 (C$_5$), 204.5 (C=O). IR (in KBr), cm^{-1}: 3346 br. (-SO$_2$NH$_2$), 3234 br. (-SO$_2$NH$_2$), 2988 w. (CH$_3$, Ar), 2938 w. (CH$_3$, Ar), 1662 v.s., 1607 s., 1568 s., 1485 m., 1386 s., 1342 s., 1238 m., 1170 s., 1058 m., 904 m., 746 m., 658 m., 597 m., 553 m. HRMS, *m/z*, calculated for $C_{18}H_{18}NO_4S^+$: 344.0952, found: 344.1006 [M + H]$^+$.

5-(4′-Chlorophenyl)-2,2-dimethyl-4-[4′-(aminosulfonyl)phenyl]-furan-3(2H)-one (**1v**) Yield 29%. Colorless crystals, m.p. 224–225 °C. ^{1}H NMR (400 MHz, 5 mg in 0.7 mL of acetone-d_6, reference: 2.05 ppm), δ, ppm: 1.53 (s, 6H, 2CH$_3$), 4.82 (s, 2H, -NH$_2$), 7.44-7.49 (m, 4H, Ar), 7.64 (d, 2H, $^3J_{HH}$ = 7.2 Hz, Ar), 7.86 (d,2H, $^3J_{HH}$ = 8.5 Hz, PhSO$_2$Me). ^{13}C NMR (75 MHz, 10 mg in 0.6 mL ofCDCl$_3$, reference: 77.00 ppm), δ, ppm: 204.51, 178.16, 140.80, 138.76, 134.76, 129.98 (2C), 129.76, 129.14 (2C), 127.68 (2C), 126.85 (2C), 112.16, 88.03, 23.31. HRMS, *m/z*, calculated for $C_{18}H_{17}ClNO_4S^+$: 378.0562, found: 378.0578 [M + H]$^+$.

5-(4′-Fluorophenyl)-2,2-dimethyl-4-[4′-(aminosulfonyl)phenyl]-furan-3(2H)-one (**1w**) Yield: 36%. Colorless crystals, m.p. 167–169 °C. ^{1}H NMR (300 MHz, 10 mg in 0.6 mL of CDCl$_3$, reference: 7.26 ppm), δ, ppm: 1.55 (s, 6H, 2CH$_3$), 5.42 (s, 2H, NH$_2$), 7.06 (t, 2H, $^3J_{HH}$ = 11.4 Hz, PhF), 7.41 (d, 2H, $^3J_{HH}$ = 8.9 Hz, PhSO$_2$NH$_2$), 7.60–7.65 (m, 2H, PhF)), 7.86 (d, 2H, *J* = 11.0 Hz, PhSO$_2$NH$_2$). ^{13}CNMR (75 MHz, 10 mg in 0.6 mL of CDCl$_3$, reference: 77.00 ppm), δ, ppm: 204.72, 178.32, 164.96 d (*J* = 255.2 Hz), 141.07, 134.64, 130.94, 130.83, 129.93, 128.46, 128.32, 126.72, 125.43(d, 2C, $^3J_{HF}$ = 3.3 Hz), 116.04 (d, 2C, $^2J_{HF}$ = 22.0 Hz), 111.80, 87.93, 23.24. HRMS, *m/z*, calculated for $C_{18}H_{17}FNO_4S^+$:362.0857, found: 362.0861 [M + H]$^+$.

3.2.4. 5-(4′-Chlorophenyl)-2,2-dimethyl-4-[4′-(methylaminosulfonyl)phenyl]-furan-3(2*H*)-one (**1vMe**)

A mixture (450 mg) of 2,2-dimethyl-4-[4′-(chlorosulfonyl)phenyl]-5-(4′-chlorophenyl)furan-3 (2*H*)-one and 2,2-dimethyl-4-[4′-chlorophenyl]-5-(phenyl)furan-3(2*H*)-one was dissolved in triethylamine (10 mL), presaturated by CH$_3$NH$_2$, and obtained by heating a 60% solution of methylamine in water and passing gas through a layer of CaCl$_2$.The resulting solution was stirred at 45 ° C for 24 h. Then the volatile products were removed on a rotary evaporator. The solid residue was applied to silica gel (3 g) and washed 3 times with a mixture of hexane-methylene chloride (1:1, 10 mL) to remove nonpolar impurities. After that, desired product **1vMe** was washed off from silica gel by ethanol. Finally, 160 mg (67%) of the product 1vMe was obtained after removing the solvent. For the purification of compound the mixtures of n-hexane/ CH$_2$Cl$_2$/ethanol (from 10:1:0.1 to 1:1:1) were used.

Colorless crystals, m.p. 204–205 °C. ^{1}H NMR (300 MHz, 15 mg in 0.6 mL of acetone-d_6, reference: 2.05 ppm), δ, ppm: 1.52 (s, 6H, 2CH$_3$), 2.57 (s, 3H, CH$_3$), 4.82 (s, 1H, NHCH$_3$), 7.48-7.52 (m, 4H, Ph) 7.66 (d, 2H, $^3J_{HH}$ = 8.7 Hz, Ar), 7.83 (d, 2H, $^3J_{HH}$ = 8.0 Hz, PhSO$_2$NHCH$_3$). ESMS, calculated for $C_{19}H_{19}ClNO_4S^+$: *m/z* = 392.0718, found: 392.0745.

3.2.5. General Method for the Synthesis of Phenanthro[9,10-b]furan-3′(2′H)-ones

A solution of 2,2-dimethyl-4-[4′-(methylthio)phenyl]-5-phenylfuran-3(2*H*)-one (500 mg, 1.61 mmol) in dichloromethane (5 mL) placed in a quartz photochemical reactor with water cooling and a volume of 50 mL and irradiated with a mercury UV lamp of 150 W during 4 h in the presence of oxygen. After completion of the reaction, the solvent was distilled off on a rotary evaporator, the mixture was applied onto a preparative chromatography plate and divided into fractions (eluent: hexane–ethyl acetate–dichloromethane). The yield of 2′,2′-dimethyl-6-(methylthio)phenanthro[9,10-b]furan-3′(2′H)-one was 70 mg (14%), colorless crystals, violet fluorescence (254 nm). The degree of conversion of the starting 3(2*H*)furanone was 90%.

2′,2′-dimethyl-3-(methylsulfonyl)-phenanthro[9,10-b]furan-3′(2′H)-one (**X-1**)

2′,2′-Dimethyl-6-(methylsulfonyl)-phenanthro[9,10-b]furan-3′(*2′H*)-one may be prepared by oxidation of corresponding sulfide as described above. Yield: 6%. Colorless crystals, m.p. 222–224 °C. ^{1}H NMR (300 MHz, 5 mg in 0.8 mL of CDCl$_3$, reference: CHCl$_3$ = 7.26 ppm), δ, ppm: 1.68 (s, 6H, 2CH$_3$), 3.06 (s, 3H, SO$_2$CH$_3$), 7.72 (td, 1H, $^3J_{HH}$ = 7.4, 1.6 Hz), 7.80 (td, 1H, $^3J_{HH}$ = 7.5, $^4J_{HH}$ = 1.5 Hz), 8.10 (dd, 1H, $^3J_{HH}$ = 7.5 Hz, $^4J_{HH}$ = 1.5 Hz), 8.32 (dd, 1H, $^3J_{HH}$ = 7.4, $^3J_{HH}$ = 5.1 Hz), 8.50 (d, 1H, J = 8.0 Hz), 8.79 (dd,1H, $^3J_{HH}$ = 7.5 Hz, $^4J_{HH}$ =1.5 Hz), 9.30 (d, 1H, $^4J_{HH}$ = 1.6 Hz). LCMS, *m/z*, calculated for C$_{19}$H$_{10}$O$_1$S$^+$: 342.0843, found: 342.0826 [M + H]$^+$.

Synthesis of Intermediate 2′,2′-dimethyl-6-(methylthio)-phenanthro[9,10-b]furan-3′(*2′H*)-one

Slightly yellow crystals, m.p. 178–180 °C. ^{1}H NMR (400 MHz, 25 mg in 0.8 mL CDCl$_3$, reference: CHCl$_3$ = 7.260 ppm), δ, ppm: 1.62 (s, 6H, 2CH$_3$), 2.64 (s, 3H, SCH$_3$), 7.60 (dd, 1H, $^3J_{HH}$ = 8.5 Hz, $^4J_{HH}$ = 1.8 Hz), 7.70 (t, 1H, $^3J_{HH}$ = 7.3 Hz), 7.90–7.84 (m, 1H), 8.33 (dd, 1H, $^3J_{HH}$ = 8.0 Hz, $^4J_{HH}$ = 1.0 Hz), 8.45 (d,1H, $^4J_{HH}$ = 1.5 Hz), 8.63 (d,1H, $^3J_{HH}$ = 8.4 Hz), 8.76 (d,1H, $^3J_{HH}$ = 8.4 Hz). LCMS, *m/z*, calculated for: C$_{19}$H$_{16}$NaO$_2$S$^+$: 331.0764, found: 331.0769 [M + Na]$^+$.

3-Fluoro-2,2-dimethyl-6-(methylsulfonyl)phenanthro[9,10-b]furan-3(2′H)-one (**X-2**)

3-Fluoro-2′, 2′-dimethyl-6-(methylsulfonyl)-phenanthro[9,10-b]furan-3′(*2′H*)-one (**X-2**) may be prepared by oxidation of corresponding sulfide as described above.

Here we present synthesis of intermediate 3-fluoro-2′, 2′-dimethyl-6-(methylthio)-phenanthro-[9,10-b]furan-3′(*2′H*)-one.

The first step: synthesis of 3-fluoro-2′,2′-dimethyl-6-(methylthio)-phenanthro[9,10-b]furan-3′(2′H)-one.

5-(4′-fluorophenyl)-2,2-dimethyl-4-[4′-(methylthio)phenyl]-)furan-3(*2H*)-one (**3, X**-2) (305 mg) dissolved in boiling *n*-hexane (30 mL) was placed in a quartz reactor and irradiated by a low-pressure UV mercury lamp (240 W, λ > 240 nm) during 2 h. After chromatography 27 mg (9%) of the product was obtained. The conversion rate was 34% and 200 mg of starting 3(*2H*)furanone was isolated.

Slightly yellow oily substance. ^{1}H NMR (300 MHz, 5 mg in 0.8 mL of CDCl$_3$, reference: CHCl$_3$ = 7.26 ppm), δ, ppm: 1.65 (s, 6H, 2CH$_3$), 2.64 (s, 3H, SCH$_3$), 7.48 (td, 1H, $^3J_{HH}$ = 7.9 Hz, $^4J_{HH}$ = 1.4 Hz), 7.66 (dd, 1H, $^3J_{HH}$ = 7.4 Hz, $^4J_{HH}$ = 1.5 Hz), 8.08 (dd, 1H, $^3J_{HH}$ = 7.5 Hz, $^4J_{HF}$ = 5.0 Hz), 8.14 (d, 1H, $^3J_{HH}$ = 7.5 Hz), 8.38 (dd, 1H, $^3J_{HH}$ = 8.0 Hz, $^4J_{HH}$ = 1.5 Hz), 8.56 (d, 1H, $^4J_{HH}$ = 1.6 Hz). LCMS, *m/z*, calculated for C$_{19}$H$_{16}$FO$_2$S$^+$: 327.0850, found: 327.0792 [M + H]$^+$.

The second step: Unreacted 5-(4′-fluorophenyl)-2,2-dimethyl-4-[4′-(methylthio)phenyl]-furan-3-(2*H*)-one (**2, X-2**)

(200 mg) dissolved in 25 mL of *n*-hexane and the irradiation procedure was repeated. A unreacted starting 3(*2H*)furanone (143 mg) was isolated and 27 mg (13.5%) of desired phenanthro[9,10-b]furanone was obtained. Total yield: 54 mg (17.7%).

3-Fluoro-2′,2′-dimethyl-6-(methylsulfonyl)-phenanthro[9,10-b]furan-3′(2′H)-one (**X-2**)

Colorless crystals, m.p. 242–244 °C. ^{1}H NMR (300 MHz, 5 mg in 0.8 mL of CDCl$_3$, reference: CHCl$_3$ = 7.26 ppm), δ, ppm: 1.65 (s, 6H, 2CH$_3$), 3.16 (s, 3H, SO$_2$CH$_3$), 7.50-7.55 (m, 1H), 8.17 (dd, 1H, $^3J_{HH}$ = 8.6 Hz, $^4J_{HH}$ = 1.7 Hz), 8.37 (dd, 1H, $^3J_{HH}$ = 10.6 Hz, $^4J_{HH}$ = 2.3 Hz), 8.43 (dd, 1H, $^3J_{HH}$ = 8.9 Hz, $^4J_{HF}$ = 5.8 Hz), 9.02 d (1H, $^4J_{HH}$ = 1.6 Hz), 9.04 (d, 1H, $^3J_{HH}$ = 6.3 Hz). LCMS, *m/z*, calculated for C$_{19}$H$_{16}$FO$_4$S$^+$: 359.0748, found: 359.0763 [M + H$^+$].

3.3. In Vitro Evaluation of COX Inhibitory Action

The COX-1 and COX-2 activities of the compounds were measured using ovine COX-1 and human recombinant COX-2 enzymes included in the "COX Inhibitor Screening Assay" kit provided by Cayman (Cayman Chemical Co., Ann Arbor, MI, USA) as reported previously [60,61]

3.4. Inhibition of the Carrageenin-Induced Edema

The anti-inflammatory activity of compounds was performed using the carrageenan mice paw oedema assay. Animal of both sex (AKR), with exception of pregnant females were housed under standard conditions, receiving only water during the experimental period. The compounds (0.01 mmol/kg body weight) were administered intraperitoneally simultaneously with the carrageenin injection. The detailed description of the whole experiment reported previously [61]. All the biological experiments were carried out in full compliance with the European Convention for the Protection of Vertebrate Animals used for Experimental or Other Scientific Purposes (ETS NO 123, Strasbourg, 03/18/1986): Strasbourg (France). European Treaty Series, № 123, March 18, 1986. 11 P. The project identification code and date of approval is 350880/3468 of 15-12-2014, University of Thessaloniki.

3.5. Cell Cultures

The antitumor activity of the compounds was tested using two different cancer cell lines, breast cancer cell line, MCF-7, which mainly expresses the COX-1 isoenzyme [43] and the tongue squamous cell carcinoma cell line, HSC3, which is known to express COX-2 isoenzyme, although COX-1 is also produced [16,20].

Malignant MCF-7 cells (human epithelial breast adenocarcinoma cell line) and HSC-3 (human oral squamous cell carcinoma cell line) cells were grown in culture (37 °C; humidified atmosphere containing 5% (v/v) CO_2) in Dulbecco's Modified Eagle's medium supplemented with 10% (v/v) fetal bovin serum (FBS), 10,000 units/mL of penicillin, 10,000 µg/mL of streptomycin, and 25 µg/mL of amphotericin B. In order to allow the continuous logarithmic phase of cell growth in culture, cells were split accordingly before reaching 80% confluency (approximately every 48–72 h) with trypsin-EDTA (0.25% w/v). According to ATCC (American Tissue Culture Collection), the MCF-7 cell express the WNT7B oncogene, the estrogen receptor, and the insulin-like growth factor binding proteins (IGFBP) BP-2, BP-4, and BP-5; addition of 0.01 mg/mL human recombinant insulin in the cell culture medium is proposed. EGF Receptors are expressed when the other growth factors' pathways are inhibited or after long term culture in the presence of Epidermal Growth Factor [62–64].

3.6. Growth Inhibition Assay

To estimate growth inhibition rate, we tested increasing concentrations (0.01–100 µM) of the compounds exposure to MCF-7 cells after appropriate serial dilutions of every compound stock solution. Each cell culture was treated appropriately, ensuring that DMSO concentration was kept below ≤ 0.2% (v/v) after every dilution. Cells were seeded in 24-well plates, at an initial density of 5×10^4 cells/mL. Cell cultures were left overnight at 37 °C to ensure appropriate cell attachment, before the compounds' addition at the specified concentrations. Cells were allowed to grow for additional 48 h before being harvested by trypsinization and subsequently counted (cell density; number of cells/mL) using a hemocytometer (Neubauer counting chamber) and optical microscopy. Cell growth was expressed as a percentage relative to that for the untreated control culture (only cells and medium, CTL). For the combination experiments, a Combination Index was measured using the equation $CI = C_A,x/Cx,_A + C_B,x/Cx,_B$ where C_A,x and C_B,x stand for the concentrations of compounds A and B respectively in the combination mixture which results in x% inhibition of cell growth, while $Cx,_A$ and $Cx,_B$ are the concentrations of compounds A and B, respectively, which alone cause x% inhibition of cell growth. A CI of less than, equal to, and more than 1 indicates synergy, additivity, and antagonism, respectively [65].

4. Conclusions

Thirteen new 4,5-diarylfuran- 3(*2H*) ones were synthesized based on molecular docking for COX inhibitory action. Their anti-inflammatory and COX-1/COX-2 inhibitory activities were evaluated in vitro. Cytotoxicity of compounds on MCS-7 and HSC-3 cancer cell lines were performed.

It was found that three compounds inhibited the MCF-7 cell growth with an IC_{50} value of 24.0 μM for **1h** and 10.0 μM for **1g** and **X-1**, while the IC_{50} values of the control compounds for gefitinib, 5-fluorouracil and celecoxib were 70.0 μM, 0.3 μM and 29.2 μM, respectively. Compound **1g** exhibited remarkable growth inhibition of the HSC-3 cells with an IC_{50} value of 7.5 μM.

The most active COX-1 selective inhibitor, **1g**, exhibited synergistic effect with both anti-neoplastic drugs with the best effect when used in combination with 5-fluorouracil (CI = 0.058). Lower synergistic effect with 5-fluorouracil was shown by the less active compound **1h** (CI = 0.533). A synergistic effect with gefitinib was found for compounds **1g** (CI = 0.780) and **X-1** (CI = 0.260), while the low activity compound **1h** exhibited antagonistic effect.

A synergistic effect of the most active COX-1 inhibitor **1g** with the known COX-2 selective inhibitor, celecoxib, was also observed (CI = 0.75), which may be indicative for the involvement of both cyclooxygenase isoenzymes in cancer cell development and the value of combined COX1/2 inhibition for best inhibitory effect.

Supplementary Materials: The following are available online, Figure S1. Structures of known COX-1/2 inhibitors (E1–E14). Figure S2. Assessment of cell growth of MCF-7 cells in culture after their treatment with increasing concentrations of the tested compounds for 48 h. Figure S3. Potential energy (NormEnergy, kcal/mol) shared by amino acids in *h*COX-2 (**a**) and *h*COX-1 (**b**) after annealing; (**c**) amino acids in *h*COX-1 with high energy values (*gray spheres*); the binding site (*yellow sphere*). Scheme S1. Synthesis of starting diaryl ketones by oxidation of substituted toluenes or benzaldehydes. For *m*-F-substituted compounds Schiemann reaction is the most rational. Table S1. Yields of the synthesis of starting diaryl ketones and their precursors according to Scheme S1.

Author Contributions: D.S. and J.M. prepared the synthesis of final compounds and their characterization by spectral analysis. Also they were involved in homology modeling and preprocessing using ICM Pro Software. 3D Molecular docking was carried out using MOE v.2014.09.01. L.-P.G., I.L. and I.S.V. contributed by performing growth inhibition assay on MCF-7 and HSC-3 cells and synergistic studies. P.E. was involved in the in vitro evaluation of COX-1/COX-2 inhibitory action and in docking studies on ovine COX-1 structure 2AYL and human COX-2 structure 5IKT and was involved in manuscript preparation. A.P. was also involved in COX-1/COX-2 evaluation. A.G. (Antonis Gavalas)was involved in antiinflammatory assay. A.G. (Athina Geronikaki)was responsible for the design of the manuscript and also performed the anti-inflammatory assay. She was responsible for the preparation of the manuscript as she is a corresponding author.

Funding: J.M. expresses his gratitude to RFBR mol_a 16-33-00059 (2015) for financial support.

Acknowledgments: Yan Ivanenkov (MIPT) for assistance in computer modeling of biological activity and Elena Bessonova (SPbSU) for mass spectrometry.

Conflicts of Interest: The authors declare no conflicts of interest.

References

1. Prasit, P.P.; Wang, Z.; Brideau, C.; Chan, C.C.; Charleson, S.E.; Cromlish, W.A.; Ethier, D.; Evans, J.F.; Ford-hutchinson, A.; Gauthier, J.Y.; et al. The discovery of rofecoxib, [MK 966, VIOXX®, 4-(4′-methylsulfonylphenyl)-3-phenyl-2(5*H*)-furanone—An orally active cyclooxygenase-2 inhibitor]. *Bioorg. Med. Chem. Lett.* **1999**, *9*, 1773–1778. [CrossRef]

2. Penning, T.D.; Talley, J.J.; Bertenshaw, S.R.; Carter, J.S.; Collins, P.W.; Docter, S.; Graneto, M.J.; Lee, L.F.; Malecha, J.W.; Miyashiro, J.M.; et al. Synthesis and Biological Evaluation of the 1.5 Diarylpyrazole Class of Cyclooxygenase-2 Inhibitors: Identification of (4-Methylphenyl)-3-(trifluoromethyl)-1H-pyrazole-1-yl]benzenesulfonamide (SC-58634, Celecoxib). *J. Med. Chem.* **1997**, *40*, 1347–1365. [CrossRef]

3. Bresalier, R.S.; Sandler, R.S.; Quan, H.; Bolognese, J.A.; Oxenius, B.; Horgan, K. Cardiovascular events associated with rofecoxib in a colorectal adenoma chemoprevention trial. *N. Engl. J. Med.* **2005**, *352*, 1092–1102. [CrossRef] [PubMed]

4. Punganuru, S.R.; Madala, H.R.; Mikelis, C.M.; Dixit, A.; Arutla, V.; Srivenugopal, K.S. Conception, synthesis, and characterization of a rofecoxib-combretastatin hybrid drug with potent cyclooxygenase-2 (COX-2) inhibiting and microtubule disrupting activities in colon cancer cell culture and xenograft models. *Oncotarget.* **2018**, *9*, 26109–26129. [CrossRef]

5. Vizirianakis, I.S.; Chatzopoulou, M.; Bonovolias, I.D.; Nicolaou, I.; Demopoulos, V.J.; Tsiftsoglou, A.S. Toward the development of innovative bifunctional agents to induce differentiation and to promote apoptosis in leukemia: Clinical candidates and perspectives. *J. Med. Chem.* **2010**, *53*, 6779–6810. [CrossRef] [PubMed]

6. Bocci, F.; Gearhart-Serna, L.; Boareto, M.; Ribeiro, M.; Ben-Jacob, E.; Devi, G.R.; Levine, H.; Onuchic, J.N.; Jolly, M.K. Toward understanding cancer stem cell heterogeneity in the tumor microenvironment. *Proc. Natl. Acad. Sci. USA* **2019**, *116*, 148–157. [CrossRef]

7. Coussens, L.M.; Werb, Z. Inflammation and cancer. *Nature* **2002**, *420*, 860–867. [CrossRef]

8. Hanahan, D.; Weinberg, R.A. Hallmarks of cancer: The next generation. *Cell* **2011**, *144*, 646–674. [CrossRef] [PubMed]

9. Bellamkonda, K.; Chandrashekar, N.K.; Osman, J.; Selvanesan, B.C.; Sayeh Savari, S.; Sjölander, A. The eicosanoids leukotriene D_4 and prostaglandin E_2 promote the tumorigenicity of colon cancer-initiating cells in a xenograft mouse model. *BMC Cancer.* **2016**, *16*, 425–439. [CrossRef]

10. Yao, L.; Liu, F.; Hong, L.; Sun, L.; Liang, S.; Wu, K.; Fan, D. The function and mechanism of COX-2 in angiogenesis of gastric cancer cells. *J. Exp. Clin. Cancer Res.* **2011**, *30*, 13–18. [CrossRef] [PubMed]

11. Diakos, C.I.; Charles, K.A.; McMillan, D.C.; Clarke, S.J. Cancer-related inflammation and treatment effectiveness. *Lancet Oncol.* **2014**, *15*, 493–503. [CrossRef]

12. Karan, D. Inflammasomes: Emerging Central Players in Cancer Immunology and Immunotherapy. *Front. Immunol.* **2018**, *9*, 3028. [CrossRef]

13. Mohamed, M.M.; Al-Raawi, D.; Sabet, S.F.; El-Shinawi, M. Inflammatory breast cancer: New factors contribute to disease etiology: A review. *J. Adv. Res.* **2014**, *5*, 525–536. [CrossRef] [PubMed]

14. Bottazzi, B.; Riboli, E.; Mantovani, A. Aging, inflammation and cancer. *Semin. Immunol.* **2018**, *40*, 74–82. [CrossRef] [PubMed]

15. Lee, C.H. Epithelial-mesenchymal transition: Initiation by cues from chronic inflammatory tumor microenvironment and termination by anti-inflammatory compounds and specialized pro-resolving lipids. *Biochem. Pharmacol.* **2018**, *158*, 261–273. [CrossRef]

16. Fujii, R.; Imanishi, Y.; Shibata, K.; Sakai, N.; Sakamoto, K.; Shigetomi, S.; Habu, N.; Otsuka, K.; Sato, Y.; Watanabe, Y.; et al. Restoration of E-cadherin expression by selective Cox-2 inhibition and the clinical relevance of epithelial-to-mesenchymal transition in head and neck squamous cell carcinoma. *J. Exp. Clin. Cancer Res.* **2014**, 33–40. [CrossRef] [PubMed]

17. Zelenay, S.; van der Veen, A.G.; Böttcher, J.P.; Snelgrove, K.J.; Rogers, N.; Acton, S.E.; Chakravarty, P.; Girotti, M.R.; Marais, R.; Quezada, S.A.; et al. Cyclooxygenase-Dependent Tumor Growth through Evasion of Immunity. *Cell* **2015**, *162*, 1257–1270. [CrossRef] [PubMed]

18. Ko, C.-Y.; Wang, W.-L.; Li, C.-F.; Jeng, Y.-M.; Chu, Y.-Y.; Wang, H.-Y.; Tseng, J.T.; Wang, J.-M. IL-18-induced interaction between IMP3 and HuR contributes to COX-2 mRNA stabilization in acute myeloid leukemia. *J. Leukoc. Biol.* **2016**, *99*, 131–141.

19. Pannunzio, A.; Coluccia, M. Cyclooxygenase-1 (COX-1) and COX-1 Inhibitors in Cancer: A Review of Oncology and Medicinal Chemistry Literature. *Pharmaceuticals* **2018**, *11*, 101. [CrossRef] [PubMed]

20. Miki, Y.; Mukae, S.; Murakami, M.; Ishikawa, Y.; Ishii, T.; Ohki, H.; Matsumoto, M.; Komiyama, K. Butyrate Inhibits Oral Cancer Cell Proliferation and Regulates Expression of Secretory Phospholipase A2-X and COX-2. *Anticancer Res.* **2007**, *27*, 1493–1502. [PubMed]

21. Ways, D.K.; Kukoly, C.A.; de Vente, J.; Hooker, J.L.; Bryant, W.O.; Posekany, K.J.; Fletcher, D.J.; Cook, P.P.; Parker, P.J. MCF-7 Breast Cancer Cells Transfected with Protein Kinase C—A Exhibit Altered Expression of Other Protein Kinase C Isoforms and Display a More Aggressive Neoplastic Phenotype. *J. Clin. Invest.* **1995**, *95*, 1906–1915. [CrossRef]

22. Rigas, B.; Goldman, I.; Levine, L. Altered eicosanoid levels in human colon cancer. *J. Lab. Clin. Med.* **1993**, *122*, 518–523.

23. Leahy, K.; Koki, A.; Masferrer, J. Role of cyclooxygenases in angiogenesis. *Curr. Med. Chem.* **2000**, *7*, 1163–1170. [CrossRef]

24. Brunelle, M.; Sartin, E.A.; Wolfe, L.G.; Sirois, J.; Doré, M. Cyclooxygenase-2 expression in normal and neoplastic canine mammary cell lines. *Vet Pathol.* **2006**, *43*, 656–666. [CrossRef]

25. Yusup, G.; Akutsu, Y.; Mutallip, M.; Qin, W.; Hu, X.; Komatsu-Akimoto, A.; Hoshino, I.; Hanari, N.; Mori, M.; Akanuma, N.; et al. A COX-2 inhibitor enhances the antitumor effects of chemotherapy and radiotherapy for esophageal squamous cell carcinoma. *Int. J. Oncol.* **2014**, *44*, 1146–1152. [CrossRef]

26. Samadder, N.J.; Neklason, D.W. Effect of Sulindac and Erlotinib vs Placebo on Duodenal Neoplasia in Familial Adenomatous Polyposis: A Randomized Clinical Trial. *JAMA* **2016**, *315*, 1266–1275. [CrossRef]

27. Shaashua, L.; Shabat-Simon, M.; Haldar, R.; Matzner, P.; Zmora, O.; Shabtai, M.; Sharon, E.; Allweis, T.; Barshack, I.; Hayman, L.; et al. Perioperative COX-2 and β-Adrenergic Blockade Improves Metastatic Biomarkers in Breast Cancer Patients in a Phase-II Randomized Trial. *Clin. Cancer Res.* **2017**, *23*, 4651–4661. [CrossRef]

28. Botti, G.; Fratangelo, F.; Cerrone, M.; Liguori, G.; Cantile, M.; Anniciello, A.M.; Scala, S.; D'Alterio, C.; Trimarco, C.; Ianaro, A.; et al. COX-2 expression positively correlates with PD-L1 expression in human melanoma cells. *J. Transl. Med.* **2017**, *15*, 16 58. [CrossRef]

29. Shin, S.S.; Noh, M.-S.; Byun, Y.J.; Choi, J.K.; Kim, J.Y.; Lim, K.M.; Ha, J.-Y.; Kim, J.K.; Lee, C.H.; Chung, S. 2,2-dimethyl-4,5-diaryl-3(2H)furanone derivatives as selective cyclo-oxygenase-2 inhibitors. *Bioorg. Med. Chem. Lett.* **2001**, *11*, 165–168. [CrossRef]

30. Shin, S.S.; Byun, Y.; Lim, K.M.; Choi, J.K.; Lee, K.W.; Moh, J.H.; Kim, J.K.; Jeong, Y.S.; Kim, J.Y.; Choi, Y.H.; et al. In vitro structure-activity relationship and in vivo studies for a novel class of cyclooxygenase-2 inhibitors: 5-aryl-2,2-dialkyl-4-phenyl-3 (2H)furanone derivatives. *J. Med. Chem.* **2004**, *47*, 792–804. [CrossRef]

31. Praveen, P.N.; Knaus, E.E. Evolution of nonsteroidal anti-inflammatory drugs (NSAIDs): Cyclooxygenase (COX) inhibition and beyond. *J. Pharm. Pharmaceut. Sci.* **2008**, *11*, 81–110.

32. Prasit, P.; Riendeau, D. Selective cyclooxygenase-2 inhibitors. *Ann. Rep. Med. Chem.* **1997**, *32*, 211–220.

33. Leval, X.D.; Julemont, F.; Delarge, J.; Sanna, V.; Pirotte, B.; Dogne, J.M. Advances in the field of COX-2 inhibition. *Expert. Opin. Ther. Patents* **2002**, *12*, 969–989.

34. Chakraborti, A.K.; Garg, S.K.; Kumar, R.; Motiwala, H.F.; Pradeep, S. Progress in COX-2 Inhibitors: A Journey So Far. *Curr. Med. Chem.* **2010**, *17*, 1563–1593. [CrossRef]

35. Rimon, G.; Sidhu, R.S.; Lauver, D.A.; Lee, J.Y.; Sharma, N.P.; Yuan, C.; Frieler, R.A.; Trievel, R.C.; Lucchesi, B.R.; Smith, W.L. Coxibs interfere with the action of aspirin by binding tightly to one monomer of cyclooxygenase-1. *Proc. Natl. Acad. Sci. USA* **2010**, *107*, 28–33. [CrossRef]

36. Wang, J.L.; Limburg, D.; Graneto, M.J.; Springer, J.; Hamper, R.J.; Liao, S.; Pawlitz, J.L.; Kurumbail, R.J.; Maziasz, T.; Talley, J.T.; et al. The novel benzopyran class of selective cyclooxygenase-2 inhibitors. Part 2: The second clinical candidate having a shorter and favorable human half-life. *Bioorg. Med. Chem. Lett.* **2010**, *20*, 7159–7163. [CrossRef]

37. Shin, S.S.; Noh, M.-S.; Byun, Y.J.; Choi, J.K.; Kim, J.K.; Lim, K.M.; Kim, J.Y.; Choi, Y.H.; Ha, J.-Y.; Lee, K.-W.; et al. 4,5-Diaryl-3(2H)furanone Derivatives as Cyclooxygenase-2 Inhibitors. U.S. Patent 6492,416, 2 December 2002.

38. Semenok, D. Selective Inhibitors of Cyclooxygenase and Method for Their Production" (**2017**). Russ. Patent RU 2631317 C1, 21 September 2017.

39. Medvedev, J.; Semenok, D.; Nikolaev, V.; Rodina, L. Method of Producing 2,2-dialkyl-4,5-diarylfuran-3(2H)-ones" (**2015**). Russ. Patent RU 2563876 C1, 21 September 2015.

40. Regitz, M.; Maas, G. *Diazo Compounds. Properties and Synthesis*; Academic Press: New York, NY, USA, 1986.

41. Medvedev, J.J.; Semenok, D.V.; Azarova, X.V.; Rodina, L.L.; Nikolaev, V.A. A New Powerful Approach to Multi-Substituted 3(2H)-Furanones via Brønsted Acid-Catalyzed Reactions of 4-Diazodihydrofuran-3-ones". *Synthesis* **2016**, *48*, 4525–4532.

42. Rodina, L.L.; Medvedev, J.J.; Galkina, O.S.; Nikolaev, V.A. Thermolysis of 4-Diazotetrahydrofuran-3-ones: Total Change of Reaction Course Compared to Photolysis. *Eur. J. Org. Chem.* **2014**, 2993–3000. [CrossRef]

43. Rodina, L.; Medvedev, Y.Y.; Moroz, P.N.; Nikolaev, V.A. Thermolysis and acid-catalyzed decomposition of 4-diazotetrahydrofuran-3-ones. A new efficient synthesis of tetrasubstituted dihydrofuran-3-ones. *Rus. J. Org. Chem.* **2012**, *48*, 602–604. [CrossRef]

44. Semenok, D.V. *Synthesis and Some Reactions of Sulfur-Containing 5,5-diaryl-4-diazodihydrofuran-3(2H)-ones-3(2H)-ones*; Skolkovo Institute of Science and Technology: Moskva, Russia, 2014. [CrossRef]

45. Nikolaev, V.A.; Galkina, O.S.; Sieler, J.; Rodina, L.L. Surprising secondary photochemical reactions observed on conventional photolysis of diazotetrahydrofuranones. *Tetrahedron Lett.* **2010**, *51*, 2713–2716. [CrossRef]

46. Moon, D.; Kim, M.; Heo, M.; Lee, J.; Choi, Y.; Kim, G. Gefitinib induces apoptosis and decreases telomerase activity in MDA-MB-231 human breast cancer cells. *Arch. Pharmacal Res.* **2009**, *32*, 1351–1360. [CrossRef]

47. Zhang, X.; Zhang, B.; Liu, J.; Liu, J.; Li, C.; Dong, W.; Fang, S.; Li, M.; Song, B.; Tang, B.; et al. Mechanisms of Gefitinib-mediated reversal of tamoxifen resistance in MCF-7 breast cancer cells by inducing ERα re-expression. *Sci. Rep.* **2015**, *5*, 7835–7842. [CrossRef]

48. Yu, L.; Chen, M.; Li, Z.; Wen, J.; Fu, J.; Guo, D.; Jiang, Y.; Wu, S.; Cho, C.-H.; Liu, S. Celecoxib Antagonizes the Cytotoxicity of Cisplatin in Human Esophageal Squamous Cell Carcinoma Cells by Reducing Intracellular Cisplatin Accumulation. *Mol. Pharm.* **2011**, *79*, 608–617. [CrossRef]

49. Liu, X.-H.; Rose, D.R. Differential Expression and Regulation of Cyclooxygenase-1 and -2 in Two Human Breast Cancer Cell Lines'. *Cancer Res.* **1996**, *56*, 5125–5127.

50. Mc Fadden, D W.; Riggs, D.R.; Jackson, B.J.; Cunningham, C. Additive effects of Cox-1 and Cox-2 inhibition on breast cancer in vitro. *Int. J. Oncol.* **2006**, *29*, 1019–1023.

51. Farrell, N.; Kiley, D.M.; Schmidt, W.; Hacker, M.P. Chemical properties and antitumor activity of complexes of platinum containing substituted sulfoxides [PtCl (R′R″SO) (diamine)]NO3. Chirality and leaving-group ability of sulfoxide affecting biological activity. *Inorg. Chem.* **1990**, *29*, 397–403. [CrossRef]

52. Suiko, M.; Maekawa, K. Synthesis and Antitumor Activity of 2-Alkanesulfinyl (or Alkanesulfonyl)-7-methyl-5H-1,3,4-thiadiazolo[3,2-a]pyrimidin-5-ones. *Agric. Biol. Chem.* **1977**, *41*, 2047–2053.

53. Tang, W.; Eisenbrand, G. Salvia miltiorrhiza Bge. *Chin. Drugs Plant Origin* **1992**, 891–902.

54. Liu, X.-Q.; Guo, Y.-Q.; Gao, W.-Y.; Zhang, T.-J.; Yan, L.-L. Two new phenanthrofurans from *Pleione bulbocodioides*. *J. Asian Nat. Prod. Res.* **2008**, *10*, 453–457. [CrossRef]

55. Prostaglandin G/H Synthase 1. Available online: http://www.uniprot.org/uniprot/P23219 (accessed on 1 May 2019).

56. Bikadi, Z.; Hazai, E. Application of the PM6 semi-empirical method to modeling proteins enhances docking accuracy of AutoDock. *J. Cheminf.* **2009**, *1*, 15–31. [CrossRef]

57. Morris, G.M.; Goodsell, D.S. Automated docking using a Lamarckian genetic algorithm and an empirical binding free energy function. *J. Comp. Chem.* **1998**, *19*, 1639–1662. [CrossRef]

58. Solis, F.J.; Wets, R.J.B. Minimization by Random Search Techniques. *MATH OPER RES.* **1981**, *6*, 19–30. [CrossRef]

59. Tsolaki, E.; Eleftheriou, P.; Kartsev, V.; Geronikaki, A.; Saxena, A.K. Application of docking analysis in the prediction and biological evaluation of the lipoxygenase inhibitory action of thiazolyl derivatives of mycophenolic acid. *Molecules* **2018**, *23*, 1621. [CrossRef]

60. Ziakas, G.N.; Rekka, E.A.; Gavalas, A.M.; Eleftheriou, P.T.; Kourounakis, P.N. New analogues of butylated hydroxytoluene asanti-inflammatory and antioxidant agents. *Bioorg. Med. Chem.* **2006**, *14*, 5616–5624. [CrossRef]

61. Lagunin, A.A.; Geronikaki, A.; Eleftheriou, P.T.; Hadjipavlou-Litina, D.I.; Filimonov, D.I.; Poroikov, V.V. Computer-aided discovery of potential anti-inflammatory thiazolidinones with dual 5-LOX/COX inhibition. *J. Med. Chem.* **2008**, *51*, 1601–1609.

62. Garcia, R.; Franklin, R.A.; McCubrey, J.A. Cell Death of MCF-7 Human Breast Cancer Cells Induced by EGFR Activation in the Absence of Other Growth Factors. *Cell Cycle* **2006**, *5*, 1840–1846.

63. Mcclelland, R.A.; Barrow, D.; Madden, T.-A.; Dutkowski, C.M.; Ppamment, J.; Knowlden, J.M.; Gee, J.M.W.; Nicholson, R.I. Enhanced Epidermal Growth Factor Receptor Signaling in MCF7 Breast Cancer Cells after Long-Term Culture in the Presence of the Pure Antiestrogen ICI 182,780 (Faslodex)*. *Endocrinology* **2001**, *142*, 2776–2788. [CrossRef]

64. Suojanen, J.; Sorsa, T.; Salo, T. Tranexamic acid can inhibit tonguesquamous cell carcinoma invasion in vitro. *Oral Dis.* **2009**, *15*, 170–175. [CrossRef]

65. Zhao, L.; Au, J.L.-S.; Wientjes, M.G. Comparison of methods for evaluating drug-drug interaction. *Front. Biosci.* **2010**, *2*, 241–249. [PubMed]

Sample Availability: Not available.

 *molecules*

Article

Synthesis of Novel 2-(Het)arylpyrrolidine Derivatives and Evaluation of Their Anticancer and Anti-Biofilm Activity

Andrey Smolobochkin [1], Almir Gazizov [1,*], Marina Sazykina [2], Nurgali Akylbekov [3], Elena Chugunova [1,4,*], Ivan Sazykin [2], Anastasiya Gildebrant [2], Julia Voronina [5], Alexander Burilov [1], Shorena Karchava [2], Maria Klimova [2], Alexandra Voloshina [1], Anastasia Sapunova [1], Elena Klimanova [6], Tatyana Sashenkova [6], Ugulzhan Allayarova [6], Anastasiya Balakina [6,7] and Denis Mishchenko [6,7]

[1] Arbuzov Institute of Organic and Physical Chemistry, FRC Kazan Scientific Center of RAS, Arbuzov str., 8, Kazan 420088, Russia
[2] Southern Federal University, Stachki Avenue, 194/2, Rostov-on-Don 344090, Russia
[3] Institute of Chemical Research and Technology of Korkyt Ata Kyzylorda State University, Aiteke bie str., 29A, Kyzylorda 120014, Kazakhstan
[4] Kazan Federal University, Kremlyovskaya str., 18, Kazan 420008, Russia
[5] N. S. Kurnakov Institute of General and Inorganic Chemistry, RAS, 31 Leninsky Av., Moscow 119991, Russia
[6] Institute of Problems of Chemical Physics RAS, Chernogolovka 142432, Russia
[7] Scientific and Educational Center in Chernogolovka of Moscow Region State University, Mytishi 141014, Russia
[*] Correspondence: agazizov@iopc.ru (A.G.); chugunova.e.a@gmail.com (E.C.);
Tel.: +7-843-272-7324 (A.G.); +7-843-272-7324 (E.C.)

Academic Editors: Carla Boga and Gabriele Micheletti
Received: 22 July 2019; Accepted: 22 August 2019; Published: 25 August 2019

Abstract: A library of novel 2-(het)arylpyrrolidine-1-carboxamides were obtained via a modular approach based on the intramolecular cyclization/Mannich-type reaction of N-(4,4-diethoxybutyl)ureas. Their anti-cancer activities both in vitro and in vivo were tested. The in vitro activity of some compounds towards M-Hela tumor cell lines was twice that of the reference drug tamoxifen, whereas cytotoxicity towards normal Chang liver cell did not exceed the tamoxifen toxicity. In vivo studies showed that the number of surviving animals on day 60 of observation was up to 83% and increased life span (ILS) was up to 447%. Additionally, some pyrrolidine-1-carboxamides possessing a benzofuroxan moiety obtained were found to effectively suppress bacterial biofilm growth. Thus, these compounds are promising candidates for further development both as anti-cancer and anti-bacterial agents.

Keywords: pyrrolidine; carboxamide; anti-tumor activity; anti-cancer activity; cytotoxicity; apoptosis; bacterial biofilm; anti-bacterial activity

1. Introduction

The pyrrolidine moiety is an important structural part of many natural alkaloids [1–4] and one of the most frequently occurring heterocyclic scaffolds in approved drugs [5]. A number of anti-cancer drugs possess an N-carboxypyrrolidine scaffold. These include both fairly old ones (e.g., dactinomycin [6], approved in 1964) and those that have appeared recently. Acalabrutinib [7,8] (approved by FDA in 2017) and larotrectinib [9,10] (approved by the FDA in 2018) may serve as illustrative examples (Figure 1). Notably, larotrectinib is the first drug to be specifically developed and approved to treat any cancer containing certain mutations, as opposed to cancers of specific tissues. It should also be emphasized that both acalabrutinib and larotrectinib contain a 2-(het)aryl-substituted pyrrolidine fragment. Considering that in the past few decades, cancer has been one of major causes of death in most countries of the world [11], a search for novel anti-cancer drug candidates among 2-(het)aryl-*N*-carboxypyrrolidine derivatives is undoubtedly a promising field of research.

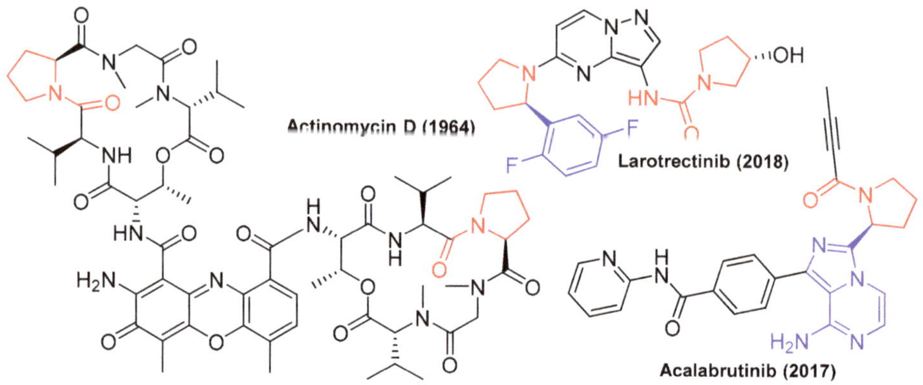

Figure 1. Approved anti-cancer drugs possessing 2-substituted *N*-carboxypyrrolidine scaffold.

Despite the increasing number of research works aimed at obtaining and studying 2-(hetaryl)pyrrolidines, the synthesis of these compounds still meets certain difficulties. Approaches to these compounds can be divided into two main groups. The first one includes the modification of an existing pyrrolidine fragment. Various cross-coupling reactions of (hetero)aromatics with appropriately substituted pyrrolidine derivatives [12–17], including oxidative [18–20] and photooxidative ones [21–24], are the most often used within this pathway. In several cases, the synthesis of enantiomerically pure 2-(het)arylpyrrolidines has been also accomplished by decarboxylative (hetero)arylation of proline derivatives [25–31]. The second approach is based on the formation of a pyrrolidine ring from acyclic precursors. Within this approach, the intermolecular [3+2] dipolar cycloaddition of activated alkenes to azomethine ylides plays a significant role [32–37]. Essential drawbacks of the abovementioned approaches are the need of expensive metal catalysts and/or harsh reaction conditions, as well as the need of the preliminary synthesis of starting compounds with appropriate functional groups and desired fragments. Hence, methods employing inexpensive, readily available reagents and catalysts and allowing simultaneous pyrrolidine ring closure and C–C$_{hetaryl}$ bond formation are of a special interest.

Earlier, we developed a metal-free approach to 2-arylsubstitued pyrrolidine derivatives based on the acid-catalyzed intramolecular cyclization of *N*-(4,4-diethoxybutyl)ureas, leading to the formation of a pyrrolinium cation. Further Mannich-type reaction of this cyclic iminium ion with various electron-rich aromatic *C*-nucleophiles allowed us to obtain a range of 2-arylpyrrolidine-1-carboxamides [38–45]. The main benefits of this approach are the usage of easily accessible starting materials and its modularity, which allows wide variability in both the aromatic moiety and substituents at the nitrogen atom (Scheme 1).

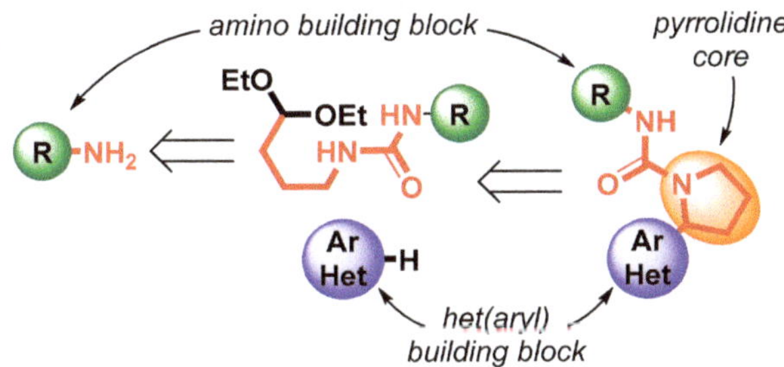

Scheme 1. Modular synthetic approach to 2-(het)arylpyrrolidine-1-carboxamides.

Herein, we report the successful extension of this approach to the synthesis of novel 2-(het)aryl-substituted pyrrolidine-1-carboxamides, as well as the evaluation of their anti-cancer activities in vivo and in vitro and studies of the inhibition of bacterial biofilm growth.

2. Results and Discussion

2.1. Chemistry

We started our research from the synthesis of initial *N*-(4,4-diethoxybutyl)ureas **1** by the previously described procedure [38,46]. Both aliphatic amines and substituted anilines were employed as the amine building block (see Scheme 1). It is well known that the ionization of an amine group is widely used for drug solubility enhancement [47,48]. Thus, *N*-(4,4-diethoxybutyl)urea **1h** possessing a dimethylamino moiety was also obtained (Table 1).

Table 1. Synthesized library of novel (het)arylpyrrolidines **5–8** [1].

		(Het)Ar Block			
	H	**5a** (93)	-	-	-
	Ph	**5b** (98)	**6b** (87)	**7b** (69)	**8b** (51)
	4-MeO-C_6H_4	**5c** (91)	**6c** (89)	**7c** (87)	**8c** (45)
R (amine block)	4-Br-C_6H_4	**5d** (95)	**6d** (95)	**7d** (75)	**8d** (68)
	4-F-C_6H_4	**5e** (91)	**6e** (76)	**7e** (58)	**8e** (51)
	n-C_6H_{13}	**5f** (72)	**6f** (44)	**7f** (47)	**8f** (34)
	cyclo-C_6H_{11}	**5g** (84)	**6g** (67)	**7g** (60)	**8g** (68)
	TFA●$(CH_2)_2NMe_2$	**5h** (75)	**6h** (57)	**7h** (77)	**8h** (73)

[1] Isolated yields are given in parentheses.

The choice of aromatic (sesamol [49,50]) and heterocyclic (4-hydroxycoumarin [51], 4-hydroxy-6-methyl-2H-pyran-2-one [52,53]) C-nucleophiles was determined by both their well-known biological activity and high reactivity in electrophilic substitution reactions. One more heterocyclic scaffold, namely, the benzofuroxan moiety, was also of interest due to its biological properties [54–56] and ability to serve as NO donor [57,58]. However, it could not be introduced to the target pyrrolidines directly due to its extremely low nucleophilicity. Thus, the phenol derivative **4** was used instead, which was obtained by the reaction of 4,6-dichloro-5-nitrobenzofuroxan **2** with 3-aminophenol **3** (Scheme 2).

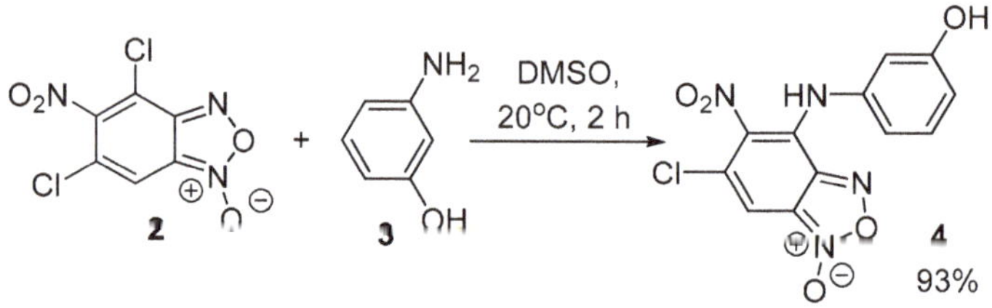

Scheme 2. Synthesis of substituted phenol **4** possessing a benzofuroxan fragment.

Next, we carried out the reaction of ureas **1a–h** with these *C*-nucleophiles in the presence of trifluoroacetic acid as catalyst (Scheme 3). As a result, a library of 28 novel 2-(het)arylpyrrolidine-1-carboxamides **5b–h**, **6b–h**, **7b–h**, **8b–h** was obtained, which included all possible combinations of amine and (het)aryl building blocks. Additionally, the non-substituted pyrrolidine-1-carboxamide **5a** was also obtained. Yields of target compounds varied from moderate to excellent (Table 1). The substitution sites of the used *C*-nucleophiles were confirmed by NMR data. The structure of compound **8f** was additionally confirmed by X-ray analysis data.

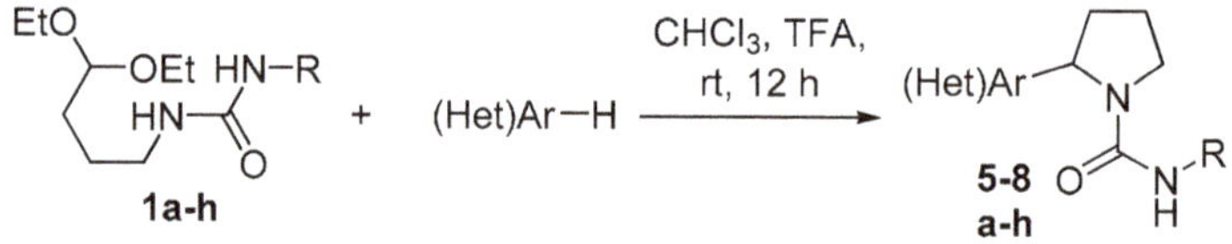

Scheme 3. Synthesis of (het)arylpyrrolidine-1-carboxamides **5–8**.

Compound **8f** crystallized with three independent molecules in the unit cell. Bond lengths, valence, and torsion angles were within the intervals typical for each bond type (Figure 2A). Molecules a, b, and c differed in the conformations of the pyrrolidine cycle (envelope in each case, however in molecules a and c atom C3 was out of the plane formed by other atoms, while in molecule b it was atom C4 and hexane substituent (Supplementary Materials, Table S1). Crystal packing consisted of centrosymmetric H-bonded dimers in which each independent molecule interacted only with its symmetric equivalent (Figure 2A, Supplementary Materials, Table S2). Dimers formed columns via π···π interactions (Supplementary Materials, Table S3). The three-dimensional system (Supplementary Materials, Figure S15) was formed by weaker CH···O (Supplementary Materials, Table S1) and CH···π interactions (Supplementary Materials, Table S4).

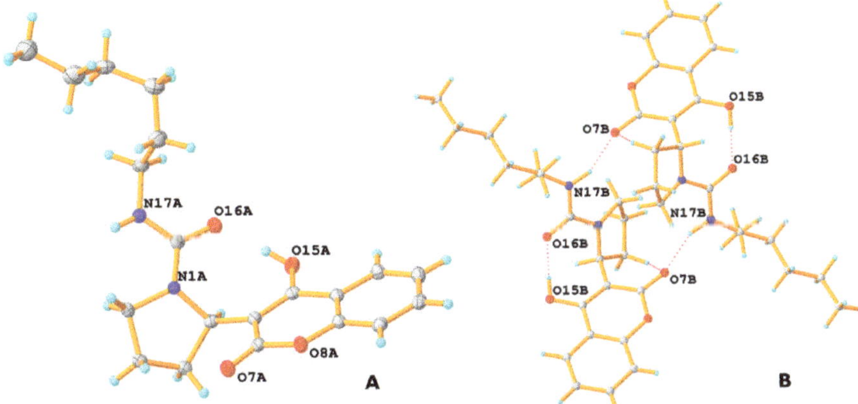

Figure 2. (**A**) Molecular structure of compound **8f** in crystal (on the example of molecule a). Ellipsoids are shown with 50% probability; (**B**) H-bonded dimer in crystal of **8f** (on the example of molecule b).

2.2. Biological Studies

2.2.1. In Vitro Studies of Anti-Cancer Activity

Cytotoxic assay. The resulting compounds were tested for cytotoxicity against normal and cancerous human cell lines at concentrations of 1–100 μM. The pyrrolidines **6b–h** containing the benzofuroxane fragment were found to be the most active, while the others (**5a–5h, 7b–7h, 8b–8h**) did not show anti-cancer activity (Table 2, Supplementary Materials, Table S5). Table 2 shows the IC_{50} data for compounds **6b–h**. It can be seen that in relation to the M-Hela cancer line, IC_{50} values for substances **6d, 6c,** and **6e** were comparable to the reference compound tamoxifen. Compound **6g** (IC_{50} −14.7 μM) was most active against cervical cancer. At the same time, with regard to the Chang's liver cell line, all test compounds were less toxic than tamoxifen.

Table 2. Cytotoxic effects of pyrrolidines **5a–6h** on the cancer and normal human cell lines [1].

Test Compound	IC_{50} (μM)		Test Compound	IC_{50} (μM)	
	Cancer Cell Line	Normal Cell Line		Cancer Cell Line	Normal Cell Line
	M-Hela	Chang Liver		M-Hela	Chang Liver
5a	>100	>100	-	-	-
5b	>100	>100	**6b**	56 ± 4.1	>100
5c	>100	>100	**6c**	26.0 ± 1.9	62 ± 4.3
5d	>100	>100	**6d**	25.5 ± 1.6	53 ± 3.8
5e	>100	>100	**6e**	26 ± 1.8	48 ± 3.0
5f	>100	>100	**6f**	47 ± 2.9	57 ± 4.3
5g	>100	>100	**6g**	14.7 ± 0.9	46 ± 2.7
5h	>100	>100	**6h**	100 ± 8.6	>100
			tamoxifen	28.0 ± 2.5	46.2 ± 3.5

[1] Three independent experiments were carried out.

Induction of apoptotic effects by test compounds. For compound **6g**, the ability to induce apoptosis in the human cancer cell line M-Hela was studied. As shown in Figure 3 (bottom), after 24 hours of treatment with compound **6g**, apoptotic effects were observed in M-Hela cells (red fluorescence). The data presented in Figure 3 show that at concentrations corresponding to the IC_{50} value, compound **6g** induced apoptosis in 25% of M-Hela cells. The top row of images in Figure 3 (control) presents images of intact M-Hela cells. It is seen that apoptotic effects were weakly expressed. The results show that the cytotoxicity of compound **6g** in M-Hela cancer cells was caused by an apoptotic pathway.

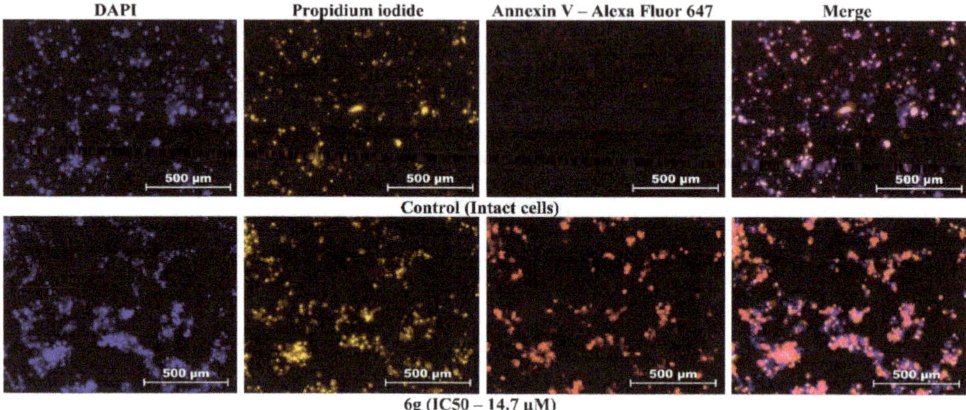

Figure 3. Images of control intact M-Hela cells (**top**) and M-Hela cells after treatment with **6g** at an IC$_{50}$ concentration of 14.7 µM (**bottom**), obtained using the Cytell Cell Imaging multifunctional system using the BioApp Automated Imaging application. Annexin V-Alexa Fluor 647 (red fluorescence) was used to detect apoptotic cells; living cells—DAPI (blue fluorescence); dead cells—propidium iodide (yellow fluorescence).

Multiplex analysis of early apoptosis markers. Next, using the MILLIPLEX® MAP 7-plex Early Phase Apoptosis Signaling kit, seven markers of early apoptosis of JNK, Bad, Bcl-2, Akt, Caspase-9, p53, and Caspase-8 were detected in M-Hela cell lysates. The median fluorescence intensity (MFI) was measured using the Luminex® system. This assay is a quick and convenient alternative to Western blot and immunoprecipitation.

Figure 4 shows that the fluorescence intensity of Caspase-8 in the experimental sample (after exposure to the test substance **6g** was two times higher than that in the control. The results suggest that apoptosis proceeded along the extrinsic pathway of activation of Caspase-8 (death is initiated by activation of the surface cell receptor), and not along the intrinsic pathway associated with the activation of Caspase-9 (fluorescence intensity at the control level) in which death occurs due to mitochondrial dysfunction. This assumption was also confirmed by the predominance of pro-apoptotic Bad proteins over anti-apoptotic Bcl-2, which are responsible for irreversible cellular damage in mitochondrial processes. At the same time, apoptosis in M-Hela cells can be induced by activating the transcription of many pro-apoptotic genes by the transcription factors AP1 (signaling pathway activated by JNK stress) and p53 (response to DNA damage).

Effects on the mitochondrial membrane potential (Δψm) by lead compounds. In confirmation of data on the course of apoptosis in cells along an external pathway not associated with dysfunctional mitochondria, the ability of the tested compounds to reduce the potential of the mitochondrial membrane (Δψm) in M-Hela culture cells was examined using the example of compound **6g**. Studies were performed using flow cytometry methods using JC-10 reagent. In normal cells with a high potential of the mitochondrial membrane, the dye JC-10 forms aggregates (J-aggregate) near the mitochondrial membranes. When the membrane potential due to the stimulation of apoptosis falls, JC-10 is evenly distributed in the cell as a monomer (J-monomer). JC-10 units in normal cells have red fluorescence, and JC-10 monomers are green. The ratio between orange-red and green fluorescence can be used to assess the onset of apoptosis. No decrease in Δψm was demonstrated using flow cytometry analysis (Figure 5). The intensity of red fluorescence after treating cells with compound **6g** in an IC$_{50}$ concentration of 14.7 µM did not actually change compared with the control. The results obtained indicate that the mechanism of action of the studied compounds is not associated with the induction of apoptosis, which proceeds along the mitochondrial pathway.

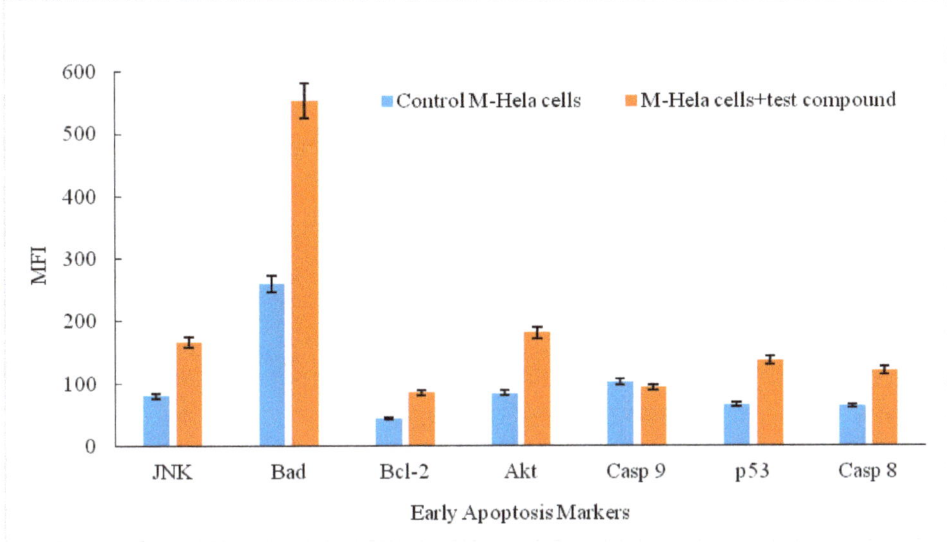

Figure 4. Multiplex analysis of early apoptosis markers in M-Hela cells treated with the test substance **6g** at an IC_{50} concentration of 14.7 µM as well as M-Hela cells untreated with the test substance (control). The median fluorescence intensity (MFI) was measured using the Luminex® system. The graph shows the mean and standard deviation values for the wells in triplicate.

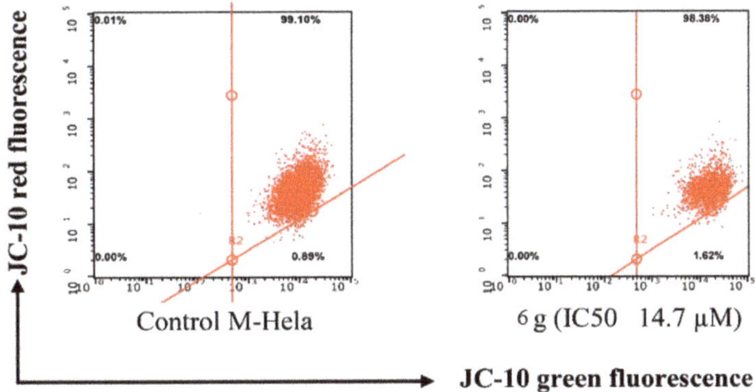

Figure 5. Flow cytometry analysis of M-Hela cells treated with compound **6g**, along with the quantification of % of cells with red aggregates. The values are presented as mean ± SD.

2.2.2. In Vivo Studies of Anti-Cancer Activity

In vivo evaluation was performed on the syngeneic P388 murine leukemia. Murine leukemia models have been an essential component of the initial drug discovery programs since the 1970s. P388 leukemia played a major role in the screening of potential antitumor agents. Today, the majority of currently used clinical drugs were first detected by the murine leukemias. These models are suitable for the initial evaluation of the antitumor activity of new compounds [59]. The compounds were administered i.p. (thus, as an intra-tumor treatment), which was believed to maximize exposure and limit pharmacokinetic influences. Due to the fact that as a solvent for parenteral administration of the substance it is permissible to use any physiologically appropriate solvent that does not cause local irritating effects such as water or saline [60], we selected only water-soluble compounds **5b**, **5h**, **6h**, and **7h** for in vivo studies.

We found that the compounds **5b**, **5h**, and **8h** had pronounced anti-leukemic activity with the number of surviving animals on day 60 of observation from 17% to 83% and increased life span (ILS) from 80% to 447% (Table 3, Table 4, and Supplementary Materials, Figure S16). Compound **5h** had the greatest anti-leukemic activity. Indeed, in the group of mice that received intraperitoneal administration of compound **5h** at a dose of 40 mg/kg/day, 1, 5, and 9 days after tumor transplantation, 83% of the animals remained alive, and ILS was 447%. In general, the studied compounds can be arranged in the following order of reducing the antitumor activity during therapy of P388 leukemia: **5h** > **8h** > **5b**. However, compounds **6h** and **7h** in the dose range and the mode of administration used did not show antitumor properties. Thus, **5b**, **5h**, and **8h** can be recommended as promising compounds for the creation of new anticancer drugs.

Table 3. Mean survival time and increased life span (ILS) of murine leukemia P388 at individual treatment with **5h**, **8h**, **5b**, **6h**, and **7h**.

Compound	Dose									
	18 mg/kg/day		26 mg/kg/day		40 mg/kg/day		55 mg/kg/day		83 mg/kg/day	
	MST ± SD [1] (days)	ILS [2] (%)	MST ± SD [1] (days)	ILS [2] (%)	MST ± SD [1] (days)	ILS [2] (%)	MST ± SD [1] (days)	ILS [2] (%)	MST ± SD [1] (days)	ILS [2] (%)
5b	27.7 ± 10.3 ***	177	26.5 ± 10.6 ***	165	26.0 ± 10.8 **	160	18.0 ± 8.4 *	80	9.7 ± 0.2	0
5h	35.7 ± 10.9 ***	276	18.5 ± 8.3 *	95	52.0 ± 8.0 ***	447	38.0 ± 9.9 ***	300	19.0 ± 8.2 **	100
6h	13.8 ± 0.6 ***	37	13.5 ± 0.5 ***	34	12.8 ± 0.7 ***	27	143.8 ± 0.2 ***	39	11.2 ± 1.6	11
7h	11.0 ± 0.6 *	0	10.3 ± 0.2	0	10.2 ± 0.2	0	10.7 ± 0.2	0	10.3 ± 0.2	0
8h	19.3 ± 8.1 *	93	20.7 ± 8.0 **	107	13.2 ± 1.6 ***	32	19.8 ± 8.1 **	98	26.2 ± 10.7 **	162

[1] MST: mean survival time; SD: standard deviation; [2] ILS%: the percentage of the median survival time (MST) of the treated group (t) to that of the control group (c). ILS% = (MSTt/MSTc) × 100. * Statistically significant increase over the control ($p < 0.05$); ** Statistically significant increase over the control ($p < 0.01$); *** Statistically significant increase over the control ($p < 0.001$).

Table 4. Effect of compounds **5b**, **5h**, **6h**, **7h**, and **8h** against i.p. implanted P388 murine tumors.

Compound	Dose									
	18 mg/kg/day		26 mg/kg/day		40 mg/kg/day		55 mg/kg/day		83 mg/kg/day	
	Survivors/Total [1]	Survival, %	Survivors/Total [1]	Survival, %	Survivors/Total [1]	Survival, %	Survivors/Total [1]	Survival, %	Survivors/Total [1]	Survival, %
Control	0/6	0	0/6	0	0/6	0	0/6	0	0/6	0
5b	2/6	33	2/6	33	2/6	33	2/6	17	0/6	0
5h	3/6	50	1/6	17	5/6	83	3/6	50	1/5	17
6h	0/6	0	0/6	0	0/6	0	0/6	0	0/5	0
7h	0/6	0	0/6	0	0/6	0	0/6	0	0/5	0
8h	1/6	17	1/6	17	1/6	0	1/6	17	1/6	33

[1] Survivors on day 60.

2.2.3. Anti-Biofilm Activity of Pyrrolidine-1-Carboxamides Possessing Benzofuroxan Moiety

Biofilms are communities of microorganisms attached to the surface or the interface in which cells are immersed in an exopolymer matrix consisting of polysaccharides, proteins, and DNA [61]. More than 90% of microorganisms occurring in nature exist in the form of biofilms [62]. Microbial biofilms are responsible for the etiology and pathogenesis of many acute and especially chronic bacterial infections in humans [63]. Among bacterial diseases in humans and animals, more than 80% are associated with the presence of stable bacterial communities enclosed in biofilms [64,65].

The possibility of interspecies community formation in biofilms along with high antibiotic resistance of many pathogenic and potentially pathogenic microorganisms make them almost invulnerable. The drug resistance of bacteria living in biofilms has increased manyfold in comparison to planktonically grown bacteria [66,67]. In this regard, the ability of pathogenic bacteria to form biofilms is a significant problem.

Moreover, in around 20% of all cases, microbial organisms are the causative agents of cancer-inducing inflammation. It is unclear if these microorganisms are causally involved in tumorigenesis, or if they benefit from the consequences of tumor growth and in turn promote tumor progression [68,69]. Furthermore, the work of Samanta et al. [70] demonstrates that the mechanisms behind advanced anti-biofilm and anticancer activities are linked to the generation of excess labile toxic reactive oxygen species (ROS). Such toxic ROS species cause the rapid oxidation and deterioration of cellular membranes. The unity of the action mechanism possibly testifies to the interconnection of anti-biofilm and anticancer activities in certain substances. Taking this into account, we also evaluated the ability of pyrrolidine-1-carboxamides **6a–e** possessing a benzofuroxan moiety and the initial benzofuroxan **4** to inhibit bacterial biofilm growth.

A staining assay was performed to estimate the extent of biofilm formation by *Vibrio aquamarinus* DSM 26054 and *Acinetobacter calcoaceticus* VKPM B-10353 under treatment with **6a, 6b, 6c, 6d, 6e**, and **4** (doses of 1×10^{-9}–1×10^{-5} M). The natural strains *A. calcoaceticus* VKPM B-10353 and *V. aquamarinus* DSM 26054 were chosen as models due to their ability to actively form biofilms and for their extreme degree of similarity to pathogenic species. The obtained biological activity results are summarized in Table 5 (see Supplementary Materials, Figures S1–S14 for additional data). The compounds were evaluated for biofilm production compared to control. The results showed that compounds exhibited a variable degree of anti-biofilm activity against *V. aquamarinus* DSM 26054 and *A. calcoaceticus* VKPM B-10353.

Table 5. Biofilm formation (%) by *Vibrio aquamarinus* DSM 26054 and *Acinetobacter calcoaceticus* VKPM B-10353 in the presence of pyrrolidine-1-carboxamides possessing benzofuroxan moiety in reference to control (control = 100%).

№	Compound	Strain	Biofilm Formation, % Compound Concentration, M				
			1×10^{-9}	1×10^{-8}	1×10^{-7}	1×10^{-6}	1×10^{-5}
1	6a	*V. aquamarinus* DSM 26054	17.70 *	10.23 *	20.34 *	16.49 *	42.65 *
		A. calcoaceticus VKPM B-10353	79.58 *	42.37 *	54.83 *	95.00	86.19
2	6b	*V. aquamarinus* DSM 26054	8.21 *	15.42 *	26.04 *	39.02 *	89.12
		A. calcoaceticus VKPM B-10353	90.41	70.79 *	73.93 *	88.44	119.69 *
3	6c	*V. aquamarinus* DSM 26054	13.54 *	14.67 *	9.22 *	18.28 *	26.65 *
		A. calcoaceticus VKPM B-10353	93.75	55.74 *	72.77 *	79.31 *	82.82
4	6d	*V. aquamarinus* DSM 26054	67.29 *	6.46 *	10.70 *	24.11 *	63.12
		A. calcoaceticus VKPM B-10353	79.16 *	46.95 *	50.00 *	95.71	89.50
5	6e	*V. aquamarinus* DSM 26054	67.29	17.96 *	11.01 *	26.02 *	39.57 *
		A. calcoaceticus VKPM B-10353	87.92 *	47.66 *	55.40 *	53.27 *	86.50
6	4	*V. aquamarinus* DSM 26054	66.58	9.31 *	11.34 *	43.56 *	96.57
		A. calcoaceticus VKPM B-10353	86.67 *	62.12 *	78.63 *	97.99	149.22*
7	Azithromycin	*V. aquamarinus* DSM 26054	103.42	102.43	99.39	103.80	81.50 *
		A. calcoaceticus VKPM B-10353	95.00	96.07	105.71	106.07	101.79

* Differences compared to the control samples are statistically significant, t criterion, $p < 0,05$; the solutions of appropriate solvent in ethanol with the same concentration were used as control in experiments with pyrrolidine-1-carboxamides; six replicates were done for each treatment and control.

Substances **6b** and **4** had different effects on *A. calcoaceticus* VKPM B-10353 biofilm formation—both suppressing and stimulating, depending on the concentration. Due to the dual nature of their action, they are not recommended for use as agents suppressing biofilm development.

For substances **6a**, **6d**, **6c**, and **6e**, the suppressive effect of different values on the formation of biofilm strains *V. aquamarinus* DSM 26054 and *A. calcoaceticus* VKPM B-10353 was registered in the range of investigated concentrations. Biofilm formation in comparison with control varied from 6.46% to 87.92%.

Biofilm formation by *A. calcoaceticus* VKPM B-10353 was less suppressed in the presence of substances than biofilm formation by *V. aquamarinus* DSM 26054. Biofilm formation varied from 42.37% to 87.92% with respect to control in the presence of the studied substances. For the *V. aquamarinus* DSM 26054 strain, it ranged from 6.46% to 67.29%.

Substances actively inhibited the growth of *A. calcoaceticus* VKPM B-10353 biofilms at the concentrations of 1×10^{-8} M and 1×10^{-7} M. The maximum inhibition of biofilms was registered under the influence of **6a** and **6d** at the concentration of 1×10^{-8} M, and biofilm formation was 42.37% and 46.95%, respectively.

The maximum inhibition of *V. aquamarinus* DSM 26054 biofilm was caused by **6a** and **6d** at the concentration of 1×10^{-8} M (biofilm formation was 10.23% and 6.46% in comparison with the control) and **6c** and **6e** at the concentration of 1×10^{-7} M (biofilm formation amounted to 9.22% and 11.01%).

Note that **6a** and **6c** actively suppressed the formation of *V. aquamarinus* DSM 26054 biofilm at the minimum concentration of 1×10^{-9} M—the preservation of biofilm was 17.7% and 13.54%, respectively.

The tested compounds were also compared with standard antibiotics azithromycin (see Supplementary Materials for additional data). Azithromycin exhibited an insignificant suppression of biofilm at high concentrations. Azithromycin suppressed the intensity of biofilm formation by *V. aquamarinus* DSM 26054 at the concentration of 1×10^{-5} M (Supplementary Materials, Figure S13). The optical density was 81.5% of the control values. The inhibitory effect of azithromycin in the studied concentrations on biofilm formation by *A. calcoaceticus* VKPM B-10353 was not detected (Supplementary Materials, Figure S14).

Taken together, pyrrolidine-1-carboxamides **6a**, **6c**, **6d**, and **6e** possessed a greater potential to suppress the formation of biofilms compared to the initial substance, as well as to the antibiotic azithromycin, and are promising as agents suppressing biofilms.

Genotoxicity and pro-oxidant characteristics of pyrrolidine-1-carboxamides possessing benzofuroxan moiety. All compounds in the same concentrations used for biofilm attenuation were also tested with *Escherichia coli* MG1655 (pRecA-lux). This strain was used for the evaluation of genotoxicity (Table 6). Bioluminescent response to DNA damage was detected for compounds **6b**, **6c**, and **6e**. The detected genotoxic effect was evaluated as medium ($I > 2$) for **6b**, **6c**, and **6e** in the concentration range of 10^{-9}–10^{-5} M, and as weak ($I < 2$) for **6b** at the concentration of 10^{-5} M. These compounds are direct mutagens. Substance **6c**, in addition, is also a promutagen. Its genotoxicity was registered under conditions of metabolic activation in the concentration range of 10^{-7}–10^{-6} M. Compounds **6a**, **6d**, and **4** do not belong to the class of DNA-damaging substances.

Table 6. Genotoxicity (induction factor, I) of the pyrrolidine-1-carboxamides possessing a benzofuroxan moiety registered with the bacterial lux-biosensor *Escherichia coli* MG1655 (pRecA-lux).

Compound	Activation [1]	Concentration of Compound, M				
		10^{-9}	10^{-8}	10^{-7}	10^{-6}	10^{-5}
6a	−	0.65 ± 0.01	0.88 ± 0.06	0.88 ± 0.03	0.61 ± 0.03	0.54 ± 0.01
	+	0.72 ± 0.04	0.86 ± 0.01	0.80 ± 0.03	0.70 ± 0.05	0.65 ± 0.06
6b	−	2.34 ± 0.06 *	2.28 ± 0.12 *	2.07 ± 0.11 *	2.15 ± 0.06 *	1.77 ± 0.04 *
	+	1.47 ± 0.04	1.31 ± 0.01	1.30 ± 0.02	1.36 ± 0.04	1.23 ± 0.02
6c	−	2.67 ± 0.07 *	2.84 ± 0.02 *	2.64 ± 0.14 *	2.82 ± 0.05 *	2.33 ± 0.04 *
	+	1.30 ± 0.01	1.37 ± 0.02	1.49 ± 0.02 *	1.54 ± 0.03 *	1.35 ± 0.02
6d	−	0.64 ± 0.03	0.70 ± 0.01	0.62 ± 0.03	0.70 ± 0.01	0.60 ± 0.02
	+	0.76 ± 0.05	0.79 ± 0.03	0.71 ± 0.05	0.71 ± 0.03	0.66 ± 0.04
6e	−	2.56 ± 0.10 *	2.74 ± 0.04 *	2.46 ± 0.02 *	2.39 ± 0.02 *	2.00 ± 0.02 *
	+	1.43 ± 0.01	1.35 ± 0.04	1.45 ± 0.02	1.47 ± 0.04	1.24 ± 0.02
4	−	1.21 ± 0.04	1.27 ± 0.07	1.14 ± 0.02	1.27 ± 0.01	1.10 ± 0.00
	+	1.19 ± 0.01	1.11 ± 0.02	1.18 ± 0.01	1.28 ± 0.04	1.11 ± 0.01

[1] Variants with metabolic activation (+S9) and without it (-S9); * difference from the control experiment are statistically significant, *t*-test; $p < 0.05$

Prooxidant characteristics (production of superoxide anion and NO) were evaluated using the biosensor *E. coli* MG1655 (pSoxS-lux) (Table 7). Compound **6b** did not possess prooxidant activity. A weak response was observed for compounds **6a** (10^{-9}–10^{-8} M), **6d** (10^{-6} M), **6c** (10^{-8}–10^{-7} M), and **6e** (10^{-7} M). On the other hand, for compound **4**, a significant effect of superoxide-anion radical or NO level increase was registered in a bacterial cell at the concentration of 1×10^{-8} M, and a weak effect was seen for concentrations 10^{-9}–10^{-8} M and 10^{-7}–10^{-5} M.

Table 7. Prooxidant activity (induction factor) of the pyrrolidine-1-carboxamides possessing a benzofuroxan moiety registered with the bacterial lux-biosensor *E. coli* MG1655 (pSoxS-lux).

Compound	Concentration of Compound, M				
	10^{-9}	10^{-8}	10^{-7}	10^{-6}	10^{-5}
6a	1.54 ± 0.13 *	1.70 ± 0.23 *	1.36 ± 0.03	1.46 ± 0.06	1.24 ± 0.08
6b	1.03 ± 0.10	1.21 ± 0.00	1.04 ± 0.04	1.24 ± 0.11	0.89 ± 0.03
6c	1.11 ± 0.03	1.54 ± 0.20 *	1.68 ± 0.20 *	1.46 ± 0.03	0.96 ± 0.09
6d	1.34 ± 0.09	1.11 ± 0.07	1.11 ± 0.01	1.60 ± 0.23 *	1.10 ± 0.07
6e	0.93 ± 0.03	1.25 ± 0.28	1.85 ± 0.15 *	1.34 ± 0.10	0.84 ± 0.04
4	1.85 ± 0.10 *	2.08 ± 0.05 *	1.74 ± 0.14 *	1.54 ± 0.07 *	1.64 ± 0.04 *

* Difference from the control experiment are statistically significant, *t*-test; $p < 0.05$.

Taken together, the most promising candidates for further studies are compounds **6a** and **6d**, for which maximum suppression of bacterial biofilm growth was observed. These compounds were found to be non-genotoxic and possessed a weak pro-oxidant activity. However, a careful study of their biological activity in eukaryotic models is still required.

3. Materials and Methods

3.1. Chemistry

IR spectra were recorded on a UR-20 spectrometer in the 400–3600 cm^{-1} range in KBr. ^{1}H-NMR spectra were recorded on a Bruker AVANCE 400 (400 MHz) spectrometer (Bruker BioSpin, Rheinstetten, Germany) with respect to the signals of residual protons of deuterated solvent (CDCl$_3$, DMSO-d$_6$). ^{13}C-NMR spectra were recorded on a Bruker Avance 600 (151 MHz) spectrometer (Bruker BioSpin, Rheinstetten, Germany) relative to signals of residual protons of deuterated solvent (CDCl$_3$, DMSO-d$_6$). Elemental analysis was performed on a CHNS-O Elemental Analyser EuroEA3028-HT-OM (EuroVector S.p.A., Milan, Italy). The melting points were determined in glass capillaries on a Stuart SMP 10 instrument. *N*-(4,4-diethoxybutyl)ureas **1a–g** were obtained as previously described [38,46].

The X-ray diffraction data for crystal of compound **8f** were collected at 150 K on a Bruker AXS Smart Apex II CCD diffractometer in the ω and φ scan modes using graphite monochromated MoKα (λ 0.71073Å) radiation. The structure was solved by direct method and refined by the full matrix least-squares using the SHELXTL program [71]. All non-hydrogen atoms were refined anisotropically. The positions of hydrogen atoms were located from the Fourier electron density synthesis and were included in the refinement in the isotropic riding model approximation. All figures were made using OLEX2 [72] and Mercury [73]. Crystallographic data for the structure reported in this paper have been deposited with the Cambridge Crystallographic Data Center (1941912, www.ccdc.ac.uk).

6-Chloro-4-((3-hydroxyphenyl)amino)-5-nitrobenzo[c][1,2,5]oxadiazole 1-oxide (**4**). To the solution of benzofuroxan 3 (0.40 g, 1.6 mmol) in DMSO (3 mL) a solution of 3-aminophenol (0.35 g, 3.2 mmol) in DMSO (3 mL) was added at room temperature. The reaction mixture was stirred at room temperature for 2 h, reagents consumption was monitored by TLC (eluent: toluene/ethylacetate, 2/1). Then the reaction mixture was poured in water (100 mL), precipitate was filtered off, washed with water, and dried. Crude product was purified by column chromatography (eluent: toluene/ethylacetate, 2/1) and recrystallized from chloroform/hexane (3/1) to give target compound **4** as dark solid. Yield 93%, m.p. 128–130 °C; IR (ν, cm^{-1}): 1563, 1628, 3094, 3320, 3447; ^{1}H-NMR (400 MHz, CHCl$_3$, δ ppm) 4.91 (s, 1H, NH), 6.73–6.75 (m, 1H, Ar-H), 6.80–6.84 (m, 2H, Ar-H), 6.92 (s, 1H, Ar-H), 7.27 (s, 1H, Ar-H), 8.49 (s, 1H, OH); ^{13}C-NMR (151 MHz, CHCl$_3$, δ ppm) 102.7, 111.8, 113.0, 114.8, 117.1, 128.4, 130.4, 130.6, 132.6, 138.8, 146.1, 156.3; Elemental analysis: calc. for C$_{12}$H$_7$ClN$_4$O$_5$ (322.5): C 44.67; H 2.19; Cl 10.99; N 17.36; found C 44.49; H 2.32; Cl 10.83; N 17.44. ESI *m/z*: [M + H]$^+$: calc. for C$_{12}$H$_8$ClN$_4$O$_5$ 323; found 323.

1-(4,4-Diethoxybutyl)-3-(2-(dimethylamino)ethyl)urea (**1h**). To a solution of N^1,N^1-dimethylethane-1,2-diamine (0.97 g, 11.0 mmol) in dichloromethane (11 mL) 1,1'-carbonyldiimidazole (2.0 g, 12.3 mmol) was added. The reaction mixture was stirred for 48 hours at room temperature. Then 4,4-diethoxybutan-1-amine (1.77 g, 11.0 mmol) was added and reaction mixture was stirred for another 48 h at room temperature. The reaction mixture was extracted with water (3 × 10 mL), the organic layer was separated, and solvent was removed in vacuum to give target compound **1h** as yellow oil. Yield 64%; ^{1}H-NMR (400 MHz, CHCl$_3$, δ ppm) 1.10 (t, 6H, *J* = 7.1 Hz, CH$_3$), 1.40–1.49 (m, 2H, CH$_2$), 1.53–1.60 (m, 2H, CH$_2$), 2.13 (s, 6H, CH$_3$), 2.31 (t, 2H, *J* = 5.9 Hz, CH$_2$), 3.04–3.11 (m, 2H, CH$_2$), 3.12–3.20 (m, 2H, CH$_2$), 3.36–3.45 (m, 2H, CH$_2$), 3.51–3.60 (m, 2H, CH$_2$), 4.39 (t, 1H, *J* = 5.6 Hz, CH), 5.37 (s, 1H,NH), 5.55 (s, 1H,NH); ^{13}C-NMR (151 MHz, CHCl$_3$, δ ppm) 15.2, 25.5, 31.1, 38.1, 40.0, 45.2, 59.3, 61.2, 102.8, 159.1.

General method for the synthesis of pyrrolidine-1-carboxamides **5–8**. To a mixture of appropriate C-nucleophile (1.61 mmol) and chloroform (5 mL), urea **1** (1.61 mmol) and trifluoroacetic acid (0.18 g, 1.61 mmol; 0.36 g, 3.22 mmol in the case of urea **1h**) were added. The reaction mixture was stirred for 24 h at room temperature, the solvent was removed in vacuum, and the residue was washed thoroughly with diethyl ether and dried in vacuum to give title compound.

2-(6-Hydroxybenzo[d][1,3]dioxol-5-yl)pyrrolidine-1-carboxamide (**5a**). Beige solid, yield 93%, m.p. 206–207 °C; IR (ν, cm^{-1}): 1534, 1637, 2986, 3174, 3294; ^{1}H-NMR (400 MHz, DMSO-d$_6$, δ ppm) 1.69–1.77 (m, 1H, CH$_2$), 1.78–1.88 (m, 2H, CH$_2$), 2.06–2.16 (m, 1H, CH$_2$), 3.29–3.36 (m, 1H, CH$_2$), 3.46–3.52 (m,

1H, CH$_2$), 4.93–4.99 (m, 1H, CH), 5.48–5.80 (br s, 2H, NH$_2$), 5.86 (dd, 2H, *J* = 4.3 Hz, 1.0 Hz, CH$_2$),), 6.41 (s, 1H, Ar-H), 6.47 (s, 1H, Ar-H); 9.07–9.86 (br s, 1H, OH). ^{13}C-NMR (151 MHz, DMSO-d_6, δ ppm) 23.8, 33.2, 46.9, 55.3, 98.4, 100.9, 106.2, 122.9, 140.1, 146.3, 149.1, 157.9; Elemental analysis: calc. for C$_{12}$H$_{14}$N$_2$O$_4$ (250): C, 57.59; H, 5.64; N, 11.19; found C, 57.70; H, 5.71; N, 11.01; ESI *m/z*: [M + H]$^+$: calc. for C$_{12}$H$_{15}$N$_2$O$_4$ 251; found 251.

2-(6-Hydroxybenzo[d][1,3]dioxol-5-yl)-N-phenylpyrrolidine-1-carboxamide (**5b**). Beige solid, yield 98%, m.p. 190–191 °C; IR (ν, cm^{-1}): 1596, 1627, 2997, 3050; ^{1}H-NMR (400 MHz, DMSO-d_6, δ ppm) 1.72–1.91 (m, 3H, CH$_2$), 2.12–2.22 (m, 1H, CH$_2$), 3.48–3.56 (m, 1H, CH$_2$), 3.72–3.80 (m, 1H, CH$_2$), 5.14–5.20 (m, 1H, CH), 5.85 (s, 2H, CH$_2$), 6.44 (s, 1H, Ar-H), 6.52 (s, 1H, Ar-H), 6.90 (t, 1H, *J* = 7.4 Hz, Ar-H), 7.20 (t, 2H, *J* = 7.8 Hz, Ar-H), 7.43 (d, 2H, *J* = 8.1 Hz, Ar-H), 7.97 (s, 1H,NH), 9.33 (s, 1H, OH); ^{13}C-NMR (151 MHz, DMSO-d_6, δ ppm) 23.7, 33.3, 47.1, 55.7, 98.2, 101.0, 106.1, 119.8, 122.1, 122.7, 128.7, 140.2, 140.9, 146.4, 148.8, 154.1; Elemental analysis: calc. for C$_{18}$H$_{18}$N$_2$O$_4$ (326): C, 66.25; H, 5.56; N, 8.58; found C, 66.31; H, 5.70; N, 8.35; ESI *m/z*: [M + H]$^+$: calc. for C$_{18}$H$_{19}$N$_2$O$_4$ 327; found 327.

2-(6-Hydroxybenzo[d][1,3]dioxol-5-yl)-N-(4-methoxyphenyl)pyrrolidine-1-carboxamide (**5c**). Beige solid, yield 91%, m.p. 112–114 °C; IR (ν, cm^{-1}): 1597, 1627, 2971, 2989, 3037; ^{1}H-NMR (400 MHz, DMSO-d_6, δ ppm) 1.72–1.92 (m, 3H, CH$_2$), 2.10–2.27 (m, 1H, CH$_2$), 3.45–3.56 (m, 2H, CH$_2$), 3.68 (s, 3H, CH$_3$), 5.12–5.22 (m, 1H, CH), 5.85 (s, 2H, CH$_2$), 6.46 (s, 1H, Ar-H), 6.52 (s, 1H, Ar-H), 6.79 (d, 2H, *J* = 9.1 Hz, Ar-H), 7.33 (d, 2H, *J* = 9.0 Hz, Ar-H), 7.85 (s, 1H,NH), 9.39 (s, 1H, OH); ^{13}C-NMR (151 MHz, DMSO-d_6, δ ppm) 23.7, 33.2, 47.0, 55.6, 55.6, 98.3, 100.9, 106.2, 114.0, 121.8, 128. 8, 133.9, 140.2, 146.4, 148.9, 154.5, 154.9; Elemental analysis: calc. for C$_{19}$H$_{20}$N$_2$O$_5$ (356): C, 64.04; H, 5.66; N, 7.86; found C, 64.21; H, 5.87; N, 7.94; ESI *m/z*: [M + H]$^+$: calc. for C$_{19}$H$_{21}$N$_2$O$_5$ 357; found 357.

N-(4-Bromophenyl)-2-(6-hydroxybenzo[d][1,3]dioxol-5-yl)pyrrolidine-1-carboxamide (**5d**). White solid, yield 95%, m.p. 164 °C; IR (ν, cm^{-1}): 1595, 1627, 2848, 2978, 3047; ^{1}H-NMR (400 MHz, DMSO-d_6, δ ppm) 1.71–1.94 (m, 3H, CH$_2$), 2.11–2.23 (m, 1H, CH$_2$), 3.46–3.57 (m, 1H, CH$_2$), 3.70–3.81 (m, 1H, CH$_2$), 5.14–5.24 (m, 1H, CH), 5.85 (s, 2H, CH$_2$), 6.44 (s, 1H, Ar-H), 6.49 (s, 1H, Ar-H), 7.37 (d, 2H, *J* = 8.8 Hz, Ar-H), 7.44 (d, 2H, *J* = 8.7 Hz, Ar-H), 8.17 (s, 1H,NH), 9.30 (s, 1H, OH); ^{13}C-NMR (151 MHz, DMSO-d_6, δ ppm) 23.7, 33.2, 47.1, 55.9, 98.2, 100.9, 106.1, 113.5, 121.6, 122.6, 131.5, 140.1, 140.4, 146.3, 148.8, 153.9; Elemental analysis: calc. for C$_{18}$H$_{17}$BrN$_2$O$_4$ (405): C, 53.35; H, 4.23; Br, 19.72; N, 6.91; found C, 53.41; H, 4.33; Br, 19.79; N, 6.79; ESI *m/z*: [M + H]$^+$: calc. for C$_{12}$H$_8$ClN$_4$O$_5$ 323; found 323; ESI *m/z*: [M + H]$^+$: calc. for C$_{18}$H$_{18}$BrN$_2$O$_4$ 406; found 406.

N-(4-Fluorophenyl)-2-(6-hydroxybenzo[d][1,3]dioxol-5-yl)pyrrolidine-1-carboxamide (**5e**). White solid, yield 91%, m.p. 184–185 °C; IR (ν, cm^{-1}): 1595, 1638, 2883, 2948, 2989, 3164; ^{1}H-NMR (400 MHz, DMSO-d_6, δ ppm) 1.72–1.83 (m, 2H, CH$_2$), 1.88–1.98 (m, 1H, CH$_2$), 2.11–2.28 (m, 1H, CH$_2$), 3.48–3.59 (m, 1H, CH$_2$), 3.68–3.78 (m, 1H, CH$_2$), 5.17–5.27 (m, 1H, CH), 6.57 (d, 2H, *J* = 7.9 Hz, CH$_2$), 6.64 (s, 1H, Ar-H), 6.85 (d, 1H, *J* = 8.2 Hz, CH$_2$), 6.99–7.07 (m, 2H, Ar-H), 7.33 (s, 1H, Ar-H), 7.40–7.48 (m, 2H, Ar-H), 8.09 (s, 1H, Ar-H), 9.72 (s, 1H,NH), 9.84 (s, 1H, OH); ^{13}C-NMR (151 MHz, DMSO-d_6, δ ppm) 23.6, 32.8, 47.1, 55.9, 102.2, 110.2, 113.7, 115.1 (d, *J* = 22.0 Hz), 121.6 (d, *J* = 7.6 Hz), 126.1, 130.5, 133.2, 137.8 (d, *J* = 124.5 Hz), 148.3, 154.2, 157.5 (d, *J* = 137.9 Hz); Elemental analysis: calc. for C$_{18}$H$_{17}$FN$_2$O$_4$ (344): C, 62.79; H, 4.98; N, 8.14; found C, 62.93; H, 5.09; N, 8.30; ESI *m/z*: [M + H]$^+$: calc. for C$_{18}$H$_{18}$FN$_2$O$_4$ 345; found 345.

N-Hexyl-2-(6-hydroxybenzo[d][1,3]dioxol-5-yl)pyrrolidine-1-carboxamide (**5f**). White solid, yield 72%, m.p. 91–93 °C; IR (ν, cm^{-1}): 1535, 1624, 2720, 2855, 2929, 3115; ^{1}H-NMR (400 MHz, CDCl$_3$, δ ppm) 0.87 (t, 3H, *J* = 6.9 Hz, CH$_3$), 1.23–1.31 (m, 6H, CH$_2$), 1.42–1.51 (m, 2H, CH$_2$), 2.04–2.13 (m, 2H, CH$_2$), 2.18–2.32 (m, 2H, CH$_2$), 3.10–3.16 (m, 1H, CH$_2$), 3.19–3.28 (m, 1H, CH$_2$), 3.40–3.54 (m, 2H, CH$_2$), 5.12–5.21 (m, 1H, CH), 5.85 (d, 2H, *J* = 12.4 Hz, CH$_2$), 6.47 (s, 1H, Ar-H), 6.61 (s, 1H, Ar-H); ^{13}C-NMR (151 MHz, CDCl$_3$, δ ppm) 14.0, 22.6, 24.9, 26.5, 30.1, 31.5, 32.7, 40.9, 46.4, 55.0, 99.9, 100.9, 105.3, 120.8, 141.0, 147.3, 150.4, 158.0; Elemental analysis: calc. for C$_{18}$H$_{26}$N$_2$O$_4$ (334): C, 64.65; H, 7.84; N, 8.38; found C, 64.75; H, 8.06; N, 8.25; ESI *m/z*: [M + H]$^+$: calc. for C$_{18}$H$_{27}$N$_2$O$_4$ 335; found 335.

N-Cyclohexyl-2-(6-hydroxybenzo[d][1,3]dioxol-5-yl)pyrrolidine-1-carboxamide (**5g**). White solid, yield 84%, m.p. 159–160 °C; IR (ν, cm^{-1}): 1528, 1624, 2720, 2855, 2929, 3113, 3403; ^{1}H-NMR (400 MHz,

DMSO-d_6, δ ppm) 1.03–1.12 (m, 2H, CH_2), 1.13–1.25 (m, 3H, CH_2), 1.47–1.66 (m, 4H, CH_2), 1.69–1.78 (m, 2H, CH_2), 1.79–1.90 (m, 2H, CH_2), 2.07–2.19 (m, 1H, CH_2), 3.31–3.43 (m, 2H, CH_2), 3.47–3.54 (m, 1H, CH_2), 4.95–5.04 (m, 1H, CH), 5.85 (d, 2H, *J* = 8.2 Hz, CH_2), 6.42 (s, 1H, Ar-H), 6.47 (s, 1H, Ar-H); ^{13}C-NMR (151 MHz, DMSO-d_6, δ ppm) 23.9, 25.2, 25.8, 33.5, 33.6, 46.7, 49.1, 55.0, 98.3, 100.9, 106.2, 122.7, 140.3, 146.5, 149.1, 156.3; Elemental analysis: calc. for $C_{18}H_{24}N_2O_4$ (332): C, 65.04; H, 7.28; N, 8.43; found C, 65.16; H, 7.42; N, 8.35; ESI *m/z*: [M + H]$^+$: calc. for $C_{18}H_{25}N_2O_4$ 333; found 333.

2-(2-(6-Hydroxybenzo[d][1,3]dioxol-5-yl)pyrrolidine-1-carboxamido)-N,N-dimethylethan-1-aminium 2,2,2-trifluoroacetate (**5h**). White solid, yield 75%, m.p. 171–172 °C; IR (ν, cm^{-1}): 1527, 1626, 2838, 2929, 3113, 3394, ^{1}H-NMR (400 MHz, DMSO-d_6, δ ppm) 1.69–1.80 (m, 2H, CH_2), 1.81–1.89 (m, 1H, CH_2), 2.04–2.09 (m, 1H, CH_2), 2.79 (s, 6H, CH_3), 3.06–3.16 (m, 2H, CH_2), 3.31–3.37 (m, 3H, CH_2), 3.53–3.60 (m, 1H, CH_2), 4.99–5.05 (m, 1H, CH), 5.85 (d, 2H, *J* = 11.0 Hz, CH_2), 6.42 (s, 1H, Ar-H), 6.44 (s, 1H, Ar-H), 9.37 (s, 1H,NH), 9.63 (s, 1H, OH); ^{13}C-NMR (151 MHz, DMSO-d_6, δ ppm) 23.4, 33.2, 35.9, 43.1, 46.7, 55.9, 57.9, 98.3, 100.9, 106.1, 117.7 (q, *J* = 300.4 Hz), 122.8, 139.9, 146.2, 148.9, 157.1, 158.6 (q, *J* = 31.0 Hz); Elemental analysis: calc. for $C_{18}H_{24}F_3N_3O_6$ (435): C, 49.66; H, 5.56; N, 9.65; found C, 49.85; H, 5.67; N, 9.80; ESI *m/z*: [M − CF$_3$CO$_2$]$^+$: calc. for $C_{16}H_{24}N_3O_4$ 322; found 322.

6-Chloro-4-((3-hydroxy-4-(1-(phenylcarbamoyl)pyrrolidin-2-yl)phenyl)amino)-5-nitrobenzo[c][1,2,5]oxadiazole 1-oxide (**6b**). Dark solid, yield 87%, m.p. 165–170 °C with decomposition; IR (ν, cm^{-1}): 753, 1383, 1559, 1627, 3073, 3336, 3396; ^{1}H-NMR (400 MHz, DMSO-d_6, δ ppm) 1.80 (m, 2H, CH_2), 1.91 (m, 1H, CH_2), 2.20 (m, 1H, CH_2), 3.53–3.54 (m, 1H, CH_2), 3.75 (m, 1H, CH_2), 5.22–5.23 (m, 1H, CH), 6.57 (d, 1H, *J* = 7.7 Hz, Ar-H), 6.64 (s, 1H, Ar-H), 6.87 (d, 1H, *J* = 8.2 Hz, Ar-H), 6.91 (t, 1H, *J* = 7.40 Hz, Ar-H), 7.20 (t, 2H, *J* = 7.9 Hz, Ar-H), 7.33 (s, 1H, Ar-H), 7.43 (d, 2H, *J* = 7.1 Hz, Ar-H), 7.99 (s, 1H, NH), 9.74 (s, 1H, NH), 9.84 (s, 1H, OH); ^{13}C-NMR (151 MHz, DMSO-d_6, δ ppm) 23.7, 33.0, 47.1, 55.8, 102.3, 110.3, 113.9, 114.3, 119.9, 122.1, 126.2, 127.1, 128.1, 128.7, 130.6, 133.3, 138.5, 140.9, 148.3, 154.2, 154.4; Elemental analysis: calc. for $C_{23}H_{19}ClN_6O_6$ (510.9): C 54.07; H 3.75; Cl 6.94; N 16.45; found C 54.18; H 3.83; Cl 6.85; N 16.32; ESI *m/z*: [M + H]$^+$: calc. for $C_{23}H_{20}ClN_6O_6$ 511.9; found 512.

6-Chloro-4-((3-hydroxy-4-(1-((4-methoxyphenyl)carbamoyl)pyrrolidin-2-yl)phenyl)amino)-5-nitrobenzo[c] [1,2,5]oxadiazole 1-oxide (**6c**). Dark solid, yield 89%, m.p. 196–202 °C with decomposition; IR (ν, cm^{-1}): 750, 1377, 1557, 1628, 3075, 3306, 3391; ^{1}H-NMR (400 MHz, DMSO-d_6, δ ppm) 1.73–1.84 (m, 2H, CH_2), 1.86–1.94 (m, 1H, CH_2), 2.12–2.23 (m, 1H, CH_2), 3.47–3.53 (m, 1H, CH_2), 3.69 (s, 3H, CH_3), 3.69–3.74 (m, 1H, CH_2), 5.16–5.23 (m, 1H, CH), 6.57 (d, 1H, *J* = 7.6 Hz, Ar-H), 6.63 (s, 1H, Ar-H), 6.78 (d, 2H, *J* = 9.02 Hz, Ar-H), 6.87 (d, 1H, *J* = 8.3 Hz, Ar-H), 7.31 (s, 1H, Ar-H), 7.33 (d, 2H, *J* = 4.8 Hz, Ar-H), 7.86 (br s, 1H, NH), 9.73 (br s, 1H, NH), 9.83 (nr s, 1H, OH); ^{13}C-NMR (151 MHz, DMSO-d_6, δ ppm) 23.7, 32.1, 32.9, 47.1, 55.6, 102.2, 107.0, 110.3, 114.0, 121.8, 121.9, 123.0, 124.5, 127.1, 130.6, 133.9, 138.5, 141.3, 146.7, 148.2, 154.5, 154.9; Elemental analysis: calc. for $C_{24}H_{21}ClN_6O_7$ (540.9): C 53.29; H 3.91; Cl 6.55; N 15.54; found C 53.38; H 3.81; Cl 6.42; N 15.67; ESI *m/z*: [M + H]$^+$: calc. for $C_{24}H_{22}ClN_6O_7$ 541.9; found 542.

4-((4-(1-((4-Bromophenyl)carbamoyl)pyrrolidin-2-yl)-3-hydroxyphenyl)amino)-6-chloro-5-nitrobenzo[c] [1,2,5]oxadiazole 1-oxide (**6d**). Dark solid, yield 95%, m.p. 166–170 °C with decomposition; IR (ν, cm^{-1}): 753, 1362, 1560, 1629, 3100, 3297, 3382; ^{1}H-NMR (400 MHz, DMSO-d_6, δ ppm) 1.74–1.84 (m, 2H, CH_2), 1.84–1.97 (m, 1H, CH_2), 2.15–2.24 (m, 1H, CH_2), 3.47–3.57 (m, 1H, CH_2), 3.68–3.77 (m, 1H, CH_2), 5.18–5.25 (m, 1H, CH), 6.56 (d, 1H, *J* = 8.2 Hz, Ar-H), 6.63 (s, 1H, Ar-H), 6.85 (d, 1H, *J* = 8.24 Hz, Ar-H), 7.33 (s, 1H, Ar-H), 7.36 (d, 2H, *J* = 8.8 Hz, Ar-H), 7.44 (t, 2H, *J* = 8.8 Hz, Ar-H), 8.19 (br s, 1H, NH), 9.71 (br s, 1H, NH), 9.83 (br s, 1H, OH); ^{13}C-NMR (151 MHz, DMSO-d_6, δ ppm) 23.6, 32.9, 47.1, 56.0, 110.2, 113.5, 113.8, 114.3, 120.8, 121.7, 126.1, 127.1, 130.6, 131.4, 132.0, 138.5, 139.5, 140.4, 148.3, 153.9, 154.4; Elemental analysis: calc. for $C_{23}H_{18}BrClN_6O_6$ (589.8): C 46.84; H 3.08; Cl 6.01; N 14.25; found C 46.92; H 3.12; Cl 6.05; N 14.34; ESI *m/z*: [M + H]$^+$: calc. for $C_{23}H_{19}BrClN_6O_6$ 590.8; found 591.

6-Chloro-4-((4-(1-((4-fluorophenyl)carbamoyl)pyrrolidin-2-yl)-3-hydroxyphenyl)amino)-5-nitrobenzo[c] [1,2,5]oxadiazole 1-oxide (**6e**). Dark solid, yield 76%, m.p. 175–177 °C with decomposition; IR (ν, cm^{-1}): 1348, 1564, 1623, 3247, 3404; ^{1}H-NMR (400 MHz, DMSO-d_6, δ ppm) 1.65–1.83 (m, 2H, CH_2), 1.86–1.98 (m, 1H, CH_2), 2.11–2.27 (m, 1H, CH_2), 3.49–3.58 (m, 1H, CH_2), 3.69–3.79 (m, 1H, CH_2), 5.16–5.25 (m, 1H, CH), 6.57 (dd, 1H, *J* = 8.2 Hz, *J* = 2.1 Hz, Ar-H), 6.64 (d, 1H, *J* = 2.1 Hz, Ar-H), 6.85 (d, 1H, *J* = 8.1

Hz, Ar-H), 6.98–7.07 (m, 2H, Ar-H), 7.33 (s, 1H, Ar-H), 7.41–7.49 (m, 2H, Ar-H), 8.09 (s, 1H, NH), 9.72 (s, 1H,NH), 9.84 (s, 1H, OH); ^{13}C-NMR (151 MHz, DMSO-d_6, δ ppm) 23.6, 32.8, 47.1, 55.9, 102.2, 110.2, 113.7, 114.2, 115.10 (d, J = 22.0 Hz), 121.6 (d, J = 7.6 Hz), 126.1, 127.1, 128.0, 130.5, 133.2, 137.2 (d, J = 2.1 Hz), 138.4, 148.3, 154.2, 157.7 (d, J = 237.9 Hz); Elemental analysis: calc. for $C_{23}H_{18}ClFN_6O_6$ (528.9): C, 52.23; H, 3.43; Cl, 6.70; N, 15.89; found C, 52.00; H, 3.61; Cl, 6.87; N, 16.07; ESI *m/z*: [M + H]$^+$: calc. for $C_{23}H_{19}ClFN_6O_6$ 529.9; found 530.

6-Chloro-4-((4-(1-(hexylcarbamoyl)pyrrolidin-2-yl)-3-hydroxyphenyl)amino)-5-nitrobenzo[c][1,2,5]oxadiazole 1-oxide (**6f**). Dark solid, yield 44%, m.p. 121–122 °C with decomposition; IR (ν, cm^{-1}): 722, 1347, 1563, 1630, 3106, 3194, 3426; ^{1}H-NMR (400 MHz, DMSO-d_6, δ ppm) 0.84 (t, 3H, J = 6.7 Hz, CH$_3$), 1.17–1.26 (m, 6H, CH$_2$), 1.31–1.38 (m, 2H, CH$_2$), 1.71–1.79 (m, 2H, CH$_2$), 1.80–1.90 (m, 1H, CH$_2$), 2.07–2.18 (m, 1H, CH$_2$), 2.90–3.02 (m, 2H, CH$_2$), 3.42–3.54 (m, 2H, CH$_2$), 4.99–5.06 (m, 1H, CH), 5.85 (br s, 1H, NH), 6.55 (d, 1H, J = 7.2 Hz, Ar-H), 6.61 (s, 1H, Ar-H), 6.79 (d, 1H, J = 8.3 Hz, Ar-H), 7.34 (s, 1H, Ar-H), 9.01 (br s, 1H, NH), 9.85 (br s, 1H, OH); ^{13}C-NMR (151 MHz, DMSO-d_6, δ ppm) 14.4, 22.5, 23.7, 26.5, 30.4, 31.5, 32.9, 46.7, 55.3, 102.3, 110.4, 113.7, 126.3, 127.1, 128.3, 130.6, 133.3, 134.9, 138.5, 148.3, 154.6, 157.0; Elemental analysis: calc. for $C_{23}H_{27}ClN_6O_6$ (518.9): C 53.23; H 5.24; Cl 6.83; N 16.19; found C 53.35; H 5.33; Cl 6.72; N 16.12; ESI *m/z*: [M + H]$^+$: calc. for $C_{23}H_{28}ClN_6O_6$ 519.9; found 520.

6-Chloro-4-((4-(1-(cyclohexylcarbamoyl)pyrrolidin-2-yl)-3-hydroxyphenyl)amino)-5-nitrobenzo[c][1,2,5]oxadiazole 1-oxide (**6g**). Dark solid, yield 67%, m.p. 170–175 °C with decomposition; IR (ν, cm^{-1}): 632, 1348, 1564, 1623, 2934, 3247, 3404; ^{1}H-NMR (400 MHz, DMSO-d_6, δ ppm) 0.99–1.23 (m, 6H, CH$_2$), 1.50–1.79 (m, 7H, CH$_2$), 2.30–2.35 (m, 1H, CH$_2$), 2.64–2.69 (m, 1H, CH$_2$), 3.27–3.33 (m, 1H, CH$_2$), 3.33–3.39 (m, 1H, CH), 4.95–5.02 (m, 1H, CH), 6.57 (d, 1H, J = 8.05 Hz, Ar-H), 6.61 (s, 1H, Ar-H), 6.82 (d, 1H, J = 8.13 Hz, Ar-H), 7.31 (s, 1H, Ar-H), 7.93 (br s, 1H, NH), 9.83 (br s, 1H, NH), 9.94 (br s, 1H, OH). ^{13}C-NMR (151 MHz, DMSO-d_6, δ ppm) 23.8, 25.3, 25.8, 33.6, 46.8, 49.2, 55.0, 65.5, 102.3, 110.4, 114.0, 126.5, 127.1, 128.7, 129.4, 130.6, 133.4, 138.7, 148.3, 154.7, 156.3; Elemental analysis: calc. for $C_{23}H_{25}ClN_6O_6$ (516.9): C 53.44; H 4.87; Cl 6.86; N 16.26; found C 53.33; H 4.75; Cl 6.79; N 16.21; ESI *m/z*: [M + H]$^+$: calc. for $C_{23}H_{26}ClN_6O_6$ 517.9; found 518.

6-Chloro-4-((4-(1-((2-(dimethylammonio)ethyl)carbamoyl)pyrrolidin-2-yl)-3-hydroxyphenyl)amino)-5-nitrobenzo[c][1,2,5]oxadiazole 1-oxide 2,2,2-trifluoroacetate (**6h**). Beige solid, yield 57%, m.p. 211–212 °C with decomposition; IR (ν, cm^{-1}): 1349, 1566, 1623, 3175, 3275, 3400; ^{1}H-NMR (400 MHz, DMSO-d_6, δ ppm) 1.70–1.79 (m, 1H, CH$_2$), 1.84–1.91 (m, 1H, CH$_2$), 2.08–2.19 (m, 1H, CH$_2$), 2.79 (s, 6H, CH$_3$), 3.07–3.13 (m, 2H, CH$_2$), 3.32–3.38 (m, 3H, CH$_2$), 3.51–3.58 (m, 1H, CH$_2$), 5.04–5.12 (m, 1H, CH), 6.54 (d, 1H, J = 7.9 Hz, Ar-H), 6.58 (s, 1H, Ar-H), 6.63 (s, 1H, Ar-H), 7.35 (s, 1H, NH), 9.76 (s, 1H,NH), 9.83 (s, 1H, OH); ^{13}C-NMR (151 MHz, DMSO-d_6, δ ppm) 23.3, 32.9, 36.0, 43.1, 46.7, 55.9, 58.0, 102.3, 110.1, 113.5, 114.3, 114.7, 117.7 (q, J = 300.3 Hz), 126.1, 127.1, 130.5, 133.4, 138.5, 140.3, 148.3, 154.5, 157.1, 158.2, 158.5 (q, J = 31.0 Hz); Elemental analysis: calc. for $C_{23}H_{25}ClF_3N_7O_8$ (619.9): C, 44.56; H, 4.06; Cl, 5.72; N, 15.82; found C, 44.79; H, 3.81; Cl, 5.87; N, 15.99; ESI *m/z*: [M − CF$_3$CO$_2$]$^+$: calc. for $C_{21}H_{25}ClN_7O_6$ 506.9; found 507.

2-(4-Hydroxy-6-methyl-2-oxo-2H-pyran-3-yl)-N-phenylpyrrolidine-1-carboxamide (**7b**). White solid, yield 69%, m.p. 190–191 °C; IR (ν, cm^{-1}): 1588, 1612, 1692, 2630, 2687, 2872, 2961; ^{1}H-NMR (400 MHz, DMSO-d_6, δ ppm) 1.76–1.87 (m, 1H, CH$_2$), 1.95–2.06 (m, 2H, CH$_2$), 2.08–2.21 (m, 1H, CH$_2$), 2.13 (s, 3H, CH$_3$), 3.50–3.65 (m, 2H, CH$_2$), 4.99–5.09 (m, 1H, CH), 5.98 (s, 1H, Ar-H), 6.88 (t, 1H, J = 7.3 Hz, Ar-H), 7.18 (t, 2H, J = 7.8 Hz, Ar-H), 7.42 (d, 2H, J = 8.1 Hz, Ar-H), 7.81 (s, 1H, NH), 11.54 (s, 1H, OH); ^{13}C-NMR (151 MHz, DMSO-d_6, δ ppm) 19.7, 25.4, 30.8, 47.4, 52.1, 100.7, 118.1, 119.4, 121.8, 128.7, 141.1, 153.6, 161.3, 163.8, 166.1; Elemental analysis: calc. for $C_{17}H_{18}N_2O_4$ (314): C, 64.96; H, 5.77; N, 8.91; found C, 65.22; H, 5.89; N, 8.80; ESI *m/z*: [M + H]$^+$: calc. for $C_{17}H_{19}N_2O_4$ 315; found 315.

2-(4-Hydroxy-6-methyl-2-oxo-2H-pyran-3-yl)-N-(4-methoxyphenyl)pyrrolidine-1-carboxamide (**7c**). White solid, yield 87%, m.p. 168–169 °C; IR (ν, cm^{-1}): 1579, 1612, 1690, 2631, 2678, 2872, 3006; ^{1}H-NMR (400 MHz, DMSO-d_6, δ ppm) 1.75–1.86 (m, 1H, CH$_2$), 1.97–2.18 (m, 3H, CH$_2$), 2.13 (s, 3H, CH$_3$), 3.46–3.59 (m, 2H, CH$_2$), 3.68 (s, 3H, CH$_3$), 4.96–5.03 (m, 1H, CH), 5.97 (s, 1H, Ar-H), 6.78 (d, 2H, J = 8.8 Hz, Ar-H), 7.31 (d, 2H, J = 9.1 Hz, Ar-H), 7.69 (s, 1H, NH), 11.54 (s, 1H, OH); ^{13}C-NMR (151 MHz,

DMSO-d_6, δ ppm) 19.7, 25.4, 30.7, 47.3, 52.1, 55.6, 100.7, 102.9, 114.0, 121.2, 134.1, 154.0, 154.7, 161.3, 163.8, 166.2; Elemental analysis: calc. for $C_{18}H_{20}N_2O_5$ (344): C, 62.78; H, 5.85; N, 8.13; found C, 62.89; H, 5.98; N, 8.06; ESI *m/z*: [M + H]$^+$: calc. for $C_{18}H_{21}N_2O_5$ 345; found 315.

N-(4-Bromophenyl)-2-(4-hydroxy-6-methyl-2-oxo-2H-pyran-3-yl)pyrrolidine-1-carboxamide (**7d**). White solid, yield 75%, m.p. 194–195 °C; IR (ν, cm^{-1}): 1578, 1612, 1679, 2654, 2815, 2877, 2989; ^{1}H-NMR (400 MHz, DMSO-d_6, δ ppm) 1.76–1.86 (m, 1H, CH$_2$), 1.92–2.04 (m, 2H, CH$_2$), 2.06–2.17 (m, 1H, CH$_2$), 2.12 (s, 3H, CH$_3$), 3.50–3.61 (m, 2H, CH$_2$), 4.97–5.05 (m, 1H, CH), 5.96 (s, 1H, Ar-H), 7.35 (d, 2H, *J* = 8.7 Hz, Ar-H), 7.43 (d, 2H, *J* = 8.5 Hz, Ar-H), 8.03 (s, 1H, NH), 11.42 (s, 1H, OH); ^{13}C-NMR (151 MHz, DMSO-d_6, δ ppm) 19.7, 25.4, 30.8, 47.4, 52.3, 100.7, 102.7, 113.1, 121.2, 131.5, 140.6, 153.3, 161.2, 163.7, 166.0; Elemental analysis: calc. for $C_{17}H_{17}BrN_2O_4$ (392): C, 51.92; H, 4.36; Br, 20.32; N, 7.12; found C, 52.09; H, 4.50; Br, 20.48; N, 7.31; ESI *m/z*: [M + H]$^+$: calc. for $C_{17}H_{18}BrN_2O_4$ 393; found 393.

N-(4-Fluorophenyl)-2-(4-hydroxy-6-methyl-2-oxo-2H-pyran-3-yl)pyrrolidine-1-carboxamide (**7e**). White solid, yield 58%, m.p. 177–178 °C; IR (ν, cm^{-1}): 1602, 1631, 1689, 2631, 2689, 2881, 3066; ^{1}H-NMR (400 MHz, DMSO-d_6, δ ppm) 1.74–1.87 (m, 1H, CH$_2$), 1.91–2.14 (m, 3H, CH$_2$), 2.12 (s, 3H, CH$_3$), 3.50–3.63 (m, 2H, CH$_2$), 4.95–5.07 (m, 1H, CH), 5.97 (s, 1H, Ar-H), 6.98–7.04 (m, 2H, Ar-H), 7.39–7.49 (m, 2H, Ar-H), 7.92 (s, 1H, NH), 11.62 (s, 1H, OH); ^{13}C-NMR (151 MHz, DMSO-d_6, δ ppm) 19.7, 25.4, 30.8, 47.3, 52.2, 100.6, 102.7, 115.1 (d, *J* = 22.0 Hz), 121.0 (d, *J* = 7.4 Hz), 137.4 (d, *J* = 2.5 Hz), 153.6, 157.5 (d, *J* = 237.6 Hz), 161.2, 163.7, 166.0; Elemental analysis: calc. for $C_{17}H_{17}FN_2O_4$ (332): C, 61.44; H, 5.16; N, 8.43; found 61.55; H, 4.89; N, 8.27; ESI *m/z*: [M + H]$^+$: calc. for $C_{17}H_{18}FN_2O_4$ 333; found 333.

N-Hexyl-2-(4-hydroxy-6-methyl-2-oxo-2H-pyran-3-yl)pyrrolidine-1-carboxamide (**7f**). White solid, yield 47%, m.p. 137–138 °C; IR (ν, cm^{-1}): 1592, 1690, 2631, 2686, 2935, 3079; ^{1}H-NMR (400 MHz, DMSO-d_6, δ ppm) 0.84 (t, 3H, *J* = 7.0 Hz, CH$_3$), 1.16–1.27 (m, 6H, CH$_2$), 1.30–1.38 (m, 2H, CH$_2$), 1.70–1.80 (m, 1H, CH$_2$), 1.95–2.10 (m, 3H, CH$_2$), 2.13 (s, 3H, CH$_3$), 2.87–2.96 (m, 1H, CH$_2$), 2.98–3.07 (m, 1H, CH$_2$), 3.31–3.35 (m, 1H, CH$_2$), 3.36–3.41 (m, 1H, CH$_2$), 4.80–4.88 (m, 1H, CH), 5.70 (s, 1H, NH), 5.96 (s, 1H, Ar-H); ^{13}C-NMR (151 MHz, DMSO-d_6, δ ppm) 14.4, 19.7, 22.5, 25.3, 26.5, 30.3, 30.4, 31.5, 35.6, 47.0, 51.9, 101.0, 103.0, 156.9, 161.4, 163.5, 167.0; Elemental analysis: calc. for $C_{17}H_{26}N_2O_4$ (322): C, 63.33; H, 8.13; N, 8.69; found C, 63.50; H, 8.31; N, 8.87; ESI *m/z*: [M + H]$^+$: calc. for $C_{17}H_{27}N_2O_4$ 323; found 323.

N-Cyclohexyl-2-(4-hydroxy-6-methyl-2-oxo-2H-pyran-3-yl)pyrrolidine-1-carboxamide (**7g**). Beige solid, yield 60%, m.p. 184–185 °C; IR (ν, cm^{-1}): 1593, 1692, 2631, 2686, 2934, 3079; ^{1}H-NMR (400 MHz, DMSO-d_6, δ ppm) 0.99–1.14 (m, 3H, CH$_2$), 1.16–1.28 (m, 2H, CH$_2$), 1.48–1.54 (m, 1H, CH$_2$), 1.55–1.66 (m, 3H, CH$_2$), 1.69–1.77 (m, 2H, CH$_2$), 1.99–2.09 (m, 3H, CH$_2$), 2.12–2.22 (m, 1H, CH$_2$), 2.14 (s, 3H, CH$_3$), 3.36–3.39 (m, 1H, CH$_2$), 3.40–3.47 (m, 1H, CH$_2$), 4.78–4.87 (m, 1H, CH), 5.36 (s, 1H, NH), 5.97 (s, 1H, Ar-H), 12.01 (s, 1H, OH); ^{13}C-NMR (151 MHz, DMSO-d_6, δ ppm) 19.7, 25.1, 25.3, 25.8, 30.3, 33.6, 47.1, 49.1, 51.6, 100.8, 102.8, 156.1, 161.7, 163.6, 167.0; Elemental analysis: calc. for $C_{17}H_{24}N_2O_4$ (320): C, 63.73; H, 7.55; N, 8.74; found C, 63.87; H, 7.76; N, 8.59; ESI *m/z*: [M + H]$^+$: calc. for $C_{17}H_{25}N_2O_4$ 321; found 321.

2-(2-(4-Hydroxy-6-methyl-2-oxo-2H-pyran-3-yl)pyrrolidine-1-carboxamido)-N,N-dimethylethan-1-aminium 2,2,2-trifluoroacetate (**7h**). White solid, yield 77%, m.p. 146–147 °C; IR (ν, cm^{-1}): 1592, 1692, 2683, 2985, 3064; ^{1}H-NMR (400 MHz, DMSO-d_6, δ ppm) 1.73–1.81 (m, 1H, CH$_2$), 1.83–1.90 (m, 1H, CH$_2$), 1.96–2.08 (m, 2H, CH$_2$), 2.12 (s, 3H, CH$_3$), 2.78 (s, 6H, CH$_3$), 3.05–3.12 (m, 2H, CH$_2$), 3.21–3.30 (m, 1H, CH$_2$), 3.32–3.39 (m, 3H, CH$_2$), 4.85–4.94 (m, 1H, CH), 5.98 (s, 1H, Ar-H), 6.22 (s, 1H, NH), 9.56 (s, 1H, OH), 11.70 (s, 1H, NH$^+$); ^{13}C-NMR (151 MHz, DMSO-d_6, δ ppm) 19.7, 25.2, 31.3, 36.0, 43.1, 46.9, 52.4, 58.1, 100.8, 102.9, 177.7 (q, *J* = 31.2 Hz), 156.8, 158.6 (q, *J* = 299.4 Hz), 160.9, 163.6, 166.2; Elemental analysis: calc. for $C_{17}H_{24}F_3N_3O_6$ (423): C, 48.23; H, 5.71; N, 9.92; found C, 48.30; H, 5.85; N, 10.14; ESI *m/z*: [M − CF$_3$CO$_2$]$^+$: calc. for $C_{15}H_{24}N_3O_4$ 310; found 310.

2-(4-Hydroxy-2-oxo-2H-chromen-3-yl)-N-phenylpyrrolidine-1-carboxamide (**8b**). Beige solid, yield 51%, m.p. 179–180 °C; IR (ν, cm^{-1}): 1595, 1616, 1695, 2853, 2930, 3075. ^{1}H-NMR (400 MHz, DMSO-d_6, δ ppm) 1.87–1.98 (m, 1H, CH$_2$), 2.13–2.27 (m, 3H, CH$_2$), 3.64–3.71 (m, 2H, CH$_2$), 5.23–5.29 (m, 1H, CH), 6.91 (t, 1H, *J* = 7.4 Hz, Ar-H), 7.19 (t, 3H, *J* = 7.9 Hz, Ar-H), 7.32–7.37 (m, 2H, Ar-H, NH), 7.42 (d, 2H, *J* = 8.0 Hz, Ar-H), 7.59 (t, 1H, *J* = 8.0 Hz, Ar-H), 7.94 (d, 1H, *J* = 7.9 Hz, Ar-H), 8.26 (s, 1H, OH); ^{13}C-NMR

(151 MHz, DMSO-d_6, δ ppm) 25.8, 29.6, 47.8, 53.2, 106.5, 116.5, 118.1, 120.2, 122.3, 124.0, 124.3, 128.7, 132.5, 140.6, 152.7, 154.9, 161.3, 162.2; Elemental analysis: calc. for $C_{20}H_{18}N_2O_4$ (350): C, 68.56; H, 5.18; N, 8.00; found C, 68.70; H, 5.40; N, 7.89; ESI *m/z*: [M + H]$^+$: calc. for $C_{20}H_{19}N_2O_4$ 351; found 351.

2-(4-Hydroxy-2-oxo-2H-chromen-3-yl)-N-(4-methoxyphenyl)pyrrolidine-1-carboxamide (**8c**). Beige solid, yield 45%, m.p. 152–153 °C; IR (ν, cm^{-1}): 1596, 1617, 1697, 2847, 2984, 3036; ^{1}H-NMR (400 MHz, DMSO-d_6, δ ppm) 1.81–2.00 (m, 1H, CH$_2$), 2.12–2.23 (m, 2H, CH$_2$), 2.24–2.34 (m, 1H, CH$_2$), 3.59–3.66 (m, 2H, CH$_2$), 3.69 (s, 3H, CH$_3$), 5.18–5.28 (m, 1H, CH), 6.79 (d, 2H, *J* = 9.1 Hz, Ar-H), 7.25–7.38 (m, 5H, Ar-H, NH), 7.59 (t, 1H, *J* = 7.0 Hz, Ar-H), 7.92 (d, 1H, *J* = 6.7 Hz, Ar-H), 8.19 (s, 1H, OH); ^{13}C-NMR (151 MHz, DMSO-d_6, δ ppm) 25.8, 29.4, 47.7, 53.2, 53.6, 106.5, 114.0, 116.4, 120.1, 122.3, 123.9, 124.3, 132.5, 133.3, 152.7, 155.1, 155.5, 161.2, 162.6; Elemental analysis: calc. for $C_{21}H_{20}N_2O_5$ (380): C, 66.53; H, 5.50; N, 7.49; found C, 66.53; H, 5.50; N, 7.49; ESI *m/z*: [M + H]$^+$: calc. for $C_{21}H_{21}N_2O_5$ 381; found 381.

N-(4-Bromophenyl)-2-(4-hydroxy-2-oxo-2H-chromen-3-yl)pyrrolidine-1-carboxamide (**8d**). Beige solid, yield 68%, m.p. 179 °C; IR (ν, cm^{-1}): 1595, 1617, 2797, 2837, 2987, 3078; ^{1}H-NMR (400 MHz, DMSO-d_6, δ ppm) 1.83–1.86 (m, 1H, CH$_2$), 2.08–2.27 (m, 3H, CH$_2$), 3.62–3.72 (m, 2H, CH$_2$), 5.23–5.31 (m, 1H, CH), 7.30–7.40 (m, 5H, Ar-H, NH), 7.43 (d, 2H, *J* = 8.90 Hz), 7.59 (t, 1H, *J* = 8.2 Hz, Ar-H), 7.94 (d, 1H, *J* = 8.3 Hz, Ar-H), 8.39 (s, 1H, OH); ^{13}C-NMR (151 MHz, DMSO-d_6, δ ppm) 25.8, 29.8, 47.8, 53.2, 106.5, 113.6, 116.5, 120.0, 121.8, 123.9, 124.3, 131.5, 132.4, 140.2, 152.7, 154.3, 161.3, 161.9; Elemental analysis: calc. for $C_{20}H_{17}BrN_2O_4$ (429): C, 55.96; H, 3.99; Br, 18.61; N, 6.53; found C, 56.14; H, 4.18; Br, 18.73; N, 6.70; ESI *m/z*: [M + H]$^+$: calc. for $C_{20}H_{18}BrN_2O_4$ 430; found 430.

N-(4-Fluorophenyl)-2-(4-hydroxy-2-oxo-2H-chromen-3-yl)pyrrolidine-1-carboxamide (**8e**). Beige solid, yield 51%, m.p. 165 °C; IR (ν, cm^{-1}): 1589, 1624, 1697, 2847, 2983, 3036, 3106; ^{1}H-NMR (400 MHz, DMSO-d_6, δ ppm) 1.92–1.98 (m, 1H, CH$_2$), 1.12–1.28 (m, 3H, CH$_2$), 3.61–3.70 (m, 2H, CH$_2$), 5.22–5.29 (m, 1H, CH), 6.99–7.05 (m, 2H, Ar-H), 7.29–7.40 (m, 3H, Ar-H, NH), 7.40–7.47 (m, 2H, Ar-H), 7.59 (t, 1H, *J* = 8.5 Hz, Ar-H), 7.93 (d, 1H, *J* = 6.9 Hz, Ar-H), 8.33 (s, 1H, OH); ^{13}C-NMR (151 MHz, DMSO-d_6, δ ppm) 25.7, 29.6, 47.7, 53.1, 106.5, 115.2 (d, *J* = 22.0 Hz), 116.4, 117.0, 121.9 (d, *J* = 7.7 Hz), 123.9, 124.2, 132.4, 136.9 (d, *J* = 2.5 Hz), 152.7, 154.8, 157.8 (d, *J* = 138.1 Hz), 161.3, 162.2; Elemental analysis: calc. for $C_{20}H_{17}FN_2O_4$ (368): C, 65.21; H, 4.65; N, 7.60; found C, 65.42; H, 4.78; N, 7.83; ESI *m/z*: [M + H]$^+$: calc. for $C_{20}H_{18}FN_2O_4$ 369; found 369.

N-Hexyl-2-(4-hydroxy-2-oxo-2H-chromen-3-yl)pyrrolidine-1-carboxamide (**8f**). Beige solid, yield 34%, m.p. 124-125 °C; IR (ν, cm^{-1}): 1554, 1614, 1687, 2858, 2929, 2953, 3075, 3374; ^{1}H-NMR (400 MHz, DMSO-d_6, δ ppm) 0.79 (t, 3H, *J* = 6.8 Hz, CH$_3$), 1.19–1.22 (m, 5H, CH$_2$), 1.30–1.35 (m, 1H, CH$_2$), 1.37–1.41 (m, 1H, CH$_2$), 1.85–1.95 (m, 1H, CH$_2$), 1.07–1.16 (m, 1H, CH$_2$), 2.20–2.17 (m, 1H, CH$_2$), 2.40–2.48 (m, 1H, CH$_2$), 2.94–3.02 (m, 2H, CH$_2$), 3.03–3.10 (m, 1H, CH$_2$), 3.31–3.39 (m, 1H, CH$_2$), 3.41–3.49 (m, 1H, CH$_2$), 5.06–5.14 (m, 1H, CH), 6.57 (s, 1H, NH); 7.29–7.33 (m, 2H, Ar-H), 7.55–7.60 (m, 1H, Ar-H), 7.87 (d, 1H, *J* = 7.8 Hz, Ar-H); ^{13}C-NMR (151 MHz, DMSO-d_6, δ ppm) 14.3, 22.5, 25.7, 26.5, 28.8, 30.1, 31.5, 40.7, 47.4, 53.4, 106.3, 116.3, 117.2, 124.1, 124.2, 132.6, 153.0, 158.8, 161.1, 164.5; Elemental analysis: calc. for $C_{20}H_{26}N_2O_4$ (358): C, 67.02; H, 7.31; N, 7.82; found C, 67.29; H, 7.55; N, 7.99; ESI *m/z*: [M + H]$^+$: calc. for $C_{20}H_{27}N_2O_4$ 359; found 359.

Crystal data: $C_{20}H_{26}N_2O_4$, M = 358.43, colorless crystal 0.12 × 0.15 × 0.15 mm^3, triclinic, space group P-1, Z = 6, a = 12.9336(12), b = 14.2803(13), c = 16.0108(14) Å, α = 70.825(2), β = 84.411(2), γ = 87.932(2)°, V = 2779.8(4) Å^3, ρ_{calc} = 1.285 g/cm^3, μ = 0.9 mm^{-1}, 33,903 reflections collected (±h, ±k, ±l), 14,791 independent (Rint 0.0849) and 7433 observed reflections [I ≥ 2σ(I)], 703 refined parameters, R = 0.0673, wR_2 = 0.1723, max. residual electron density was 0.667 (−0.528) eÅ^{-3}.

N-Cyclohexyl-2-(4-hydroxy-2-oxo-2H-chromen-3-yl)pyrrolidine-1-carboxamide (**8g**). Beige solid, yield 68%, m.p. 166–167 °C; IR (ν, cm^{-1}): 1549, 1616, 1694, 2475, 2853, 2935, 3075, 3373; ^{1}H-NMR (400 MHz, DMSO-d_6, δ ppm) 1.01–1.10 (m, 1H, CH$_2$), 1.14–1.26 (m, 4H, CH$_2$), 1.48–1.56 (m, 1H, CH$_2$), 1.59–1.67 (m, 2H, CH$_2$), 1.68–1.78 (m, 2H, CH$_2$), 1.84–1.94 (m, 1H, CH$_2$), 2.06–2.16 (m, 1H, CH$_2$), 2.20–2.29 (m, 1H, CH$_2$), 2.39–2.48 (m, 1H, CH$_2$), 3.35–3.50 (m, 3H, CH$_2$, CH), 5.04–5.13 (m, 1H, CH), 6.26 (s, 1H, OH), 7.30–7.35 (m, 2H, Ar-H), 7.59 (td, 1H, *J* = 7.8 Hz, *J* = 1.6 Hz, Ar-H) 7.88 (dd, 1H, *J* = 8.3 Hz, *J* = 1.6 Hz, Ar-H); ^{13}C-NMR (151 MHz, DMSO-d_6, δ ppm) 25.4, 25.7, 28.7, 33.3, 33.5, 47.5, 49.9, 53.4, 106.2, 116.3,

117.2, 124.1, 124.2, 132.6, 153.0, 158.1, 161.1, 164.5; Elemental analysis: calc. for $C_{20}H_{24}N_2O_4$ (356): C, 67.40; H, 6.79; N, 7.86; found C, 67.54; H, 6.89; N, 7.73; ESI *m/z*: [M + H]$^+$: calc. for $C_{20}H_{25}N_2O_4$ 357; found 357.

2-(2-(4-Hydroxy-2-oxo-2H-chromen-3-yl)pyrrolidine-1-carboxamido)-N,N-dimethylethan-1-aminium 2,2,2-trifluoroacetate (**8h**). Beige solid, yield 73%, m.p. 147–148 °C; IR (v, cm^{-1}): 1544, 1615, 1684, 2718, 2876, 2957, 3038, 3368; ^{1}H-NMR (400 MHz, DMSO-d_6, δ ppm) 1.83–1.94 (m, 1H, CH$_2$), 2.08–2.22 (m, 2H, CH$_2$), 2.24–2.35 (m, 1H, CH$_2$), 2.78 (s, 6H, CH$_3$), 3.06–3.15 (m, 2H, CH$_2$), 3.26–3.33 (m, 1H, CH$_2$), 3.35–3.41 (m, 1H, CH$_2$), 3.43–3.49 (m, 2H, CH$_2$), 5.08–5.16 (m, 1H, CH), 6.75 (s, 1H, NH), 7.29–7.38 (m, 2H, Ar-H), 7.59 (t, 1H, *J* = 8.2 Hz, Ar-H) 7.92 (d, 1H, *J* = 7.9 Hz, Ar-H) 9.62 (s, 1H, OH); ^{13}C-NMR (151 MHz, DMSO-d_6, δ ppm) 25.6, 29.7, 36.0, 43.0, 47.2, 53.3, 57.4, 106.2, 116.2, 116.9 (q, *J* = 199.9 Hz), 117.4, 124.1, 124.2, 132.4, 152.9, 157.9, 158.7 (q, *J* = 32.1 Hz), 161.3, 163.5; Elemental analysis: calc. for $C_{20}H_{24}F_3N_3O_6$ (458): C, 52.29; H, 5.27; N, 9.15; found C, 52.48; H, 5.16; N, 8.97; ESI *m/z*: [M + H]$^+$: calc. for $C_{18}H_{24}N_3O_4$ 346; found 346.

3.2. Biological Studies

3.2.1. In Vitro Studies of Anti-Cancer Activity

Cytotoxicity assay. Cytotoxic effects of the test compounds on human cancer and normal cells were estimated by means of the multifunctional Cytell Cell Imaging system (GE Health Care Life Science, Sweden) using the Cell Viability Bio App which precisely counts the number of cells and evaluates their viability from fluorescence intensity data. Two fluorescent dyes that selectively penetrate the cell membranes and fluoresce at different wavelengths were used in the experiments. A low-molecular-weight 4′,6-diamidin-2-phenylindol dye (DAPI) is able to penetrate the intact membranes of living cells and color nuclei in blue. The high-molecular-weight propidium iodide dye penetrates only dead cells with damaged membranes, staining them in yellow. As a result, living cells are painted in blue and dead cells are painted in yellow. DAPI and propidium iodide were purchased from Sigma. The M-Hela clone 11 human, epithelioid cervical carcinoma, strain of Hela, clone of M-Hela from the Type Culture Collection of the Institute of Cytology (Russian Academy of Sciences) and Chang liver cell line (Human liver cells) from N. F. Gamaleya Research Center of Epidemiology and Microbiology were used in the experiments. The cells were cultured in a standard Eagle's nutrient medium manufactured at the Chumakov Institute of Poliomyelitis and Virus Encephalitis (PanEco company) and supplemented with 10% fetal calf serum and 1% nonessential amino acids. The cells were plated into a 96-well plate (Eppendorf) at a concentration of 100,000 cells/mL, 150 µL of medium per well, and cultured in a CO$_2$ incubator at 37 °C. Twenty-four hours after seeding the cells into wells, the compound under study was added at a preset dilution, 150 µL to each well. The dilutions of the compounds at concentrations of 1–100 µM were prepared immediately in nutrient media; 5% DMSO (which does not induce the inhibition of cells at this concentration) was added for better solubility. The experiments were repeated three times. Intact cells cultured in parallel with experimental cells were used as a control.

Induction of Apoptotic Effects by test compounds. Cell Culture. M-Hela cells at 1×10^6 cells/well in a final volume of 2 mL were seeded into six-well plates. After 24 hours of incubation, a solution of the test compound **6g** was added to the wells at the concentration studied.

Cytell Cell Imaging System Assay. M-Hela cells were plated into a 24-well plate (Eppendorf) at a concentration of 1×10^6 cells/mL, 500 µL of medium per well, and cultured in a CO$_2$ incubator at 37 °C. Twenty-four hours after seeding the cells into wells the compound was added at a preset dilution, 500 µL to each well. The dilutions of compound **6g** were prepared immediately in nutrient media; 5% DMSO (which did not induce the inhibition of cells at this concentration) was added for better solubility. Evaluation of apoptotic effects was performed with the help of multifunctional system Cytell Cell Imaging, using Cell Viability BioApp and Automated Imaging BioApp applications. The annexin V-Alexa Fluor 647 apoptosis detection kit, DAPI and propidium iodide purchased from Sigma.

Multiplex analysis of early apoptosis markers. M-Hela cells were incubated for 24 hours with the test substance. Cells were lysed in MILLIPLEX® MAP Lysis buffer containing protease inhibitors. Twenty micrograms of total protein of each lysate diluted in MILLIPLEX® MAP Assay Buffer 2 was analyzed according to the analysis protocol (the lysate was incubated at 4 °C overnight). The mean fluorescence intensity (MFI) was detected using the Luminex® system, MERCK, USA.

Flow Cytometry Assay. Mitochondrial membrane potential. Cells were harvested at 2000 rpm for 5 min and then washed twice with ice-cold PBS, followed by resuspension in JC-10 (10 µg/mL) and incubation at 37 °C for 10 min. After the cells were rinsed three times and suspended in PBS, the JC-10 fluorescence was observed by flow cytometry (Guava easy Cyte 0I IT, Guava Technologies Inc., Hayward, CA, USA).

Statistical analysis. The experiments were repeated three times. The cytometric results were analyzed by the Cytell Cell Imaging multifunctional system using the Cell Viability BioApp and Apoptosis BioApp application. The data in the tables and graphs are given as the mean ± standard error.

3.2.2. In Vivo Studies of Anti-Cancer Activity

Animals. In vivo experiments were performed using the BDF_1 hybrid male mice of 22–24 g weight. The experimental animals were caged in a standard vivarium in 12 h light conditions with free access to food and water. All manipulations with the animals were performed in accordance with the solutions of the Commission on Bioethics of the Institute of Problems of Chemical Physics, Russian Academy of Sciences (IPCP RAS).

Anti-tumor activity. The tumors were transplanted intraperitoneally (i.p.) in accordance with a standard procedure inoculum: 10^6 tumor cells in isotonic solution of NaCl, V = 0.2 cm^3 (leukemia P388) [60]. Original compounds were injected intraperitoneally as aqueous solution. Doses from 18 to 83 mg/kg/day and the mode of administration on days 1, 5, and 9 after transplantation were used. In each experiment, a single group of tumor-bearing animals not injected with the compounds served as the control group. Each group consisted of six mice. The animals were observed daily for survival for a minimum of 60 days. The efficacy of the therapy against leukemia (defined as increase in lifespan—ILS) was assessed as the percentage of the median survival time (MST) of the treated group (t) to that of the control group (c): ILS(%) = (MSTt/MSTc) × 100.

Statistics. The experiments were carried out in triplicate. The data are presented in the form X ± SD (mean ± standard deviation). The significance of the differences between the groups was assessed using Student's *t*-test. Values of $p < 0.05$ were considered statistically significant. The data were processed statistically using GraphPad Prism.

3.2.3. Bacterial Biofilm Formation Inhibitory Activity

Bacterial strains and cultivation conditions. For the detection of biofilms, formation strains *Vibrio aquamarinus* DSM 26054 and *Acinetobacter calcoaceticus* VKPM B-10353 were used. These strains form biofilms, making them useful for studying biofilms.

E. coli MG1655 (pRecA-lux) was used for the evaluation of the genotoxicity of the synthesized compounds. The biosensor with the PrecA promotor fixes the presence of the factors causing damage of DNA in a cell [74]. The biosensor *E. coli* MG1655 (pSoxS-lux) was used for the evaluation of prooxidant activity. The biosensor with the PsoxS promoter fixes the production of superoxide anion and NO [75]. Bioluminescent strains were obtained by the transformation of *E. coli* MG1655 by hybrid plasmids pRecA-lux, pSoxS-lux. The gene cassette luxCDABE *Photorhabdus luminescens* under the control PrecA promoters was used in this biosensor. This plasmid was created on the basis of pBR322 and contained a selective marker of ampicillin resistance (Amp gene). The strains were kindly furnished by Manukhov I.V., Federal State Unitary Enterprise "GosNIIGenetika").

The bacterial strains *Acinetobacter calcoaceticus* VKPM B-10353, *E. coli* MG1655 (pRecA-lux), and *E. coli* MG1655 (pSoxS-lux) were cultivated in Luria–Bertani (LB) medium [76] under constant shaking to early exponential phase at 37 °C. Cells were used immediately for stress induction tests. One hundred micrograms of ampicillin per milliliter were added into LB medium at cultivation of *E. coli* MG1655 (pRecA-lux) and *E. coli* MG1655 (pSoxS-lux). Strain *V. aquamarinus* DSM 26054 was grown in LB medium supplemented with 3% NaCl.

Chemicals. All of the chemicals used were of analytical grade. Crystal violet and *N*-methyl-*N*-nitro-*N*-nitrosoguanidine were obtained from Sigma-Aldrich (USA). Ampicillin was obtained from Sintez (Russia). Azithromycin was obtained from Farmstandart (Russia). Test solutions were prepared in deionized water immediately before the tests. Rat liver microsomal enzymes (S9 fraction) were from Moltox (USA).

The test compounds were dissolved in DMSO to the concentration of 1×10^{-2} M. Then, they were diluted with ethanol. The control solutions were analogous dilutions of DMSO in ethanol. The tested compounds were also compared with the standard antibiotic azithromycin. Azithromycin was dissolved in DMSO to the concentration of 5×10^{-3} M and then diluted with deionized water.

Biosensors assay procedure. The detailed protocol of toxicity testing by means of a bacterial lux-biosensors is described in the article [77].

Calculation. The criterion of toxic influence was bioluminescence intensity change of the test object in the researched sample in comparison with the control sample.

The induction factor (I) was defined as the relation of luminescence intensity of a lux-biosensor suspension containing tested sample (L_c) to the luminescence intensity of a lux-biosensor control suspension (L_k): $I = L_c/L_k$.

If at significant differences from control induction factor values were ≤ 2, the detected genotoxic effect was evaluated as "weak", if they were in the range from 2 to 10 as "medium", and above 10 as "strong". All the experiments were carried out three times independently.

Difference reliability of bioluminescence in experiment from control value was estimated by *t*-criterion with the help of Excel software. The conclusions about sample toxicity were made at $p < 0.05$.

Test system for evaluation of biofilms production. To quantify the formation of biofilms, the crystal violet assay was used, with some modifications [78]. The necessary concentrations of the test compounds were prepared as described above.

V. aquamarinus DSM 26054 was cultivated for 24 h in LB medium supplemented with 3% NaCl in the Innova 40R shaker incubator (New Brunswick Scientific, USA) at 25 °C and 200 rpm. *A. calcoaceticus* VKPM B-10353 was cultivated for 24 h in LB medium in the Innova 40R shaker incubator (New Brunswick Scientific, Enfield, CT, USA) at 30 °C and 200 rpm. Then, the suspensions of the daily culture of *V. aquamarinus* DSM 26054 and *A. calcoaceticus* VKPM B-10353 were diluted with LB medium supplemented with 3% NaCl to the density of 1×10^8 cells/mL.

The resulting suspension (180 µL) was added to the wells of a polystyrene microplate (Nuova Aptaca, Canelli, Italy). To some of the wells, 20 µL of the test substances at various concentrations were added. Since solvents used could also influence the biofilm formation, 20 µL of the appropriate solvent was added to the other part of the wells at same dilutions (control). Six replicates were done for each treatment and control. The microplate was covered with a lid and wrapped with Parafilm (Bemis Company, Inc., Oshkosh, WI, USA).

After incubation at 25 °C for 72 h, biofilms were stained. The contents in the wells were removed by means of a dispenser. The wells were then carefully washed three times with 250 µL of sterile saline. The microplates were shaken to remove all non-adherent bacteria. Biofilms were fixed with 200 µL of 96% ethanol for 15 min. After the microplates had dried in air, 200 µL of 0.5% crystal violet was introduced into the wells. After 10 min, the dye was removed. The excess dye was removed by washing with water three times. After the microplates were air-dried, the dye in the wells bound to biofilms was dissolved with 200 µL of 96% ethanol. The extraction level (absorption) of crystal violet by ethanol was measured after 60 minutes at a wavelength of 570 nm using a FLUOstar Omega microplate reader (BMG Labtech, Offenburg, Germany) in optical density units (OD570). The intensity of biofilm formation directly corresponds to the intensity of staining of the contents of the wells with the dye. Biofilm formation was determined by the difference between the mean OD readings obtained in the presence of compounds and the control.

Each experiment was performed in triplicate. The values were expressed as mean + SD. Student's t-test was used to compare these values. Differences were considered statistically significant at $p < 0.05$.

4. Conclusions

In conclusion, a series of novel 2-(het)arylpyrrolidine-1-carboxamides were obtained via a modular approach based on intramolecular cyclization/Mannich-type reaction of N-(4,4-diethoxybutyl)ureas. Their anti-cancer activities were tested both in vitro and in vivo. A pyrrolidine derivative possessing a *cyclo*-hexyl substituent in the carboxamide moiety and a benzofuroxan fragment in the pyrrolidine ring was determined as the most active in the in vitro assay. Notably, its activity towards M-Hela tumor cell lines was found to be twice that of reference drug tamoxifen. At the same time, its cytotoxicity towards normal Chang liver cells did not exceed tamoxifen's toxicity. The obtained results indicate that the death of M-Hela cells presumably occurs via an apoptotic pathway due to activation of the surface cell receptors and not due to mitochondrial dysfunction. In the in vivo studies, water-soluble compounds possessing N-(2-(dimethylamino)ethyl)pyrrolidine-1-carboxamide scaffold and either a heterocyclic (hydroxycoumarine) or aromatic (sesamol) substituent in the pyrrolidine core were proven to be the most effective. The number of surviving animals on day 60 of observation ranged from 17% to 83% and increased life span (ILS) ranged from 80% to 447%. Additionally, compounds possessing a benzofuroxan moiety were found to effectively suppress bacterial biofilm growth, and thus are promising candidates for further development as anti-bacterial agents.

Supplementary Materials: The following are available online at http://www.mdpi.com/1420-3049/24/17/3086/s1, Figures S1–S14 (Anti-biofilm activity data), Figure S15, Tables S1–S4 (X-ray data); Figure S16 (In vivo anti-cancer activity data), copies of NMR spectra of all synthesized compounds.

Author Contributions: A.S.—investigation (chemistry), N.A.—investigation (chemistry), A.G.—supervision (chemistry), conceptualization, writing—original draft, E.C.—writing—review & editing, funding acquisition, A.B.—project administration, I.S., A.G., S.K., M.Z.—investigation (anti-biofilm activity), M.S.—supervision (anti-biofilm activity), J.V.—investigation (X-ray study), A.V.—supervision (in vitro anti-cancer studies), A.S.—investigation (in vitro anti-cancer studies), E.K., T.S., U.A., A.B.—investigation (in vivo anti-cancer studies), D.M.—supervision (in vivo anti-cancer studies).

Funding: The reported study was funded by Russian Foundation for Basic Research (RFBR) according to the research project № 18-33-20023 and as part of the government assignment for the FRC Kazan Scientific Center of RAS. The study of anti-biofilm activity was funded by the Ministry of Education and Science of the Russian Federation (grant N 6.2379.2017/PCh), President of the Russian Federation (grant N NSh-3464.2018.11). The study of in vivo anti-cancer activity was funded by the Ministry of Education and Science of the Russian Federation (N 0089-2019-0014, 0089-2019-0016).

Acknowledgments: The authors are grateful to the Assigned Spectral-Analytical Center of FRC Kazan Scientific Center of RAS for technical assistance in research. The X-ray measurements were performed using shared experimental facilities supported by IGIC RAS state assignment.

Conflicts of Interest: The authors declare no conflicts of interest.

References

1. Talapatra, S.K.; Talapatra, B. Hygrine, hygroline, and cuscohygrine (ornithine-derived alkaloids) BT—Chemistry of plant natural products: Stereochemistry, conformation, synthesis, biology, and medicine. In *Chemistry of Plant Natural Products*; Talapatra, S.K., Talapatra, B., Eds.; Springer: Berlin/Heidelberg, Germany, 2015; pp. 725–732. ISBN 978-3-642-45410-3.

2. Carroll, F.I. Epibatidine analogs synthesized for characterization of nicotinic pharmacophores—A Review. *Heterocycles* **2009**, *79*, 99–120. [CrossRef]

3. Robertson, J.; Stevens, K. Pyrrolizidine alkaloids. *Nat. Prod. Rep.* **2014**, *31*, 1721–1788. [CrossRef] [PubMed]

4. Michael, J.P. Simple indolizidine and quinolizidine alkaloids. In *Alkaloids: Chemistry and Biology*; Academic Press: Cambridge, MA, USA, 2016; Volume 75, pp. 1–498. ISBN 9780128034347.

5. Taylor, R.D.; MacCoss, M.; Lawson, A.D.G. Rings in drugs. *J. Med. Chem.* **2014**, *57*, 5845–5859. [CrossRef] [PubMed]

6. Hollstein, U. Actinomycin. Chemistry and mechanism of action. *Chem. Rev.* **1974**, *74*, 625–652. [CrossRef]

7. Byrd, J.C.; Harrington, B.; O'Brien, S.; Jones, J.A.; Schuh, A.; Devereux, S.; Chaves, J.; Wierda, W.G.; Awan, F.T.; Brown, J.R.; et al. Acalabrutinib (ACP-196) in relapsed chronic lymphocytic leukemia. *N. Engl. J. Med.* **2016**, *374*, 323–332. [CrossRef]

8. Wu, J.; Zhang, M.; Liu, D. Acalabrutinib (ACP-196): A selective second-generation BTK inhibitor. *J. Hematol. Oncol.* **2016**, *9*, 21. [CrossRef]

9. DuBois, S.G.; Laetsch, T.W.; Federman, N.; Turpin, B.K.; Albert, C.M.; Nagasubramanian, R.; Anderson, M.E.; Davis, J.L.; Qamoos, H.E.; Reynolds, M.E.; et al. The use of neoadjuvant larotrectinib in the management of children with locally advanced TRK fusion sarcomas. *Cancer* **2018**, *124*, 4241–4247. [CrossRef]

10. Ziegler, D.S.; Wong, M.; Mayoh, C.; Kumar, A.; Tsoli, M.; Mould, E.; Tyrrell, V.; Khuong-Quang, D.-A.; Pinese, M.; Gayevskiy, V.; et al. Brief Report: Potent clinical and radiological response to larotrectinib in TRK fusion-driven high-grade glioma. *Br. J. Cancer* **2018**, *119*, 693–696. [CrossRef]

11. Jemal, A.; Bray, F.; Center, M.M.; Ferlay, J.; Ward, E.; Forman, D. Global cancer statistics. *CA Cancer J. Clin.* **2011**, *61*, 69–90. [CrossRef]

12. Le, C.; Liang, Y.; Evans, R.W.; Li, X.; MacMillan, D.W.C. Selective sp3 C–H alkylation via polarity-match-based cross-coupling. *Nature* **2017**, *547*, 79–83. [CrossRef]

13. Deng, X.; Lei, X.; Nie, G.; Jia, L.; Li, Y.; Chen, Y. Copper-catalyzed cross-dehydrogenative N 2 -coupling of NH—1,2,3-Triazoles with N, N—Dialkylamides: N—Amidoalkylation of NH—1,2,3-Triazoles. *J. Org. Chem.* **2017**, *82*, 6163–6171. [CrossRef] [PubMed]

14. Angibaud, P.R.; Querolle, O.A.G.; Berthelot, D.J.-C.; Meyer, C.; Willot, M.P.V.; Meerpoel, L. Preparation of quinoxaline and pyridopyrazine derivatives as pI3Kβ inhibitors. Intnl. Patent Appl. WO 2017060406, 13 April 2017.

15. Barlaam, B.; Cosulich, S.; Degorce, S.; Ellston, R.; Fitzek, M.; Green, S.; Hancox, U.; Lambert-van der Brempt, C.; Lohmann, J.-J.; Maudet, M.; et al. Discovery of a series of 8-(1-phenylpyrrolidin-2-yl)-6-carboxamide-2-morpholino-4H-chromen-4-one as PI3Kβ/δ inhibitors for the treatment of PTEN-deficient tumours. *Bioorg. Med. Chem. Lett.* **2017**, *27*, 1949–1954. [CrossRef] [PubMed]

16. Mu, X.; Shibata, Y.; Makida, Y.; Fu, G.C. Control of vicinal stereocenters through nickel-catalyzed alkyl-alkyl cross-coupling. *Angew. Chemie Int. Ed.* **2017**, *56*, 5821–5824. [CrossRef] [PubMed]

17. Park, Y.; Schindler, C.S.; Jacobsen, E.N. Enantioselective Aza-Sakurai Cyclizations: Dual role of thiourea as H-bond donor and Lewis base. *J. Am. Chem. Soc.* **2016**, *138*, 14848–14851. [CrossRef] [PubMed]

18. He, J.; Dhakshinamoorthy, A.; Primo, A.; Garcia, H. Iron nanoparticles embedded in graphitic carbon matrix as heterogeneous catalysts for the oxidative C–N coupling of aromatic N–H compounds and amides. *ChemCatChem* **2017**, *9*, 3003–3012. [CrossRef]

19. Dian, L.; Zhang-Negrerie, D.; Du, Y. Transition metal-free oxidative cross-coupling C(sp 2)–C(sp 3) bond formation: Regioselective C-3 alkylation of coumarins with tertiary amines. *Adv. Synth. Catal.* **2017**, *359*, 3090–3094. [CrossRef]

20. Panda, S.; Coffin, A.; Nguyen, Q.N.; Tantillo, D.J.; Ready, J.M. Synthesis and Utility of Dihydropyridine Boronic Esters. *Angew. Chemie Int. Ed.* **2016**, *55*, 2205–2209. [CrossRef]

21. Xie, J.; Rudolph, M.; Rominger, F.; Hashmi, A.S.K. Photoredox-controlled mono- and Di-multifluoroarylation of C(sp 3)–H bonds with aryl fluorides. *Angew. Chemie Int. Ed.* **2017**, *56*, 7266–7270. [CrossRef]

22. Kamijo, S.; Kamijo, K.; Murafuji, T. Synthesis of alkylated pyrimidines via photoinduced coupling using benzophenone as a mediator. *J. Org. Chem.* **2017**, *82*, 2664–2671. [CrossRef]

23. Vega, J.A.; Alonso, J.M.; Méndez, G.; Ciordia, M.; Delgado, F.; Trabanco, A.A. Continuous flow α-Arylation of N, N—Dialkylhydrazones under visible-light photoredox catalysis. *Org. Lett.* **2017**, *19*, 938–941. [CrossRef]

24. Ahneman, D.T.; Doyle, A.G. C–H functionalization of amines with aryl halides by nickel-photoredox catalysis. *Chem. Sci.* **2016**, *7*, 7002–7006. [CrossRef] [PubMed]

25. Shaw, M.H.; Shurtleff, V.W.; Terrett, J.A.; Cuthbertson, J.D.; MacMillan, D.W.C. Native functionality in triple catalytic cross-coupling: Sp3 C-H bonds as latent nucleophiles. *Science* **2016**, *352*, 1304–1308. [CrossRef] [PubMed]

26. Shih, Y.-C.; Wang, J.-S.; Hsu, C.-C.; Tsai, P.-H.; Chien, T.-C. Identification of reactive intermediates for the decarbonylative reaction of 1-alkylprolines. *Synlett* **2016**, *27*, 2841–2845.

27. Cheng, W.-M.; Shang, R.; Fu, Y. Photoredox/Brønsted acid co-catalysis enabling decarboxylative coupling of amino acid and peptide redox-active esters with N-heteroarenes. *ACS Catal.* **2017**, *7*, 907–911. [CrossRef]

28. Lipp, B.; Nauth, A.M.; Opatz, T. Transition-metal-free decarboxylative photoredox coupling of carboxylic acids and alcohols with aromatic nitriles. *J. Org. Chem.* **2016**, *81*, 6875–6882. [CrossRef] [PubMed]

29. Luo, J.; Zhang, J. Donor—Acceptor fluorophores for visible-light-promoted organic synthesis: Photoredox/Ni dual catalytic C (sp 3)–C (sp 2) cross-coupling. *ACS Catal.* **2016**, *6*, 873–877. [CrossRef]

30. Lovett, G.H.; Sparling, B.A. Decarboxylative anti-michael addition to olefins mediated by photoredox catalysis. *Org. Lett.* **2016**, *18*, 3494–3497. [CrossRef]

31. Daudelet, D.; Daïch, A.; Rigo, B.; Lipka, E.; Gautret, P.; Homerin, G.; Claverie, C.; Rousseau, J.; Abuhaie, C.-M.; Ghinet, A. Impact of functional groups on the copper-initiated N-arylation of 5-functionalized pyrrolidin-2-ones and their vinylogues. *Synthesis (Stuttg)* **2016**, *48*, 2226–2244. [CrossRef]

32. Yamashita, Y.; Nam, L.C.; Dutton, M.J.; Yoshimoto, S.; Kobayashi, S. Catalytic asymmetric endo-selective [3 + 2] cycloaddition reactions of Schiff bases of α-aminophosphonates with olefins using chiral metal amides. *Chem. Commun. (Cambridge, United Kingdom)* **2015**, *51*, 17064–17067. [CrossRef]

33. Yamashita, Y.; Imaizumi, T.; Guo, X.-X.; Kobayashi, S. Chiral Silver Amides as Effective Catalysts for Enantioselective [3 + 2] Cycloaddition Reactions. *Chem. Asian J.* **2011**, *6*, 2550–2559. [CrossRef]

34. Yamashita, Y.; Guo, X.-X.; Takashita, R.; Kobayashi, S. Chiral silver amide-catalyzed enantioselective [3 + 2] cycloaddition of α-aminophosphonates with olefins. *J. Am. Chem. Soc.* **2010**, *132*, 3262–3263. [CrossRef] [PubMed]

35. Dondas, H.A.; Durust, Y.; Grigg, R.; Slater, M.J.; Sarker, M.A.B. XYZH systems as potential 1,3-dipoles. Part 62: 1,3-dipolar cycloaddition reactions of metallo-azomethine ylides derived from α-iminophosphonates. *Tetrahedron* **2005**, *61*, 10667–10682. [CrossRef]

36. Feng, B.; Chen, J.-R.; Yang, Y.-F.; Lu, B.; Xiao, W.-J. A highly enantioselective copper/phosphoramidite-thioether-catalyzed diastereodivergent 1,3-dipolar cycloaddition of azomethine ylides and nitroalkenes. *Chem. A Eur. J.* **2018**, *24*, 1714–1719. [CrossRef] [PubMed]

37. Ponce, A.; Alonso, I.; Adrio, J.; Carretero, J.C. Stereoselective Ag-catalyzed 1,3-dipolar cycloaddition of activated trifluoromethyl-substituted azomethine ylides. *Chem. A Eur. J.* **2016**, *22*, 4952–4959. [CrossRef] [PubMed]

38. Gazizov, A.S.; Smolobochkin, A.V.; Burilov, A.R.; Pudovik, M.A. Interaction of 2-naphthol with γ-ureidoacetals. A new method for the synthesis of 2-arylpyrrolidines. *Chem. Heterocycl. Compd.* **2014**, *50*, 707–714. [CrossRef]

39. Gazizov, A.S.; Smolobochkin, A.V.; Voronina, J.K.; Burilov, A.R.; Pudovik, M.A. Acid-catalyzed ring opening in 2-(2-hydroxynaphthalene-1-yl)-pyrrolidine-1-carboxamides: Formation of dibenzoxanthenes, diarylmethanes, and calixarenes. *Tetrahedron* **2015**, *71*, 445–450. [CrossRef]

40. Gazizov, A.S.; Kharitonova, N.I.; Smolobochkin, A.V.; Syakaev, V.V.; Burilov, A.R.; Pudovik, M.A. Facile synthesis of 2-(2-arylpyrrolidin-1-yl)pyrimidines via acid-catalyzed reaction of N-(4,4-diethoxybutyl)pyrimidin-2-amine with phenols. *Monat. Chem.* **2015**, *146*, 1845–1849. [CrossRef]

41. Gazizov, A.S.; Burilov, A.R.; Pudovik, M.A. Reactions of polyhydric phenols with nitrogen-containing acetals in the synthesis of polyphenols and heterocyclic compounds. *Russ. Chem. Bull.* **2016**, *65*, 2143–2150. [CrossRef]

42. Smolobochkin, A.V.; Gazizov, A.S.; Syakaev, V.V.; Anikina, E.A.; Burilov, A.R.; Pudovik, M.A. Synthesis of 2-arylpyrrolidine-1-carboxamides via acid-catalyzed reaction of (4,4-diethoxybutyl)ureas with 3-aminophenol. *Monat. Chem.* **2017**, *148*, 1433–1438. [CrossRef]

43. Smolobochkin, A.V.; Gazizov, A.S.; Anikina, E.A.; Burilov, A.R.; Pudovik, M.A. Acid-catalyzed reaction of phenols with N-(4,4-diethoxybutyl) sulfonamides—A new method for the synthesis of 2-aryl-1-sulfonylpyrrolidines. *Chem. Heterocycl. Compd.* **2017**, *53*, 161–166. [CrossRef]

44. Gazizov, A.S.; Smolobochkin, A.V.; Anikina, E.A.; Voronina, J.K.; Burilov, A.R.; Pudovik, M.A. Acid-catalyzed intramolecular cyclization of N-(4,4-diethoxybutyl) sulfonamides as a novel approach to the 1-sulfonyl-2-arylpyrrolidines. *Synth. Commun.* **2017**, *47*, 44–52. [CrossRef]

45. Smolobochkin, A.V.; Gazizov, A.S.; Burilov, A.R.; Pudovik, M.A. Synthesis of functionalized diarylbutane derivatives by the reaction of 2-methylresorcinol with γ-ureidoacetals. *Russ. J. Gen. Chem.* **2015**, *85*, 1779–1782. [CrossRef]

46. Gazizov, A.S.; Smolobochkin, A.V.; Voronina, Y.K.; Burilov, A.R.; Pudovik, M.A. New method of synthesis of 2-arylpyrrolidines: Reaction of resorcinol and its derivatives with γ-ureidoacetals. *Arkivoc* **2014**, *2014*, 319–327.

47. Huang, L. Impact of solid state properties on developability assessment of drug candidates. *Adv. Drug Deliv. Rev.* **2004**, *56*, 321–334. [CrossRef] [PubMed]

48. Corrigan, O.I. Salt forms: Pharmaceutical aspects. In *Encyclopedia of Pharmaceutical Technology*; Swarbrick, J., Ed.; Informa Healthcare: New York, NY, USA, 2006; pp. 3177–3187.

49. Kim, J.Y.; Choi, D.S.; Jung, M.Y. Antiphoto-oxidative activity of sesamol in methylene blue- and chlorophyll-sensitized photo-oxidation of oil. *J. Agric. Food Chem.* **2003**, *51*, 3460–3465. [CrossRef]

50. Fukuda, Y.; Nagata, M.; Osawa, T.; Namiki, M. Contribution of lignan analogues to antioxidative activity of refined unroasted sesame seed oil. *J. Am. Oil Chem. Soc.* **1986**, *63*, 1027–1031. [CrossRef]

51. Zhang, M.-Z.; Zhang, R.-R.; Wang, J.-Q.; Yu, X.; Zhang, Y.-L.; Wang, Q.-Q.; Zhang, W.-H. Microwave-assisted synthesis and antifungal activity of novel fused osthole derivatives. *Eur. J. Med. Chem.* **2016**, *124*, 10–16. [CrossRef]

52. Molodtsov, V.; Fleming, P.R.; Eyermann, C.J.; Ferguson, A.D.; Foulk, M.A.; McKinney, D.C.; Masse, C.E.; Buurman, E.T.; Murakami, K.S. X-ray crystal structures of *Escherichia coli* RNA polymerase with switch region binding inhibitors enable rational design of squaramides with an improved fraction unbound to human plasma protein. *J. Med. Chem.* **2015**, *58*, 3156–3171. [CrossRef]

53. Fang, Z.; Liao, P.-C.; Yang, Y.-L.; Yang, F.-L.; Chen, Y.-L.; Lam, Y.; Hua, K.-F.; Wu, S.-H. Synthesis and biological evaluation of polyenylpyrrole derivatives as anticancer agents acting through caspases-dependent apoptosis. *J. Med. Chem.* **2010**, *53*, 7967–7978. [CrossRef]

54. Chugunova, E.; Akylbekov, N.; Shakirova, L.; Dobrynin, A.; Syakaev, V.; Latypov, S.; Bukharov, S.; Burilov, A. Synthesis of hybrids of benzofuroxan and N-, S-containing sterically hindered phenols derivatives. Tautomerism. *Tetrahedron* **2016**, *72*, 6415–6420. [CrossRef]

55. Chugunova, E.A.; Voloshina, A.D.; Mukhamatdinova, R.E.; Serkov, I.V.; Proshin, A.N.; Gibadullina, E.M.; Burilov, A.R.; Kulik, N.V.; Zobov, V.V.; Krivolapov, D.B.; et al. The study of the biological activity of amino-substituted benzofuroxans. *Lett. Drug Des. Discov.* **2014**, *11*, 502–512. [CrossRef]

56. Chugunova, E.A.; Sazykina, M.A.; Gibadullina, E.M.; Burilov, A.R.; Sazykin, I.S.; Chistyakov, V.A.; Timasheva, R.E.; Krivolapov, D.B.; Goumont, R. Synthesis, genotoxicity and uv-protective activity of new benzofuroxans substituted by aromatic amines. *Lett. Drug Des. Discov.* **2013**, *10*, 145–154. [CrossRef]

57. Schiefer, I.T.; VandeVrede, L.; Fa', M.; Arancio, O.; Thatcher, G.R.J. Furoxans (1,2,5-oxadiazole- N -oxides) as novel NO mimetic neuroprotective and procognitive agents. *J. Med. Chem.* **2012**, *55*, 3076–3087. [CrossRef] [PubMed]

58. Gasco, A.; Fruttero, R.; Sorba, G.; Di Stilo, A.; Calvino, R. NO donors: Focus on furoxan derivatives. *Pure Appl. Chem.* **2004**, *76*, 973–981. [CrossRef]

59. Teicher, B.A. (Ed.) *Tumor Models in Cancer Research*; Humana Press: Totowa, NJ, USA, 2011; ISBN 978-1-60761-967-3.

60. Mironov, A.N.; Bunyatyan, N.D. (Eds.) *Handbook for Preclinical Drug Trials*; Ministry of Public Health and Social Development of the RF, FGBU Scientific Centre for the Expert Evaluation of Medicinal Products: Moscow, Russia, 2012.

61. Rabin, N.; Zheng, Y.; Opoku-Temeng, C.; Du, Y.; Bonsu, E.; Sintim, H.O. Biofilm formation mechanisms and targets for developing antibiofilm agents. *Future Med. Chem.* **2015**, *7*, 493–512. [CrossRef] [PubMed]

62. Rittmann, B.E. Biofilms, active substrata, and me. *Water Res.* **2018**, *132*, 135–145. [CrossRef] [PubMed]

63. Hobley, L.; Harkins, C.; MacPhee, C.E.; Stanley-Wall, N.R. Giving structure to the biofilm matrix: An overview of individual strategies and emerging common themes. *FEMS Microbiol. Rev.* **2015**, *39*, 649–669. [CrossRef] [PubMed]

64. Hall, M.R.; McGillicuddy, E.; Kaplan, L.J. Biofilm: Basic principles, pathophysiology, and implications for clinicians. *Surg. Infect. (Larchmt).* **2014**, *15*, 1–7. [CrossRef]

65. Jamal, M.; Ahmad, W.; Andleeb, S.; Jalil, F.; Imran, M.; Nawaz, M.A.; Hussain, T.; Ali, M.; Rafiq, M.; Kamil, M.A. Bacterial biofilm and associated infections. *J. Chin. Med. Assoc.* **2018**, *81*, 7–11. [CrossRef]

66. Peng, J.-S.; Tsai, W.-C.; Chou, C.-C. Inactivation and removal of *Bacillus cereus* by sanitizer and detergent. *Int. J. Food Microbiol.* **2002**, *77*, 11–18. [CrossRef]

67. Goldberg, J. Biofilms and antibiotic resistance: A genetic linkage. *Trends Microbiol.* **2002**, *10*, 264. [CrossRef]

68. Elinav, E.; Nowarski, R.; Thaiss, C.A.; Hu, B.; Jin, C.; Flavell, R.A. Inflammation-induced cancer: Crosstalk between tumours, immune cells and microorganisms. *Nat. Rev. Cancer* **2013**, *13*, 759–771. [CrossRef] [PubMed]

69. Rao, C.; Verma, A.; Gupta, P.; Vijayakumar, M. Anti-inflammatory and anti-nociceptive activities of Fumaria indica whole plant extract in experimental animals. *Acta Pharm.* **2007**, *57*, 491–498. [CrossRef] [PubMed]

70. Samanta, A.; Podder, S.; Kumarasamy, M.; Ghosh, C.K.; Lahiri, D.; Roy, P.; Bhattacharjee, S.; Ghosh, J.; Mukhopadhyay, A.K. Au nanoparticle-decorated aragonite microdumbbells for enhanced antibacterial and anticancer activities. *Mater. Sci. Eng. C* **2019**, *103*, 109734. [CrossRef] [PubMed]

71. Sheldrick, G.M. *SHELXTL v.6.12, Structure Determination Software Suite;* Bruker AXS: Madison, WI, USA, 2000.

72. Dolomanov, O.V.; Bourhis, L.J.; Gildea, R.J.; Howard, J.A.K.; Puschmann, H. OLEX2: A complete structure solution, refinement and analysis program. *J. Appl. Crystallogr.* **2009**, *42*, 339–341. [CrossRef]

73. Macrae, C.F.; Bruno, I.J.; Chisholm, J.A.; Edgington, P.R.; McCabe, P.; Pidcock, E.; Rodriguez-Monge, L.; Taylor, R.; van de Streek, J.; Wood, P.A. Mercury CSD 2.0—New features for the visualization and investigation of crystal structures. *J. Appl. Crystallogr.* **2008**, *41*, 466–470. [CrossRef]

74. Vollmer, A.C.; Belkin, S.; Smulski, D.R.; Van Dyk, T.K.; LaRossa, R.A. Detection of DNA damage by use of *Escherichia coli* carrying recA′::lux, uvrA′::lux, or alkA′::lux reporter plasmids. *Appl. Environ. Microbiol.* **1997**, *63*, 2566–2571.

75. Lushchak, V.I. Adaptive response to oxidative stress: Bacteria, fungi, plants and animals. *Comp. Biochem. Physiol. Part C Toxicol. Pharmacol.* **2011**, *153*, 175–190. [CrossRef]

76. Maniatis, T.; Fritsch, E.F.F.; Sambrook, J.; Fritsch, E.F.F.; Maniatis, T. *Molecular Cloning: A Laboratory Manual;* Cold Spring Harbor Laboratory Press: New York, NY, USA, 1982; ISBN 0-87969-136-0.

77. Chugunova, E.; Boga, C.; Sazykin, I.; Cino, S.; Micheletti, G.; Mazzanti, A.; Sazykina, M.; Burilov, A.; Khmelevtsova, L.; Kostina, N. Synthesis and antimicrobial activity of novel structural hybrids of benzofuroxan and benzothiazole derivatives. *Eur. J. Med. Chem.* **2015**, *93*, 349–359. [CrossRef]

78. Stepanović, S.; Vuković, D.; Dakić, I.; Savić, B.; Švabić-Vlahović, M. A modified microtiter-plate test for quantification of staphylococcal biofilm formation. *J. Microbiol. Methods* **2000**, *40*, 175–179. [CrossRef]

Sample Availability: Samples of the compounds **4**, **5–8** are available from the authors.

 *molecules*

Article

Design, Synthesis and Cancer Cell Growth Inhibition Evaluation of New Aminoquinone Hybrid Molecules

Andrea Defant and Ines Mancini *

Laboratory of Bioorganic Chemistry, Department of Physics, University of Trento, Via Sommarive 14, 38123 Trento, Italy; andrea.defant@unitn.it

* Correspondence: ines.mancini@unitn.it; Tel.: +39-461-281-548

Academic Editors: Carla Boga and Gabriele Micheletti
Received: 4 June 2019; Accepted: 12 June 2019; Published: 14 June 2019

Abstract: Molecular hybridization has proven to be a successful multi-target strategy in the design and development of new antitumor agents. Based on this rational approach, we have planned hybrid molecules containing covalently linked pharmacophoric units, present individually in compounds acting as inhibitors of the cancer protein targets tubulin, human topoisomerase II and ROCK1. Seven new molecules, selected by docking calculation of the complexes with each of the proteins taken into consideration, have been efficiently synthesized starting from 2,3-dichloro-1,4-naphtoquinone or 6,7-dichloro-5,8-quinolinquinone. By screening the full National Cancer Institute (NCI) panel, including 60 human cancer cell lines, four molecules displayed good and sometimes better growth inhibition GI_{50} than the ROCK inhibitor Y-27632, the Topo II inhibitor podophyllotoxin and the tubulin inhibitor combretastatin A-4. The relative position of *N,N* heteroatoms in the structures of the tested compounds was crucial in affecting bioactivity and selectivity. Furthermore, compound **3** (2-(4-(2-hydroxyethyl)piperazin-1-yl)-3-(3,4,5-trimethoxyphenoxy)naphthalene-1,4-dione) emerged as the most active in the series, showing a potent and selective inhibition of breast cancer BT-549 cells (GI_{50} < 10 nM).

Keywords: hybrid molecules; quinolinequinones; naphtoquinones; 3,4,5-trimethoxyphenyl group; docking calculation; antitumor activity; cytotoxicity

---

1. Introduction

Cancer is one of the prominent causes of death worldwide, and there is an urgent need to introduce new therapeutic agents due to resistance and severe side effects shown by the currently available drugs. Using molecules to act on different tumor targets simultaneously has shown a higher therapeutic potential than single-target chemotherapy. In the recent rational design of anticancer drugs, molecular hybridization is a promising approach. A hybrid molecule usually contains two or more pharmacophore scaffolds present in single therapeutically active agents, connected together in a new single structure by covalent bonds. The selection of these moieties is based on the strategy of combining structures and pharmacological activities of known drugs and bioactive natural or synthetic compounds [1]. The aim of this strategy is to obtain new therapeutic agents that are able not only to reduce undesirable side effects of the parent drugs, but also to display a modified selectivity profile, a higher affinity and a better therapeutic effect than the administration of a combination of the single-target drugs [2]. Hybrid molecules have demonstrated more favorable pharmacokinetic and pharmacodynamic parameters, in addition to dual or multiple modes of action, due to their ability to inhibit more than one biological target [1]. It is therefore understandable that the investigation of new hybrid anticancer drugs has recently become of great therapeutic interest.

Tubulin is a protein composed of microtubules which are the main components of the cellular cytoskeleton. They play a pivotal role in proliferation, migration and mitosis. Molecules able to bind this protein interfere with microtubule polymerization and depolymerization, inducing cell cycle arrest and leading to apoptosis in cancer cells. A variety of molecules have been proven to act efficiently as tubulin inhibitors, but their therapeutic use is limited by toxicity and development of resistance. Molecular hybridization has been used with good results to target this protein [3].

Topoisomerases (Topo I and Topo II) are enzymes that change the topological state of DNA through the breaking and rejoining of DNA strands. They have a relevant role in replication, recombination, transcription and preservation of genome stability. A series of currently used anticancer drugs and molecules in clinical trials act as inhibitors of these enzymes, stabilizing the DNA-Topo complex by intercalation between DNA base pairs. Therefore, inhibition of human topoisomerases is a promising target in the development of new antitumor agents [4].

Rho-associated kinases, known as two isoforms, ROCK1 and ROCK2, have been shown to induce stress fiber formation, cancer cell migration and metastasis. ROCK protein expression is elevated in several types of cancer. In a series of both in vitro and in vivo studies, advantages have been demonstrated by blocking these proteins, obtaining results especially in reducing tumor growth and in preventing metastases. The data support the potential of ROCKs as targets against tumors [5].

We report here on the efficient synthesis of seven new amino-quinone derivatives, designed as hybrid molecules that combine structural units present in inhibitors of tubulin, topoisomerase II and ROCK as tumor targets, and selected by docking calculations as ligands of these proteins. The evaluation on the growth inhibition of cancer cells by the in vitro NCI screening has been related to the structural features of these hybrid molecules.

2. Results and Discussion

2.1. In Silico Molecular Modeling

The 3,4,5-trimethoxyphenyl (TMP) unit is a structural feature of anti-neoplastic molecules. It is the case of the natural phenols combretastatins, whose biological activities have been attributed to the presence of this moiety's ability to target tubulin by the inhibition of microtubule formation. Other representative examples are given by colchicine and podophyllotoxin (Figure 1) [6], which make this pharmacophore a peculiar structural motif among the effective tubulin inhibitors studied in the last decades, with a special value in the design of new antitumor drugs. In addition, podophyllotoxin derivatives acting as topoisomerase II inhibitors are drugs commonly used in clinical oncology [7]. MPT0B214, also showing the TMP unit (Figure 1), inhibits tubulin polymerization and induces apoptosis through mitochondria-mediated pathways [8]. Naphthoquinone and quinolinequinones cores are other representative chemical scaffolds for the development of antitumor agents, which are present in structures of natural and synthetic bioactive products [9,10]. Their molecular mechanism includes reactive oxygen species (ROS) generation mediated by naphtoquinone oxidoreductase 1 (NQO1) bioreduction [11]. Quinolinequinone is a structural unit also present in 7-chloro-6-piperidinyl-quinoline-5,8-dione (PT-262, Figure 1), a synthetic molecule showing an effective inhibition of ROCK kinase activities [12].

Based on these evidences, the structures considered in this molecular hybridization design show a quinone unit with X, Y as CH or *N*, substituted by both a linear or cyclic amine and a 3,4,5 trimethoxyphenyl ether (Figure 1). A very good overlap has been observed for the energy-minimized structures of PT-262 and podophyllotoxin in cases where quinolinequinone is substituted by piperidine as a cyclic amine (corresponding to **1b** in Scheme 1) in the planned molecule, showing an *N,N*-anti configuration (as in PT-262 (Figure S1)).

In silico screening of the interactions with the target proteins tubulin, human topoisomerase II β and human ROCK1 (Table S1) via docking calculation permitted the selection of the new molecules **1a–c**, **2a–c** and **3** (Scheme 1). Their drug-likeness was computationally predicted regarding the physico-chemical properties relevant for the development of a drug. By using the free web tool SwissADME, a series of descriptors were calculated, taking into account lipophilicity, size, polarity, solubility, flexibility and unsaturation [13]. All the designed molecules showed a behavior that respected the parameters necessary for good bioavailability (Table S2).

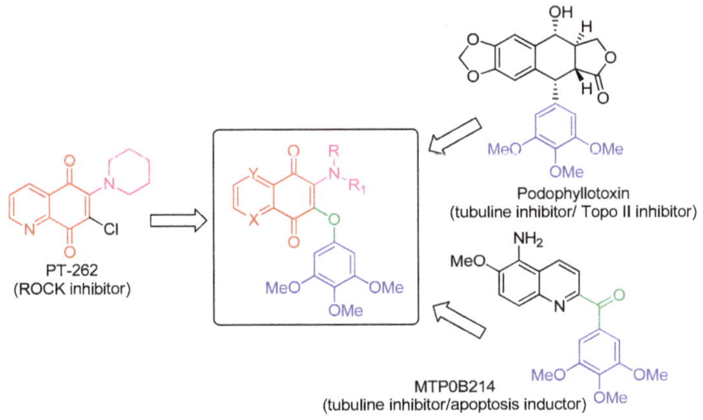

Figure 1. Design strategy of the target hybrid molecules.

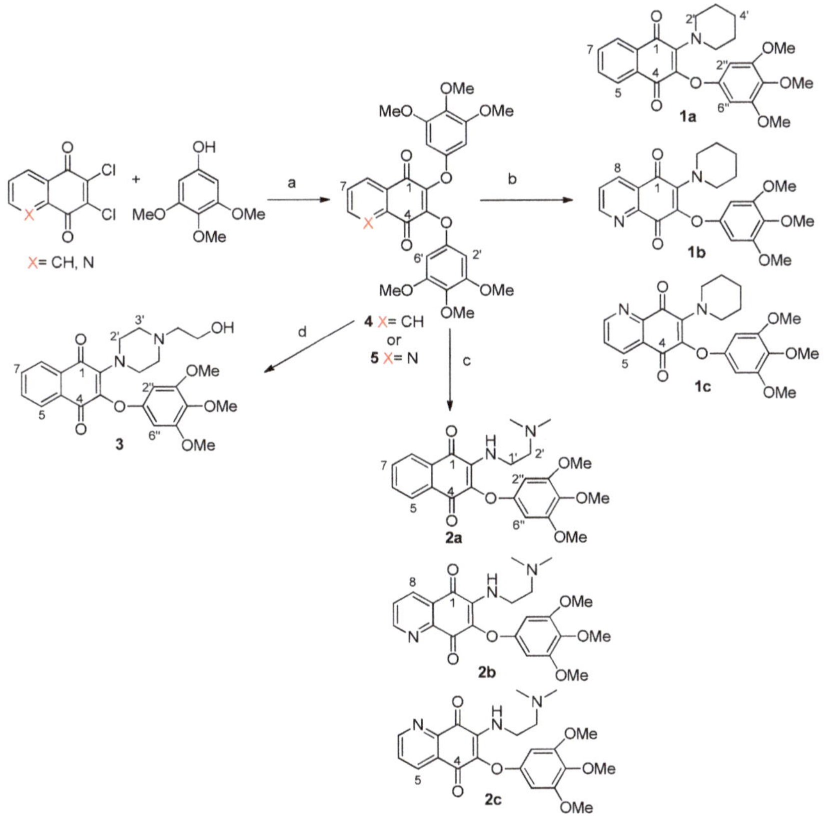

Scheme 1. Synthesis of molecules **1a–c**, **3a–c** and **3**. Reagents and conditions: (a) K_2CO_3, DMSO, r.t. 48 h; (b) piperidine, CH_2Cl_2, r.t. 24 h; (c) *N1,N1*-dimethylethane-1,2-diamine, CH_2Cl_2, r.t. 24 h; (d) 2-(piperazin-1-yl)ethan-1-ol, CH_2Cl_2, r.t. 24 h. Arbitrary numbering is for convenience.

2.2. Synthesis of Compounds 1a–c, 2a–c and 3

The desired products were easily accessible using the common precursors **4** and **5**, by the reaction carried out at room temperature in DMSO in the presence of potassium carbonate of 3,4,5-trimethoxyphenol with 2,3-dicholoro-1,4-naphtoquinone or 6,7-dichloro-5,8-quinolinequinone respectively (Scheme 1), via a proposed mechanism involving a Michael addition followed by chloride elimination [14]. The following treatment of compound **4** with piperidine, *N,N*dimethylethane-1,2 diamine or 2-(piperazin-1-yl)ethan-1-ol in dichloromethane at room temperature provided the products **1a**, **2a** and **3**, respectively, with global yields in the range 77% ÷ 79%, as evaluated after chromatographic purification of the products. It has been reported that similar quinones with one chlorine and one alkoxyl unit preferentially react by replacing the latter group [15,16], and therefore the use of the symmetric disubstituted 3,4,5-trimethoxyphenoxyl reagent was essential. Through a similar procedure, the products **1b**, **1c**, **2b** and **2c** were obtained starting from the quinolinequinone **5**. Pure regioisomers **1b** and **1c** with a 40:60 ratio resulted from chromatographic separation on silica gel by elution with dichloromethane/methanol/triethylamine 95:5:0.1. Similarly, **2b** and **2c** were isolated in a 42:58 ratio. Structural assignments of the regioisomer pairs **1b/2b** and **1c/2c** were based on long-range hetero-correlations observed by HMBC experiments. In detail, taking as references the data obtained for similar isomers [17]: (*i*) $^3J(^1H, {}^{13}C)$ couplings of the signal at 8.35 ppm for H-8 with 183.2 ppm for C(1)=O in **1b**, and 8.36 ppm with 182.0 ppm for **2b**, supported the *N,N*-anti configuration; (*ii*) $^3J(^1H, {}^{13}C)$ couplings of the signal at 8.34 ppm for H-5 with 177.1 ppm for C(1)=O in **1c**, and 8.38 ppm with 176.3 ppm for **2c**, supported the *N,N*-syn configuration (Scheme 1).

2.3. Biological Evaluation

Compounds **1a–c**, **2a–c** and **3** were subjected to in vitro growth percentage activity on the NCI full panel containing 60 human cancer cell lines at a single dose at 10 μM. From this evaluation compounds **1a**, **1c** and **2a** were considered inactive (Table 1), whereas **1b**, **2b**, **2c** and **3** were selected for further investigation at five concentration levels. Table 2 reports GI_{50} data (defined as the concentration values of the molecules inhibiting the growth of cancer cells by 50%) in comparison with the values reported in the NCI database for the ROCK inhibitor Y-27632, the Topo II inhibitor podophyllotoxin and the tubulin inhibitor combretastatin A-4 [18]. The GI_{50} values of the new tested compounds resulted in 10 nM ÷ 10 μM range.

Table 1. Mean dose percent values by one-dose assay (10^{-5} M) of all the tested compounds in the full NCI 60 cell panel.

Compound	Mean Growth Percent	Activity
1a	93.74	Inactive
1b	20.92	Active
1c	67.49	Inactive
2a	78.44	Inactive
2b	−4.33	Active
2c	12.79	Active
3	17.88	Active

Table 2. Inhibition of in vitro human cancer cell lines by compounds **1a**, **2b**, **2c** and **3**, in comparison with combretastatin A-4, podophyllotoxin and Y-27632 taken as reference compounds.

	Cytotoxicity GI_{50} (μM)						
	1b	2b	2c	3	Y-27632	Podophyllotoxin	Combretastatin A-4
Cell lines							
Leukemia							
CCRF-CEM	2.98	2.21	2.24	2.48	31.6	0.01	0.251
HL-60(TB)	1.76	0.811	1.43	1.66	100	0.01	0.01
K-562	0.354	0.659	2.90	2.19	100	-	0.316
MOLT-4	3.58	3.07	2.48	2.35	100	0.01	0.501
RPMI-8226	0.486	1.92	2.23	2.03	100	0.01	0.063
SR	1.89	3.17	2.13	2.87	25.1	0.01	1.99
Non-Small Cell							
Lung Cancer							
A549/ATCC	0.482	3.43	1.87	13.8	100	0.0126	0.020
HOP-92	10.4	1.88	2.01	1.46	1.26	0.0316	0.100
NCI-H226	0.506	16.67	2.08	2.92	100	0.01	0.251
NCI-H23	2.11	2.25	1.76	2.22	100	0.01	0.040
NCI-H322M	6.27	2.65	7.04	1.23	100	0.01	0.063
NCI-H460	2.77	3.11	1.65	3.13	100	0.01	0.050
NCI-H522	1.47	1.46	1.39	0.571	100	0.01	0.032
Colon Cancer							
COLO 205	2.79	1.10	2.44	1.80	100	0.01	6.31
HCC-2998	5.57	1.62	4.71	12.4	100	0.0126	0.158
HCT-116	0.513	0.277	1.85	1.44	100	0.01	0.079
HCT-15	0.442	2.07	2.78	1.56	100	0.0126	0.040
HT29	3.74	2.36	3.77	2.27	100	0.01	6.31
KM12	4.59	2.91	3.34	4.39	100	0.01	0.063
SW-620	1.37	0.318	1.55	1.79	100	0.01	0.063
CNS Cancer							
SF-268	1.94	2.39	2.87	1.73	63.1	0.01	0.063
SF-295	0.465	9.88	3.76	5.36	100	0.01	0.032
SF-539	0.434	1.78	1.85	2.48	100	0.01	0.025
SNB-19	0.450	2.00	2.86	3.41	100	0.01	0.025
SNB-75	0.334	1.89	1.39	1.68	10	0.01	1.259
U251	0.440	2.00	2.12	3.10	100	0.01	0.079
Melanoma							
LOX IMVI	1.01	0.508	1.76	1.81	100	0.01	0.050
MALME-3M	2.73	0.967	2.23	3.99	100	50.1	0.631
M14	0.552	1.84	3.07	1.96	100	0.0126	0.100
MDA-MB-435	1.11	1.72	1.53	1.76	100	0.01	0.010
SK-MEL-2	4.16	2.03	2.01	2.01	100	0.016	0.050
SK-MEL-28	1.35	-	2.00	2.34	100	0.01	5.012
SK-MEL-5	1.95	1.70	1.55	1.81	100	0.01	0.013
UACC-257	0.846	1.78	1.56	3.56	100	0.01	0.063
UACC-62	2.53	3.17	1.60	1.79	100	0.01	0.040
Ovarian Cancer							
OVCAR-3	1.77	0.364	2.02	1.02	79.4	0.0126	0.051
OVCAR-4	1.31	0.937	1.53	1.63	100	0.016	1.995
OVCAR-5	0.439	2.35	2.19	2.60	100	0.251	3.981
OVCAR-8	0.463	0.386	2.13	2.91	100	0.01	0.079
NCI/ADR-RES	1.06	7.40	5.95	3.19	100	0.01	0.063
SK-OV-3	3.28	11.6	7.49	5.31	100	0.01	0.251
Renal Cancer							
786-0	2.03	2.46	2.04	1.93	100	0.016	0.631
A498	0.640	2.17	2.09	2.04	100	0.01	0.100
ACHN	1.25	1.81	2.36	1.87	100	0.01	0.199
CAKI-1	0.595	2.10	2.51	3.59	100	0.1	0.251
RXF 393	0.694	2.00	2.83	1.55	100	0.01	0.398
SN12C	0.829	2.81	2.02	3.11	100	0.016	0.251
TK-10	3.27	3.63	3.91	4.97	10	0.0316	3.162
UO-31	1.18	1.80	1.29	1.61	100	0.016	1.000
Prostate Cancer							
PC-3	0.914	2.43	2.72	3.19	100	0.01	0.010
DU-145	2.60	3.43	4.19	-	100	0.01	0.013

Table 2. *Cont.*

	Cytotoxicity GI$_{50}$ (µM)						
	1b	**2b**	**2c**	**3**	**Y-27632**	**Podophyllotoxin**	**Combretastatin A-4**
Breast Cancer							
MCF7	0.509	0.304	1.20	1.15	100	0.01	0.010
MDA-MB-231/ATCC	4.14	2.55	1.81	2.00	100	0.01	0.016
HS 578T	0.415	3.06	2.84	2.69	100	0.01	0.010
BT-549	0.819	2.68	4.96	<0.01	100	0.01	0.020
T-47D	2.53	1.67	1.98	0.510	100	79.4	50.12
MDA-MB-468	1.77	0.133	0.285	1.37	100	0.01	0.079
MGM [a]	1.29	1.82	2.24	2.24	91.6	1.52	1.52

[a] MGM (mean graph medium) value as average GI50 (µM) over all cell lines investigated.

Submicromolar activities were observed for compound **1b** (Table 2), in particular against central nervous system (CNS) cancer for most of the tested cell lines (Figure 2a). Compound **2b** showed the highest inhibitions on MDA-MB-468 breast cancer cells (GI$_{50}$ = 0.133 µM) and on both HCT-116 and SW-620 colon cell lines (GI$_{50}$ 0.277 and 0.318 µM, respectively, Table 2). Notably, the relative structural position of *N,N* heteroatoms is crucial in affecting bioactivity and selectivity. In fact, compared with **2b**, its *N,N*-syn regioisomer **2c** displayed a lower activity corresponding to a growth inhibition in the micromolar range on the same cancer cell lines. A more pronounced disparity was observed for the pair **1b/1c**, where the *N,N*-syn isomer **1c** showed no activity. This is evident taking into account the values observed for all the synthesized compounds reported in Table 1 when expressed as the mean dose percentage of single high doses (10^{-5} M) from the full NCI 60 cell panel. Moreover, the structures lacking of the N atom on quinone unit as 1a and 2a provided inactive compounds (Table 1), whereas product 3 having additional heteroatoms on the C-2 substituent (specifically an NCH$_2$CH$_2$OH moiety) when compared with the inactive 1a, emerged as the most active molecule. Product **3** displayed submicromolar inhibition and was about 160 and 100 times more active than podophyllotoxin and combretastatin A-4, respectively, on T-47D breast cancer lines (Table 2). Furthermore, **3** emerged as the most promising in the series, with a selective GI$_{50}$ value lower than 10 nM on BT-549 breast cancer cell line (Figure 2b).

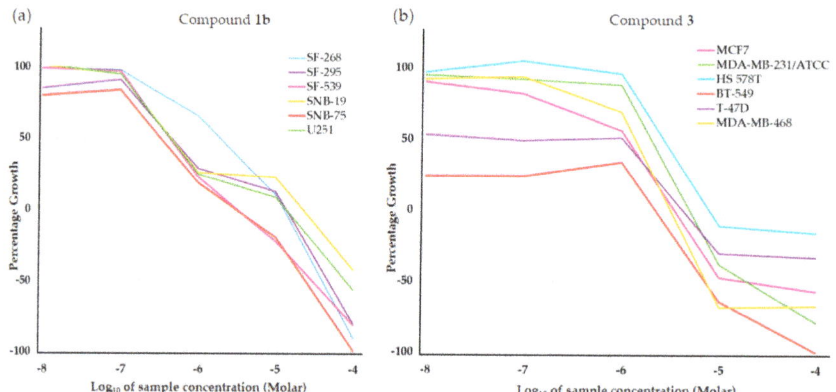

Figure 2. Dose response curves at NCI fixed protocol: (**a**) for compound **1b** on the indicated central nervous system (CNS) cancer cell lines, and (**b**) for compound **3** on the indicated breast cancer cell lines. The curves were obtained at five concentrations (log dilutions from 10^{-4} to 10^{-8} M). The concentrations of each compound, which inhibited 50% of cancer cell growth (GI$_{50}$), were deduced and are the values reported in Table 1.

3. Materials and Methods

3.1. Chemistry

3.1.1. General

All reagents were purchased from Sigma Aldrich and used without further purification. Preparation of 6,7-Dichloroquinoline-5,8-dione followed a reported procedure [14,19]. The reaction yields were calculated for the products after chromatographic purification. Thin layer chromatography (TLC): Merck silica gel F_{254} or reversed phase Merck RP-18 F_{254}, with visualization using UV light. Flash chromatography (FC): Merck Si 15–25 µm. Preparative thin layer chromatography (PLC): 20 × 20 cm Merck Kieselgel 60 F_{254} 0.5-mm plates. NMR spectra were recorded on a Bruker-Avance 400 spectrometer using a 5-mm BBI probe ^{1}H at 400 MHz and ^{13}C at 100 MHz in CDCl$_3$ (relative to δ_H 7.25 and δ_C 77.00 ppm), with δ values in ppm and J values in Hz; assignments were supported by heteronuclear single quantum correlation (HSQC) and heteronuclear multiple bond correlation (HMBC) experiments. Electrospray ionization (ESI)-MS mass spectra were recorded using a Bruker Esquire-LC spectrometer by direct infusion of a methanol solution (source temperature 300 °C, drying gas N$_2$, 4 L min^{-1}, scan range m/z 100 ÷ 1000). Electron ionization (EI) mass spectra (m/z; rel%) and high resolution (HR)-EI data were recorded with a Kratos-MS80 mass spectrometer, heating at 213 °C for **2a–c**, at 276 °C for **1a–c** and at 417 °C for compound **3**, using a home-built computerized acquisition software.

3.1.2. Typical Reaction Procedure for Precursors **4** and **5**

Anhydrous potassium carbonate (414 mg, 3.00 mmol, 1.5 equiv) was added to a solution of 2,3-dichloro-1,4-naphtoquinone (227 mg, 1.00 mmol, 1.0 equiv) or 6,7-dichloro-5,8-quinolinquinone (228 mg, 1.00 mmol, 1.0 equiv) and 3,4,5-trimethoxyphenol (384 mg, 2.00 mmol, 2.00 equiv) in 2.5 mL of dry DMSO, and the reaction mixture was stirred at room temperature for 48 h. The mixture was decanted to remove inorganic salt, and partitioned between dichloromethane/water (X3). The combined organic extracts were washed with water, dried over anhydrous Na$_2$SO$_4$ and concentrated in vacuo to give a solid that was purified by silica gel FC eluting with hexane/EtOAc (from 9:1 to 6:4 v/v) for **4**, and with dichloromethane/methanol/triethylamine (96:4:0.1 v/v) for **5**.

2,3-Bis(3,4,5-trimethoxyphenoxy)naphthalene-1,4-dione (**4**). TLC (hexane: EtOAc = 1:1 v/v): R_f = 0.50. Light orange solid. Yield: 87%. ^{1}H-NMR (400 MHz, CDCl$_3$) δ 8.15 and 7.79 (two m, 2H each, H-5/H-8 and H-6/H-7), 6.11 (s, 4H, H-2′, H-6′, H-2′and H-6′), 3.75 (s, 6H, -OCH$_3$), 3.72 (s, 12H, -OCH$_3$). ^{13}C-NMR (100 MHz, CDCl$_3$) δ 180.4 (C=O), 153.7, 134.7, 134.5, 130.7, 126.5, 94.4, 60.7 (-OCH$_3$), 56.0(-OCH$_3$). Significant HMBC correlations: 8.15 ppm with 180.4 ppm; 7.79 ppm with 130.7 ppm. ESI(+)-MS: m/z 545 [M + Na]$^+$. HRMS(EI) calcd. for C$_{28}$H$_{26}$O$_{10}$, 522.15260, found 522.15248.

6,7-Bis(3,4,5-trimethoxyphenoxy)quinoline-5,8-dione (**5**). TLC (CH$_2$Cl$_2$/MeOH = 95:5 v/v, with two drops of Et$_3$N): R_f = 0.90. Brown solid. Yield: 93%. ^{1}H-NMR (400 MHz, CDCl$_3$) δ 9.11 (br d, J = 4.9 Hz, 1H, H-6), 8.49 (d, J = 7.5 Hz, 1H, H-8), 7.75 (dd, J = 7.5, 4.9 Hz 1H, H-7), 6.12 and 6.08 (two s, 2H each, H-2′, H-6′, H-2′and H-6′), 3.76 (s, 6H, -OCH$_3$), 3.73 (s, 12H, -OCH$_3$). ^{13}C-NMR (100 MHz, CDCl$_3$) δ 179.7 (C=O), 178.8 (C=O), 153.6, 153.5, 152.4, 146.7, 144.7, 135.5, 134.6, 127.8, 127.7, 94.0, 60.2 (-OCH$_3$), 55.5 (-OCH$_3$). Significant HMBC correlations: 9.11 ppm with 146.7 ppm; 8.49 ppm with 179.7 ppm; 7.75 ppm with 127.7 ppm; 6.12 and 6.10 ppm with 152.4, 146.7 and 134.6 ppm; 3.76 ppm with 134.6 ppm; 3.73 ppm with 152.4 ppm ESI(+)-MS: m/z 524 [M + H]$^+$, 545 [M + Na]$^+$, 562 [M + K]$^+$. HRMS(EI) calcd. for C$_{27}$H$_{25}$NO$_{10}$, 523.14785, found 523.14778.

3.1.3. Typical Reaction Procedure for the Synthesis of Compounds **1a–c**, **2a–c** and **3**

A mixture of compound **4** or **5** (52.2 mg, 0.10 mmol, 1.0 equiv) and the suitable amine (0.20 mmol, 2.0 equiv) in 2 mL of anhydrous dichloromethane was stirred at room temperature for 24 h. The solvent was removed in vacuo and the residue was purified by PLC eluting with hexane/EtOAc 1:1 (v/v) for **1a** and CH_2Cl_2/MeOH/Et_3N 95:5:0.1 (v/v) for the other compounds.

2-(Piperidin-1-yl)-3-(3,4,5-trimethoxyphenoxy)naphthalene-1,4-dione (**1a**). TLC (hexane: EtOAc = 1:1 v/v): Rf = 0.74. Red solid. Yield: 91%. ^{1}H-NMR (400 MHz, $CDCl_3$) δ 8.02 and 7.65 (two m, 2H each, H-5/H-8 and H-6/H-7), 6.18 (s, 2H, H-2′ and H-6′), 3.77 (s, 9H, three –OCH_3), 3.43 (br s, 4H, H-2′and H-6′), 1.65 (br s, 6H, H-3′, H-4′and H-5′). ^{13}C-NMR (100 MHz, $CDCl_3$) δ 184.1 and 178.4 (two C=O), 154.3, 142.2, 134.2, 133.3, 132.3, 126.2, 96.2, 56.2 (-OCH_3), 51.4, 28.2. Significant HMBC correlations: 8.02 ppm with 184.1 and 178.4 ppm; 7.65 ppm with 132.3 ppm. ESI(+) MS: m/z 462 [M + K]$^+$, 446[M + Na]$^+$, 424 [M + H]$^+$. HRMS(EI) calcd. for $C_{24}H_{25}NO_6$, 423.16819, found 423.16810.

6-(Piperidin-1-yl)-7-(3,4,5-trimethoxyphenoxy)quinoline-5,8-dione (**1b**). TLC (CH_2Cl_2/MeOH = 95:5 v/v, with two drops of Et_3N): R_f = 0.93. Violet solid. Yield: 40% (as single regioisomer). ^{1}H-NMR (400 MHz, $CDCl_3$) δ 8.97 (br d, J = 4.9 Hz, 1H, H-6), 8.35 (br d, J = 7.9 Hz, 1H, H-8), 7.59 (m, 1H, H-7), 6.20 (s, 2H, H-2′ and H-6′), 3.79 and 3.77 (two s, 9H, OCH_3), 3.45 (m, 4H, H-2′ and H-6′), 1.65 (m, 6H, H-3′, H-4′and H-5′). ^{13}C-NMR (100 MHz, $CDCl_3$) δ 183.2 (C=O), 154.9, 154.0, 147.3, 145.4, 139.7, 135.7, 133.7, 128.5, 126.1, 91.8, 56.7, 51.4, 29.2, 24.0. Significant HMBC correlations: 8.35 ppm with 183.2 and 128.5 ppm; 7.59 ppm with 128.5 ppm; 6.20 ppm with 139.8 ppm; 3.45 ppm with 145.4 and 29.2 ppm. ESI(+)-MS: m/z 447 [M + Na]$^+$. HRMS(EI) calcd. for $C_{23}H_{24}N_2O_6$, 424.16344, found 424.16329.

7-(Piperidin-1-yl)-6-(3,4,5-trimethoxyphenoxy)quinoline-5,8-dione (**1c**). TLC (CH_2Cl_2/MeOH = 95:5 v/v, with two drops of Et_3N): R_f = 0.90. Violet solid. Yield: 60% (as single regioisomer). ^{1}H-NMR (400 MHz, $CDCl_3$) δ 8.93 (br d, J = 4.8 Hz, 1H, H-7), 8.34 (br d, J = 7.9 Hz, 1H, H-5), 7.61 (m, 1H, H-6), 6.19 (s, 2H, H-2′ and H-6′), 3.78 (s, 9H, OCH_3), 3.50 (br s, 4H, H-2′ and H-6′), 1.69 (m, 6H, H-3′, H-4′ and H-5′). ^{13}C-NMR (100 MHz, $CDCl_3$) δ 182.4 and 177.1 (two C=O), 154.4 154.1, 151.6, 151.1, 147.7, 134.2, 133.5, 127.9, 92.9, 56.2, 52.6, 26.9, 25.0. Significant HMBC correlations: 8.93 ppm with 151.1 and 127.9 ppm; 8.34 ppm with 177.1 and 151.1 ppm; 7.61 ppm with 154.4 ppm; 6.19 ppm with 154.1 and 133.5 ppm. ESI(+)-MS: m/z 463 [M + K]$^+$, 447 [M + Na]$^+$, 425 [M + H]$^+$. HRMS(EI) calcd. for $C_{23}H_{24}N_2O_6$, 424.16344, found 424.16328.

2-((2-(Dimethylamino)ethyl)amino)-3-(3,4,5-trimethoxyphenoxy)naphthalene-1,4-dione (**2a**). TLC (CH_2Cl_2/MeOH = 95:5 v/v, with two drops of Et_3N): R_f = 0.30. Orange solid. Yield: 90%. ^{1}H-NMR (400 MHz, $CDCl_3$) δ 8.05 (d, J = 7.5 Hz, 2H, H-5 and H-8), 7.70 and 7.61 (two t, J = 7.7 Hz, 2H, H-6 and H-7), 6.40 (br s, 1H, NH), 6.22 (s, 2H, H-2′ and H-6′), 3.77 (s, 6H, two –OCH_3), 3.75 (s, 3H, –OCH_3), 3.56 (q, J = 5.5 Hz, 2H, H-1′), 2.49 (t, J = 5.5 Hz, 2H, H-2′), 2.22 (s, 6H, N(CH_3)$_2$). ^{13}C-NMR (100 MHz, $CDCl_3$) δ 182.4 and 177.8 (C=O), 155.1, 134.2, 133.3, 132.1, 129.8, 126.4, 92.6, 58.9 (C-2′), 55.9 (-OCH_3), 44.6 (N(CH_3)$_2$), 41.7 (C-1′). Significant HMBC correlations: 8.05 ppm with 182.4 and 177.8 ppm; 7.65 ppm with 132.3 ppm; 7.70 ppm with 134.2 ppm; 7.61 ppm with 132.1 ppm; 6.22 ppm with 155.1 and 133.3 ppm. ESI(+)-MS: m/z 449 [M + Na]$^+$, 427 [M + H]$^+$. HRMS(EI) calcd. for $C_{23}H_{26}N_2O_6$, 426.17909, found 426.17896.

6-((2-(Dimethylamino)ethyl)amino)-7-(3,4,5-trimethoxyphenoxy)quinoline-5,8-dione (**2b**). TLC (CH_2Cl_2/MeOH = 95:5 v/v, with two drops of Et_3N): R_f = 0.28. Orange solid. Yield: 42% (as single regioisomer). ^{1}H-NMR (400 MHz, $CDCl_3$) δ 8.99 (br d, J = 4.8 Hz, 1H, H-6), 8.36 (br d, J = 7.9 Hz, 1H, H-8), 7.55 (dd, J = 7.9, 4.8 Hz, 1H, H-7), 6.46 (br s, 1H, NH), 6.23 (s, 2H, H-2′and H-6′), 3.78 and 3.77 (two s, 9H, three –OCH_3), 3.57 (m, 2H, H-1′), 2.49 (br t, J = 5.3 Hz, 2H, H-2′,), 2.22 (s, 6H, N(CH_3)$_2$). ^{13}C-NMR (100 MHz,$CDCl_3$) δ 182.0 (C=O), 176.1 (C=O), 154.6, 154.4, 148.6, 136.9, 133.7, 130.1, 123.8, 93.2, 57.8 (-OCH_3), 57.7 (C-2′), 44.5 (-N(CH_3)$_2$), 41.1(C-1′). Significant HMBC correlations: 8.99 ppm with 123.8 ppm; 8.36 ppm with 182.0, 154.4 and 148.6 ppm; 7.55 ppm with 154.4 and 130.1 ppm; 6.24 ppm with 154.6 and 133.7 ppm; 3.78 and 3.77 ppm with 154.6 and 133.7 ppm; 2.49 ppm with 41.1 ppm; 2.23 ppm with 57.7 ppm. ESI(+)-MS: m/z 450 [M + Na]$^+$, 428 [M + H]$^+$. HRMS(EI) calcd. for $C_{22}H_{25}N_3O_6$, 427.17434, found 427.17446.

7-((2-(Dimethylamino)ethyl)amino)-6-(3,4,5-trimethoxyphenoxy)quinoline-5,8-dione (**2c**). TLC (CH_2Cl_2/MeOH = 95:5 v/v, with two drops of Et_3N): R_f = 0.31. Orange solid. Yield: 58% (as single regioisomer).

^{1}H-NMR (400 MHz, CDCl$_3$) δ 8.92 (br d, J = 4.6 Hz, 1H, H-7), 8.38 (br d, J = 7.9 Hz, 1H, H-5), 7.62 (dd, J = 7.8, 4.5 Hz, 1H, H-6), 6.53 (br s, 1H, NH), 6.22 (s, 2H, H-2' and H-6'), 3.78 (s, 9H, three –OCH$_3$), 3.61 (m, 2H, H-1'), 2.50 (br t, J = 5.7 Hz, 2H, H-2'), 2.22 (s, 6H, N(CH$_3$)$_2$). ^{13}C-NMR (100 MHz, CDCl$_3$) δ 180.9 (C=O), 176.3 (C=O), 155.0, 153.2, 151.4, 147.2, 135.1, 133.6, 132.2, 127.0, 93.2, 58.5 (C-2'), 56.5 (-OCH$_3$), 44.7 (-N(CH$_3$)$_2$), 41.2 (C-1'). Significant HMBC correlations: 8.92 ppm with 147.2, 135.1 and 127.0 ppm; 8.38 ppm with 180.9 (small), 176.3, 153.2 and 147.2 ppm; 7.62 ppm with 153.2, 135.1 and 132.2 ppm; 6.22 ppm with 155.0 and 133.6 ppm; 3.78 ppm with 155.0 and 133.6 ppm. ESI(+)-MS: m/z 450 [M + Na]$^+$. HRMS(EI) calcd. for C$_{22}$H$_{25}$N$_3$O$_6$, 427.17434, found 427.17427.

2-(4-(2-Hydroxyethyl)piperazin-1-yl)-3-(3,4,5-trimethoxyphenoxy)naphthalene-1,4-dione (**3**). TLC (CH$_2$Cl$_2$/MeOH = 95:5 v/v, with two drops of Et$_3$N): R$_f$ = 0.30. Red solid. Yield: 88%. ^{1}H-NMR (400 MHz, CDCl$_3$) δ 8.01 and 7.66 (two m, 2H each, H-5/H-8 and H-6/H-7), 6.17 (s, 2H, H-2' and H-6'), 3.77 (s, 9H, OCH$_3$), 3.60 (m, 2H, CH$_2$OH), 3.52 (m, 4H, H-2'), 2.59 (m, 4H, H-2'), 2.54 (m, 2H, NCH$_2$). ^{13}C-NMR (100 MHz, CDCl$_3$) δ 183.7 and 178.6 (C=O), 154.0, 133.6, 133.5, 132.3, 131.7, 126.2, 92.6, 56.6, 55.9 (OCH$_3$), 54.2, 49.8. Significant HMBC correlations: 8.01 ppm with 183.7 and 178.6 ppm; 7.66 ppm with 132.3 and 131.7 ppm; 6.17 ppm with 154.0 and 133.5 ppm. ESI(+)-MS: m/z 491 [M + Na]$^+$, 469 [M + H]$^+$. HRMS(EI) calcd. for C$_{25}$H$_{28}$N$_2$O$_7$, 468.18965, found 468.18951.

3.2. Computational Analysis

Calculations were carried out using Autodock Vina 1.1.2 [20], adopting a reported procedure [21]. The structures of human ROCK 1 (PDB ID: 2ETK), human topoisomerase II β (PDB ID: 3QX3) and tubulin (PDB ID: 5JCB) were determined by X-ray crystallography with a resolution of 2.9, 2.2 and 2.3 Å respectively. For the docking calculation, a grid box of 16 × 16 × 24 Å in x, y, z directions was created with a spacing of 1.00 Å, and centered at x = 51.964, y = 101.296, z = 29.213 for 2ETK; a grid box of 22 × 18 × 18 Å in x, y, z directions was created with a spacing of 1.00 Å, and centered at x = 32.884, y = 95.413, z = 50.785 for 3QX3; and a grid box of 18 × 28 × 26 Å in x, y, z directions was created with a spacing of 1.00 Å, centered at x = −13.614, y = 9.720, z = 20.911 for 5JCB. Results were expressed as energy associated to each ligand–enzyme complex in terms of Gibbs free energy values (Table S1). The visual ligand–enzyme interactions were displayed using Discovery Studio Visualizer v.19.1.0.18287 [22]. ADME predictions were performed using the online server Swiss-ADME [23].

3.3. Biological Evaluation

The synthesized compounds were evaluated for their in vitro activity against cancer cell lines by the National Cancer Institute (NCI-USA) following its anticancer drug development program based on automated sulforhodamine blue (SRB) cytotoxicity assay. The screening was a two-stage process, where after a first evaluation was carried out against the full panel of cell lines at a single dose of 10 μM, with the compounds exhibiting significant growth inhibition being tested at five concentration levels [24].

4. Conclusions

Based on the known benefits of considering a molecular hybridization approach in the design and development of new antitumor agents, we have planned new hybrid molecules containing pharmacophoric units, which present individually in compounds acting as inhibitors of the cancer protein targets tubulin, human topoisomerase II and ROCK1. Docking calculation of the complexes with each protein allowed us to select seven molecules, structurally characterized by a naphtoquinone or quinolinequinone moiety, and substituted by both a cyclic or functionalized amine and a 3,4,5-trimethoxyphenyl group.

The evaluation of human cancer cell inhibition by the seven synthetic compounds provided a qualitative structure–activity relationship study. What is more, compound **3** emerged as the most active in the series, displaying a selective nanomolar inhibition of breast cancer BT-549 cells. According to these promising findings and their easily accessible synthesis, the molecules herein reported are worthy of further biological investigation.

Supplementary Materials: The following are available online. Table S1: Energy data from docking calculation by Autodock Vina for **1a–c**, **2a–c** and **3**, in comparison with original and reference ligands; Table S2: ADME prediction of **1a–c**, **2a–c**, **3** and reference compounds evaluated by on-line Server Swiss-ADME; Figure S1: Overlapping of the energy minimized structures **1b**, PT-262 and podophyllotoxin; Figures S2–S23: NMR spectra of compounds **1a–c**, **2a–c**, **3**, **4** and **5**.

Author Contributions: Conceptualization, A.D.; software, A.D.; investigation, A.D.; writing—original draft preparation, A.D and I.M.; writing—review and editing, I.M., supervision, I.M., funding acquisition, I.M.

Funding: This research received no external funding.

Acknowledgments: The authors thank the National Cancer Institute (NCI) at Bethesda, USA for the antitumor screening tests of the compounds. We are grateful to Adriano Sterni, university of Trento for mass spectra recording.

Conflicts of Interest: The authors declare no conflict of interest.

References

1. Abbot, V.; Sharma, P.; Dhiman, S.; Noolvi, M.N.; Patel, H.M.; Bhardwaj, V. Small hybrid heteroaromatics: Resourceful biological tools in cancer research. *RSC Adv.* **2017**, *7*, 28313–28349. [CrossRef]

2. Kerru, N.; Singh, P.; Koorbanally, N.; Raj, R.; Kumar, V. Recent advances (2015–2016) in anticancer hybrids. *Eur. J. Med. Chem.* **2017**, *142*, 179–212. [CrossRef] [PubMed]

3. Nepali, K.; Sharma, S.; Sharma, M.; Bedi, P.M.; Dhar, K.L. Rational approaches, design strategies, structure activity relationship and mechanistic insights for anticancer hybrids. *Eur. J. Med. Chem.* **2014**, *77*, 422–487. [CrossRef] [PubMed]

4. Pommier, Y.; Sun, Y.; Huang, S.N.; Nitiss, J.L. Roles of eukaryotic topoisomerases in transcription, replication and genomic stability. *Nat. Rev. Mol. Cell Biol.* **2016**, *17*, 703–721. [CrossRef] [PubMed]

5. Morgan-Fisher, M.; Wewer, U.M.; Yoneda, A. Regulation of ROCK activity in cancer. *J. Histochem. Cytochem.* **2013**, *61*, 185–198. [CrossRef] [PubMed]

6. Li, L.; Jiang, S.; Li, X.; Liu, Y.; Su, J.; Chen, J. Recent advances in trimethoxyphenyl (TMP) based tubulin inhibitors targeting the colchicine binding site. *Eur. J. Med. Chem.* **2018**, *151*, 482–494. [CrossRef] [PubMed]

7. Hartmann, J.T.; Li, H.P. Camptothecin and podophyllotoxin derivatives inhibitors of topoisomerase I and II—Mechanisms of Action, Pharmacokinetics and Toxicity Profile. *Drug Saf.* **2006**, *29*, 209–230. [CrossRef] [PubMed]

8. Chiang, N.J.; Lin, C.I.; Liou, J.P.; Kuo, C.C.; Chang, C.Y.; Chen, L.T.; Chang, J.Y. A novel synthetic microtubule inhibitor, mpt0b214 exhibits antitumor activity in human tumor cells through mitochondria-dependent intrinsic pathway. *PLoS ONE* **2013**, *8*, e58953. [CrossRef] [PubMed]

9. Defant, A.; Guella, G.; Mancini, I. Synthesis and in vitro cytotoxicity evaluation of novel naphthindolizinedione derivatives. *Arch. Pharm. Chem. Life Sci.* **2007**, *340*, 147–153. [CrossRef] [PubMed]

10. Defant, A.; Guella, G.; Mancini, I. Synthesis and *in-vitro* cytotoxicity evaluation of novel naphtindolizinedione derivatives, part II: Improved activity for aza-analogues. *Arch. Pharm. Chem. Life Sci.* **2009**, *342*, 80–86. [CrossRef] [PubMed]

11. Qiu, H.Y.; Wang, P.F.; Lin, H.Y.; Tang, C.Y.; Zhu, H.L.; Yang, Y.H. Naphthoquinones: A continuing source for discovery of therapeutic antineoplastic agents. *Chem. Biol. Drug Des.* **2018**, *91*, 681–690. [CrossRef] [PubMed]

12. Tsai, C.C.; Liu, H.F.; Hsu, K.C.; Yang, J.M.; Chen, C.; Liu, K.K.; Hsu, T.S.; Chao, J.I. 7-Chloro-6-piperidin-1-yl-quinoline-5,8-dione (PT-262), a novel ROCK inhibitor blocks cytoskeleton function and cell migration. *Biochem. Pharmacol.* **2011**, *81*, 856–865. [CrossRef] [PubMed]

13. Daina, A.; Michielin, O.; Zoete, V. SwissADME: A free web tool to evaluate pharmacokinetics, druglikeness and medicinal chemistry friendliness of small molecules. *Sci. Rep.* **2017**, *7*, 42717. [CrossRef] [PubMed]

14. Defant, A.; Guella, G.; Mancini, I. Regioselectivity in the multi-component synthesis of indolizinoquinoline-5,12-dione derivatives. *Eur. J. Org. Chem.* **2006**, 4201–4210. [CrossRef]

15. Silver, R.F.; Holmes, H.L. Synthesis of some 1,6-naphthoquinones and reactions relating to their use in the study of bacterial growth inhibition. *Can. J. Chem.* **1968**, *46*, 1859–1864. [CrossRef]

16. Egleton, J.E.; Thinnes, C.C.; Seden, P.T.; Laurieri, N.; Lee, S.P.; Hadavizadeh, K.S.; Measures, A.R.; Jones, A.M.; Thompson, S.; Varney, A.; et al. Structure–activity relationships and colorimetric properties of specific probes for the putative cancer biomarker human arylamine N-acetyltransferase 1. *Bioorg. Med. Chem.* **2014**, *22*, 3030–3054. [CrossRef] [PubMed]

17. Defant, A.; Rossi, B.; Viliani, G.; Guella, G.; Mancini, I. Metal-assisted regioselectivity in nucleophilic substitutions: A study by Raman spectroscopy and density functional theory calculations. *J. Raman Spectrosc.* **2010**, *41*, 1398–1403. [CrossRef]

18. NCI. Available online: https://dtp.cancer.gov/dtpstandard/cancerscreeningdata/index.jsp (accessed on 30 May 2019).

19. Shaikh, I.A.; Johnson, F.; Grollman, A.P. Streptonigrin.1. Structure-activity relationship among simple bicyclic analogues. Rate dependence of DNA on quinolone reduction potential. *J. Med. Chem.* **1986**, *29*, 1329–1340. [CrossRef] [PubMed]

20. Trott, O.; Olson, A.J. AutoDock Vina: Improving the speed and accuracy of docking with a new scoring function, efficient optimization, and multithreading. *J. Comp. Chem.* **2010**, *31*, 455–461. [CrossRef] [PubMed]

21. Bosco, B.; Defant, A.; Messina, A.; Incitti, T.; Sighel, D.; Bozza, A.; Ciribilli, Y.; Inga, A.; Casarosa, S.; Mancini, I. Synthesis of 2,6-diamino-substituted purine derivatives and evaluation of cell cycle arrest in breast and colorectal cancer cells. *Molecules* **2018**, *23*, 1996. [CrossRef] [PubMed]

22. Systèmes, D. *BIOVIA, Discovery Studio Modeling Environment, Release 2019*; Dassault Systèmes: San Diego, CA, USA, 2019.

23. Swiss ADME. Available online: http://www.swissadme.ch/ (accessed on 28 April 2019).

24. NCI (NIH). Available online: https://dtp.cancer.gov/discovery_development/nci-60/methodology.htm (accessed on 30 May 2019).

Sample Availability: Samples of the compounds **1a–c**, **2a–c** and **3** are available from the authors.

Article

Synthesis of Novel Structural Hybrids between Aza-Heterocycles and Azelaic Acid Moiety with a Specific Activity on Osteosarcoma Cells

Gabriele Micheletti [1,*], Natalia Calonghi [2,*], Giovanna Farruggia [2,3], Elena Strocchi [1], Vincenzo Palmacci [1], Dario Telese [1], Silvia Bordoni [1], Giulia Frisco [2] and Carla Boga [1]

[1] Department of Industrial Chemistry 'Toso Montanari', Alma Mater Studiorum University of Bologna Viale Del Risorgimento, 4 402136 Bologna, Italy; elena.strocchi@unibo.it (E.S.); vincenzo.palmacci@studio.unibo.it (V.P.); dario.telese2@unibo.it (D.T.); silvia.bordoni@unibo.it (S.B.); carla.boga@unibo.it (C.B.)

[2] Department of Pharmacy and Biotechnology, University of Bologna, Via Irnerio 48,40126 Bologna, Italy; giovanna.farruggia@unibo.it (G.F.); giulia.frisco2@unibo.it (G.F.)

[3] National Institute of Biostructures and Biosystems, Viale delle Medaglie d'Oro, 305, 00136 Rome, Italy

[*] Correspondence: gabriele.micheletti3@unibo.it (G.M.); natalia.calonghi@unibo.it (N.C.); Tel.: +39-051-2093641 (G.M.); +39-051-2091231 (N.C.)

Academic Editors: Miriam Rossi and Roman Dembinski
Received: 24 November 2019; Accepted: 16 January 2020; Published: 18 January 2020

Abstract: Nine compounds bearing pyridinyl (or piperidinyl, benzimidazolyl, benzotriazolyl) groups bound to an azelayl moiety through an amide bond were synthesized. The structural analogy with some histone deacetylase inhibitors inspired their syntheses, seeking new selective histone deacetylase inhibitors (HDACi). The azelayl moiety recalls part of 9-hydroxystearic acid, a cellular lipid showing antiproliferative activity toward cancer cells with HDAC as a molecular target. Azelayl derivatives bound to a benzothiazolyl moiety further proved to be active as HDACi. The novel compounds were tested on a panel of both normal and tumor cell lines. Non-specific induction of cytotoxicity was observed in the normal cell line, while three of them induced a biological effect only on the osteosarcoma (U2OS) cell line. One of them induced a change in nuclear shape and size. Cell-cycle alterations are associated with post-transcriptional modification of both H2/H3 and H4 histones. In line with recent studies, revealing unexpected HDAC7 function in osteoclasts, molecular docking studies on the active molecules predicted their proneness to interact with HDAC7. By reducing side effects associated with the action of the first-generation inhibitors, the herein reported compounds, thus, sound promising as selective HDACi.

Keywords: cancer; pyridine; pyrimidine; 9-hydroxystearic acid; azelaic acid; osteosarcoma; molecular docking

---

1. Introduction

Nitrogen-containing heterocycles are an important class of compounds of wide interest toward many applied fields, such as material, agrochemical, and medicinal chemistry. One of the simplest aza-heteroaromatics is pyridine (**1**, Figure 1), first isolated by Anderson in 1851 [1], which plays a key role both in organic chemistry (as solvent and reactant) and in biological systems, since its nucleus belongs to many coenzymes and vitamins.

Insertion of a further aza-moiety within the six-membered pyridine structure gives 1,2-diazines, 1,3-diazines, or 1,4-diazines, whose ancestors are pyridazine (**2**), pyrimidine (**3**), and pyrazine (**4**), respectively (Figure 1).

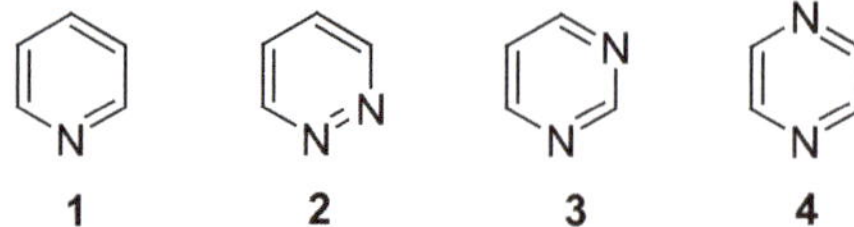

Figure 1. Structures of pyridine (**1**), pyridazine (**2**), pyrimidine (**3**), and pyrazine (**4**).

Most of them are found in plants, animals, insects, marine organisms, and microorganisms. The effectiveness in the biological field lies in the key-core structure of nucleotides and coenzymes and in their role as pharmaceutical drugs, including anticancer agents. Some examples (Figure 2) are (i) imitanib, a tyrosine kinase inhibitor used in some kinds of chronic myelogenous leukemia (CML), acute lymphocytic leukemia (ALL), and gastrointestinal stromal tumors (GIST) [2,3]; (ii) mocetinostat, a histone deacetylase inhibitor for the treatment of a variety of cancers, including follicular lymphoma, Hodgkin's lymphoma, and acute myelogenous leukemia [4]; (iii) berzosertib (VE-822), an ATR kinase inhibitor, which proved to be effective in vivo in the treatment of non-small-cell lung and pancreatic cancer [5,6].

Figure 2. Structures of imatinib, mocetinostat, and berzosertib.

Pentatomic-aza-heteroaromatics bearing two different heteroatoms, such as thiazoles or imidazoles, together with their benzocondensated derivatives, are also found in many natural compounds and are of great importance as bioactive compounds [7,8]. For a long time, our interest focused on the synthesis and reactivity of thiazoles and benzothiazole derivatives [9–15]. Recently, we performed a S_NAr reaction between 2-aminobenzothiazole derivatives and 7-chloro-4,6-dinitrobenzofuroxan, and we exploited the action of the products toward the natural strain in *Vibrio* genus and different bacterial lux-biosensors [16]. Furthermore, we explored the activity shown by the azelayl scaffold connected to the 2-aminobenzothiazolyl moiety, disclosing that some of the synthesized hybrid systems behave as histone deacetylase inhibitors (HDACi) [17]. The choice to bind the benzothiazolyl group to the –$(CH_2)_7$COOMe chain through an amide bond was inspired by the following circumstances: (i) the above reported carbon chain constitutes a moiety of the endogenous cellular lipid 9-hydroxystearic acid (9-HSA) [18], with antiproliferative activity against cancer cells, including human colon cancer [19–21] and osteosarcoma [22,23]; (ii) 9-HSA, as well as the methyl ester [24], acts as a histone deacetylase inhibitor (HDACi) [25–27]; (iii) the structure of the designed compounds is analogous to the well-known vorinostat molecule [28], where a methyl ester replaced a hydroxamic acid group. Based on the above considerations, we planned to synthesize similar novel derivatives with an azelayl scaffold bound through an amide bond to pyridine, 1,3-diazine, benzimidazol, and benzotriazol moieties. All the novel compounds were tested on five cell lines. For the compounds that showed promising half maximal inhibitory concentration (IC$_{50}$) values, further experiments and in silico studies were run to predict whether the molecular target might be HDACi, as in the case of benzothiazolyl derivatives. Herein, we report the results obtained.

2. Results and Discussion

2.1. Chemistry

The two series of novel heterocyclic derivatives **4a–c** and **5a–c** were synthesized through a Schotten Bauman type reaction (Scheme 1), by reacting acyl chloride of the mono methyl azelate (**1**, synthesized from oxalyl chloride and mono methyl ester of azelaic acid) and aminopyridines **2a–c** or aminopyrimidines **3a–c** (Scheme 1).

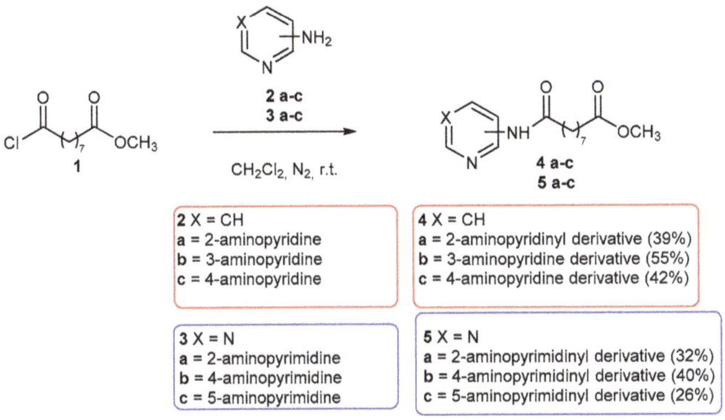

2 X = CH	**4 X = CH**
a = 2-aminopyridine	**a** = 2-aminopyridinyl derivative (39%)
b = 3-aminopyridine	**b** = 3-aminopyridine derivative (55%)
c = 4-aminopyridine	**c** = 4-aminopyridine derivative (42%)

3 X = N	**5 X = N**
a = 2-aminopyrimidine	**a** = 2-aminopyrimidinyl derivative (32%)
b = 4-aminopyrimidine	**b** = 4-aminopyrimidinyl derivative (40%)
c = 5-aminopyrimidine	**c** = 5-aminopyrimidinyl derivative (26%)

Scheme 1. Reactions of methyl 9-chloro-9-oxononanoate (**1**) and aminopyridines **2a–c** or aminopyrimidines **3a–c**.

The reactions were carried out in anhydrous dichloromethane under nitrogen atmosphere, by using two equivalents of amine reagent to remove the hydrochloric acid formed during the reaction course. All products were purified on a silica gel column and fully characterized. They were recovered in not optimized yields ranging from 20% to 55%; in some cases, the mono methyl azelate was recovered, likely due to a certain amount of hydrolyzed acyl chloride before the amidation reaction.

Concerning the reaction with 2-aminopyrimidine (**3a**), it is worth noting that, in addition to the mono acyl derivative **5a**, product **6** is formed from the attack of the amino group of **3a** on two molecules of acyl chloride **1** (Scheme 2).

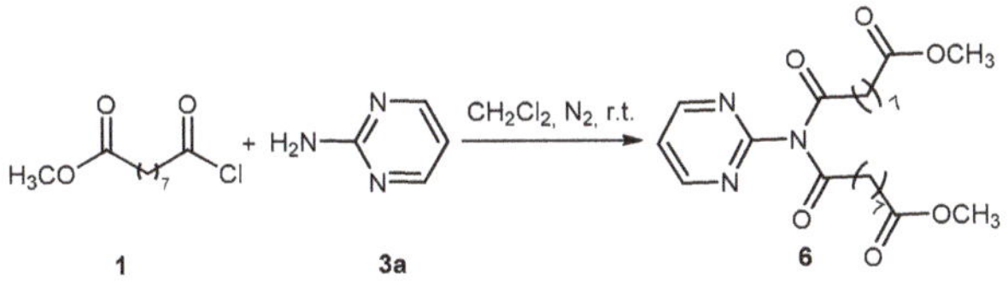

Scheme 2. Double attack of methyl 9-chloro-9-oxononanoate (**1**) to 2-aminopyrimidine **3a**.

To the best of our knowledge, such a reaction is not reported in the literature so far. This might be due to the stronger basicity of the amino group of the 2-aminopyrimidine (**3a**) with respect to that of the other isomers, as supported by comparing the pKa values of **3a** and **3c** (20.5 and 18.4, respectively) [29].

Adopting the same strategy, the by-reactions between **1** and benzimidazole (**7a**) or benzotriazole (**7b**) afford the azelaic derivatives **8a** and **8b**, respectively (Scheme 3).

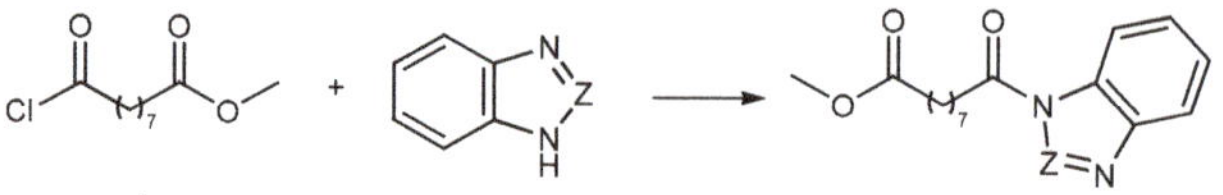

Scheme 3. Synthesis of benzimidazole and benzotriazole derivatives **8a** and **8b**.

All the above compounds underwent biological tests to assess their activity toward four cancer cells lines, U2OS (human osteosarcoma), HT29 (human colon adenocarcinoma), PC3 (human prostatic carcinoma), and IGROV1 (human ovarian carcinoma), as well as a normal human adult fibroblast cell line (see Section 2.2).

2.2. Biological Activity

2.2.1. In Vitro Effects on Cell Viability

Cell lines included in the evaluation of toxicity profiles were malignant U2OS, HT29, PC3, and IGROV1, and a normal human adult fibroblast cell line HDFa. IC_{50} values of the drugs were calculated using Prism, fitted by means of sigmoidal fit and listed in Table 1.

Table 1. Half maximal inhibitory concentration (IC_{50}) of compounds in different cell lines after 48 h of treatment (μM). (n.a. = non active; DMSO = dimethyl sulfoxide).

Compound	Solvent	U2OS IC_{50} (μM)	HDFa IC_{50} (μM)	HT29 IC_{50} (μM)	PC3 IC_{50} (μM)	IGROV1 IC_{50} (μM)
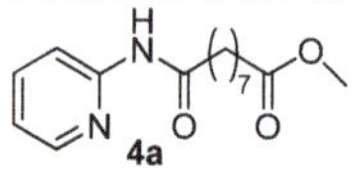 **4a**	DMSO 60 mM	>100	n.a.	>100	n.a.	n.a.
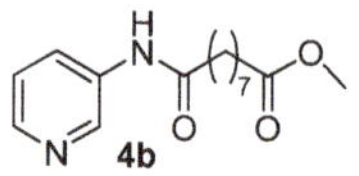 **4b**	DMSO 60 mM	>100	n.a.	>100	n.a.	n.a.
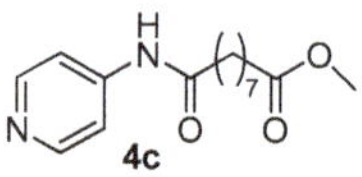 **4c**	DMSO 60 mM	>100	n.a.	n.a.	n.a.	n.a.
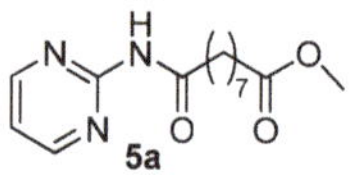 **5a**	DMSO 60 mM	50	n.a.	n.a	n.a	n.a
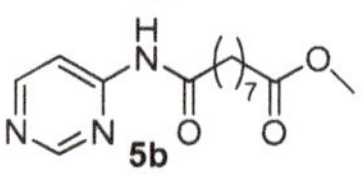 **5b**	DMSO 60 mM	>100	n.a.	n.a.	n.a.	>100
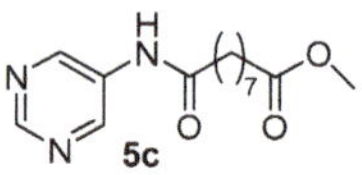 **5c**	DMSO 60 mM	n.a.	n.a.	>100	>100	>100
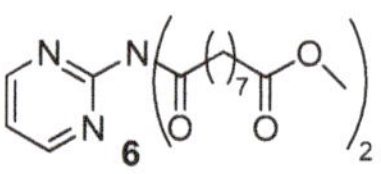 **6**	DMSO 60 mM	35	n.a.	>100	n.a.	n.a.
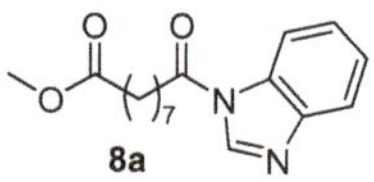 **8a**	DMSO 60 mM	50	n.a.	>100	>100	n.a.
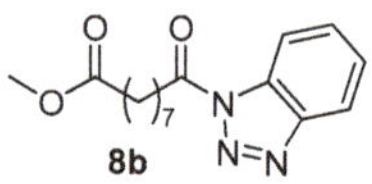 **8b**	DMSO 60 mM	n.a.	n.a.	n.a.	n.a.	>100

IC$_{50}$ analysis showed that all compounds had no effect on the normal cell line. Among tumor cell lines, only U2OS was sensitive to some compounds, such as **5a, 6**, and **8a**.

The chemotherapeutic drugs used in the treatment of osteosarcoma belong to two different classes: the DNA intercalating agents and the histone deacetylase inhibitors. Cisplatin, doxorubicin, and 5-fluorouracile show an IC$_{50}$ of 1.67 µM, 0.5 µM, and 0.3 µM, respectively, in the U2OS cell line [30,31]. These effective and powerful compounds are associated with two important problems: (1) their mode of action is linked to their ability to crosslink with the purine bases on the DNA, thereby interfering with DNA repair mechanisms, causing DNA damage, and subsequently inducing death in cancer and normal cells; (2) they frequently induce multidrug resistance.

Some studies demonstrated that HDACIs, such as SAHA (SuberAniloHydroxamic Acid), sodium butyrate, and trichostatin A (TSA), are able to inhibit human osteosarcoma growth at different IC$_{50}$ values of 5 µM, 5 µM, and 0.5 µM, respectively [32,33]. Our compounds show higher IC$_{50}$, but they offer the important advantage of not inducing cytotoxic effects in normal cells.

2.2.2. Effects on Cell Proliferation

To gain some insight into the biological effects of these novel derivatives, the most active compounds on U2OS cell lines, 5a, **6**, and 8a were subjected to additional studies. To infer whether their effects were due to interference with cell-cycle progression, DNA profiles of cultured cells were examined by flow cytometry. While compound **5a** and **8a** caused a slight accumulation in the synthesis (S) phase, compound **6** induced an increase in the gap 0/gap 1 (G0/G1) fraction, with concomitant decrease in the G2/mitosis (M) phase. Interestingly, compound **6** enlarged cell nucleus size, as shown in Figure 3, where the propidium iodide (PI) fluorescence, indicating the amount of DNA per cell, is plotted against the forward scatter (FS) signal, which is proportional to the size of the analyzed particle. Again, the mean channel of FS distribution was very similar for control cells, **5a**, and **8a** treated samples (730, 735, and 780, respectively), while sample **6** showed a mean channel of the FS distribution of 905. As clear from Figure 3, the nuclei in all cell-cycle phases were affected.

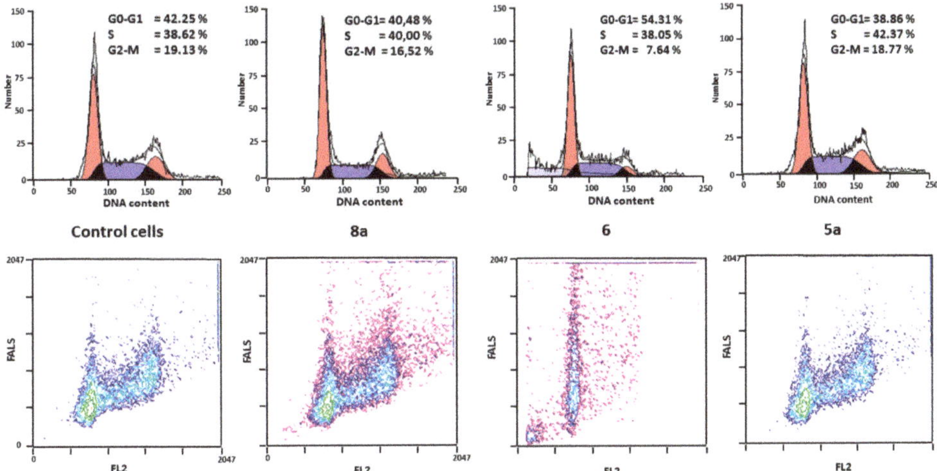

Figure 3. Flow cytometric assays of U2OS cells treated with compounds **5a, 6,** and **8a**, at the concentrations reported. In the upper row, the DNA distribution of a typical experiment in control and treated cells; the lower row indicates the biparametric cytograms of the DNA distribution (FL2) versus forward scatter (FAL signal. The compounds analyzed are reported between the two rows.

Nuclear size and shape are currently not only considered connected to the changes in DNA content and structure during cell-cycle progression, but it is also well known that alteration of nuclear morphology could be due to differentiation [34], pathologies as cancer [35,36], or to senescence [37].

Nuclear size could also be affected by drugs, such as intercalating drugs [37] or etoposide [38]. Interestingly, the fatty palmitic acid increases nuclear size in chick embryonic cardiomyocytes [39], as well as in mouse fibroblasts [40]. The mechanism of these effects is not disclosed, even if it was hypothesized that the lipid, or its metabolites, could interact with DNA, thereby altering the structure of the nucleus, or that they activate cellular metabolism, thereby increasing protein synthesis [39].

2.2.3. Effect on Histone Acetylation

To identify acetylated histones, we analyzed the nuclear cell lysates by Western blot using a 15% polyacrylamide gel electrophoresis. After electrophoresis, the proteins were transferred to a nitrocellulose membrane and then immunoblotted with an anti-acetyl lysine monoclonal antibody. As shown in Figure 4, the antibody could detect the accumulation of acetylated proteins induced in U2OS cells by treatment with compounds **6**, **5a**, and **8a**. Differences in the density of bands are thought to reflect differences in protein acetylation levels. Histone acetylation signals were quantified by densitometry and normalized on histone H1. Histones H2/H3 acetylation increased by 28% upon treatment with **6** for 6 h with no effect on H4 acetylation, whereas compound **8a** induced hyperacetylation by 30% on histone H4 only. Interestingly, compound **5a** caused a post-transcriptional modification of both H2/H3 and H4 histones, increasing acetylation by 89% and 22%, respectively.

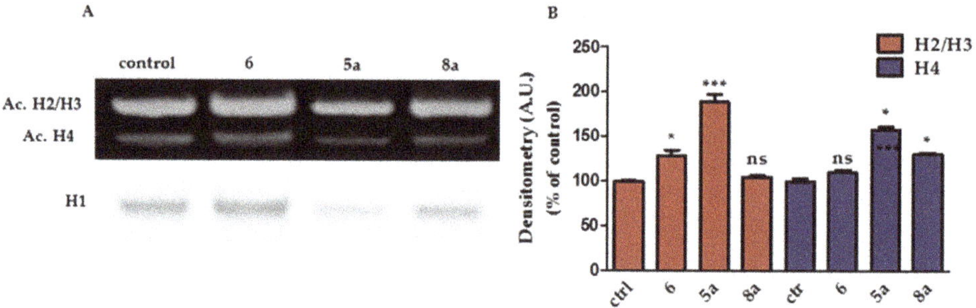

Figure 4. Effect of compounds **5a**, **6**, and **8a** on histone acetylation levels. (**A**) Cell nuclear extracts were prepared and subjected to Western blot analysis for acetylated (Ac) histones H2/H3 and H4. Histone H1 was used as loading control. A representative experiment is shown, which was repeated three times. (**B**) Densitometric analysis of the bands (mean ± SD; $n = 3$) is shown. * $p \leq 0.05$, *** $p \leq 0.01$ vs. control.

These effects induced by the above compounds on histone acetylation are not associated with events involved in apoptotic death. The analysis of the nucleus labeling with Hoechst 33,342 shows indeed no morphological alterations typical of apoptotic cell death (Figure 5). However, the compounds induced a state of nuclear alteration, as shown in Figure 5, where it is possible to see how the compounds **5a** and **8a** caused a dim staining of nuclei, while compound **6** induced an increase in nuclear size, as observed by the flow cytometric assay.

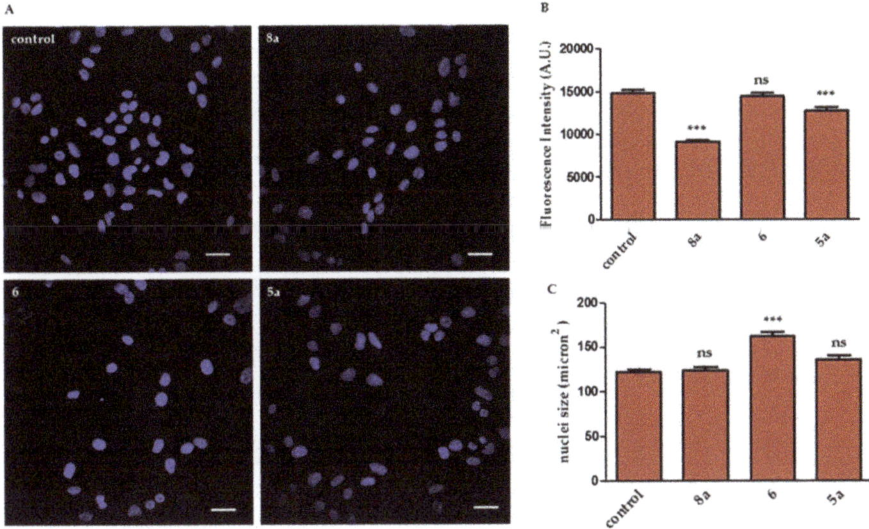

Figure 5. Confocal microscopy of Hoechst 33,342 nuclear staining in control and treated U2OS cells. (**A**) U2OS cells were treated with compounds **5a**, **6**, or **8a** for 24 h, and then stained with Hoechst 33,342. Representative images are shown (scale bar = 20 μm). (**B**) Image densitometry measured using ImageJ (*** $p < 0.01$ with respect to control). (**C**) Cell size was estimated by Hoechst 33,342 staining by image densitometry using ImageJ (*** $p < 0.01$ with respect to control).

2.3. Docking Evaluation

To verify the biochemical data observed for **5a**, **6**, and **8a** species and to explore the binding mode and the potential affinity toward the catalytic Zn^{2+} active site of some HDACs, the synthesized inhibitors were docked to the following structures involved in the histone deacetylation pathway: (*i*) HDAC1, HDAC2, HDAC3, and HDAC8, belonging to the class I HDACs; (*ii*) HDAC4 and HDAC7, belonging to the class II HDACs.

The target macromolecules were selected by analyzing the action modes of some antitumor drugs pertaining to the category of HDACs. Recent studies revealed an unexpected function for HDAC7 in osteoclasts different from the function of HDAC3. Suppression of HDAC7 enhances osteoclast formation, whereas overexpression of HDAC7 impairs osteoclast formation, indicating that HDAC7 represses osteoclast differentiation [41]. Moreover, HDAC7 was found to be overexpressed in pancreatic cancer and acute lymphoblastic leukemia and was reported as insensitive to its previously designated HDACi, trichostatin A (TSA) [42,43].

Our benzimidazolyl (**8a**) and pyrimidinyl (**5a,6**) derivatives were found to have the minimum requirements for tight binding at the pocket Zn^{2+} active site of the HDAC domain whether in class I or II examined macromolecules. In the literature, compounds similar to the studies herein showed evidence for Zn complex formation [44], and the selected target proteins for docking evaluation of the proposed compounds were retrieved from the Protein Data Bank (PDB; http://www.rcsb.org/pdb/) and were as follows: HDAC1 (PDB: 5ICN), HDAC2 (PDB: 4LXZ), HDAC3 (PDB: 4A69, HDAC8 (PDB: 4QA3), HDAC4 (PDB: 2VQM), and HDAC7 (PDB: 3C0Z). Binding modes and binding affinities of the evaluated compounds within the catalytic Zn^{2+} pocket site of all the selected HDACs were calculated using the Autogrid 4.0 and Autodock 4.2 programs [45]. The ADT 1.6.1 package was used for visualizing the results (for the preparation of the molecules and the target macromolecules, and for the procedure of molecular docking, see Section 3).

Analysis of the Binding Mode

The binding pose of the evaluated compounds and their interactions in the active binding site of the selected HDACs were analyzed. The predicted binding free energy (BE, which includes intermolecular energy and torsional free energy) was used as a ranking criterion. The conformation with the lowest ranking docked binding energy (BE) was considered as the "best" docking result.

Satisfactory docking results with good affinity results were observed for the three compounds (**5a**, **6**, and **8a**) whether for the evaluated class I or class II HDACs, as well as virtual constants of inhibition (Ki) at micromolar concentration. The compounds docked and perfectly fitted into the active pocket site of the examined HDAC2, HDAC8, and HDAC7, as indicated from binding free energy reported in Table 2 and from analysis of the ligand binding pose inside the active binding sites.

Table 2. Docking results, expressed in term of binding free energy (ΔG, kcal/mol) and calculated inhibition constant (Ki, nM or μM).

	HDAC Class I				HDAC Class II	
	HDAC1-5ICN	**HDAC2-4LXZ**	**HDAC3-4A69**	**HDAC8-4QA3**	**HDAC4-2VQM**	**HDAC7-3C0Z**
BE_ΔG (mol_5a)	−5.80	−6.90	−5.47	−8.05	−6.03	−7.26
ΔG (mol_6)	−6.36	−7.24	−5.92	−8.14	−7.61	−7.00
ΔG (mol_8a)	−6.83	−7.73	−6.38	−8.63	−6.70	−7.61
Ki (mol_5a)	55.93 μM	8.77 μM	98.23 μM	1.26 μM	37.85 μM	4.78 μM
Ki (mol_6)	21.78 μM	4.94 μM	45.52 μM	1.07 μM	2.62 μM	7.37 μM
Ki (mol_8a)	9.92 μM	2.17 μM	21.08 μM	472.15 nM	12.29 μM	2.62 μM

BE_ΔG: binding energy (kcal/mol); Ki: calculated inhibition constant.

Along the evaluated compounds, the best docking results were observed for the class I HDAC8 and class II HDAC7; the best poses of compounds **5a**, **6**, and **8a** into the active binding site had an estimated binding free energy (ΔG) of −8.05, −8.14, and −8.63 kcal/mol, respectively (for HDAC8), and −7.26, −7.00, and −7.61 kcal/mol, respectively (for HDAC7). The best calculated corresponding inhibition constants (Ki) related to HDAC8 were found to be 1.26 μM, 1.07 μM, and 472.15 nM, respectively, for **5a**, **6**, and **8a** compounds, while they were 4.78 μM, 7.37 μM, and 2.62 μM, for the same compounds in the case of HDAC7. Similar results for ΔG of −7.24 and −7.73 kcal/mol were observed with HDAC2 for compounds **6** and **8a** with the corresponding best calculated inhibition constants (Ki) equal to 4.94 μM and 2.17 μM, respectively. Compound **6** also showed similarly good results with the other examined HDAC4 (ΔG of −7.61 kcal/mol and inhibition constant (Ki) equal to 2.62 μM).

The structures of HDAC enzymes are characterized by the active Zn^{2+} ion at the pocket bottom, the hydrophobic channel reaching the active Zn^{2+} ion, and the surface rim at the entrance of the pocket. Thus, to assess the precise binding pose of the synthesized compounds in the binding pocket of HDAC1 (PDB code: 5ICN), HDAC2 (PDB code 4LXZ), HDAC3 (PDB code 4A69), HDAC8 (PDB code 4QA3), HDAC4 (PDB code 2VQM), and HDAC7 (PDB code: 3C0Z), we performed a molecular docking study.

This docking study acknowledged the inhibitory potential of **5a**, **6**, and **8a** against both HDAC8 and HDAC7 isoforms.

The docking results showed that these inhibitors are able to coordinate to the zinc ion. The in silico docking study displayed different poses of synthesized compounds for HDAC8 and HDAC7 isoforms.

All the compounds (**5a**, **6**, and **8a**) are characterized by one (**5a** and **8a**) or two (**6**) terminal methyl ester groups at the end of the long aliphatic chain, connected to the heterocyclic moiety by an amide bond, and they exhibit slightly better values of binding free energies to class I HDAC8, (Table 2).

For the sake of clarity, in Figure 6, we report only the docking pose of the molecule **8a** (showing the best Ki inhibition) in the binding site of HDAC8 and HDAC7, belonging to class I and II HDACs. In the Supplementary Materials, the figures related to the following docking poses are reported: for molecule **8a** in HDAC2 (Figure S28), for molecules **5a** and **6** in the binding site of HDAC8, HADC7, and HDAC2 (Figures S29 and S30, respectively), and for all the evaluated compounds in the HDAC8 and HDAC7 active binding sites (Figure S31).

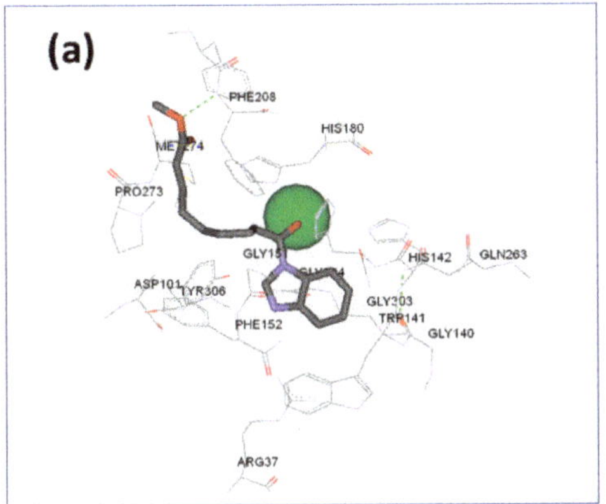
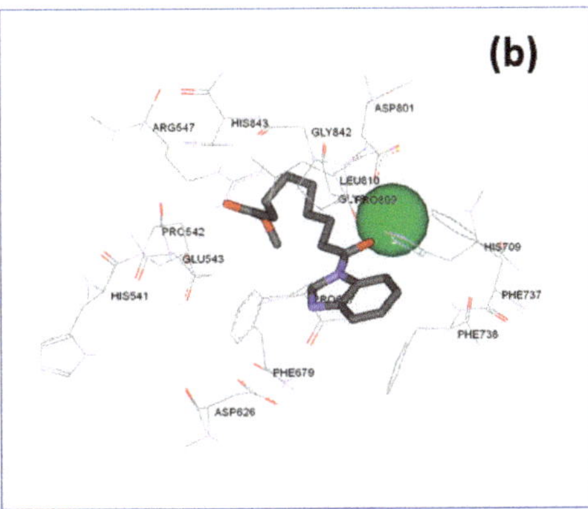

Figure 6. Molecular docking analysis of compound **8a** (showing best *Ki* inhibition docking results) with HDAC8 and HDAC7 isoforms. Docking poses of compound **8a** in the binding site of (**a**) HDAC8 structure (Protein Data Bank identifier (PDB ID): 4QA3) and (**b**) HDAC7 structure (PDB ID: 3C0Z). In both pictures, the zinc ion is represented as a green sphere.

The terminal ester group of compound **8a** and of one lateral chain of **6** were accommodated into the active pocket found in the HDAC7 isoform, while **5a** had the pyrimidine group inserted in the bottom of the pocket. The benzimidazole moiety of **8a** and the pyrimidinyl group of compound **5a** fit perfectly into the active pocket in HDAC8 differently from the HDAC7 isoform, while the orientation of compound **6** showed a terminal ester group of one lateral chain into the pocket. The docking pose of **5a**, **6**, and **8a** with HDAC8 indicated that the oxygen atom of the amide group of all compounds formed a potential hydrogen bonding interaction with the NH^+ of His180 at the rim of the pocket.

On the other hand, the pose of all the compounds to HDAC7 indicated that the oxygen atom of the amide group formed a potential hydrogen bonding interaction with the NH^+ of His709 at the rim of the pocket. A comprehensive overview is reported in the Supplementary Materials (Figure S31).

Figure 7 reports hydrogen bond and chelation interactions in the active site of complexes between **8a** and HDAC8-4QA3 (**a**) and **8a** with HDAC7-3C0Z (**b**); amino-acid residues within 5 Å are indicated without a dashed line.

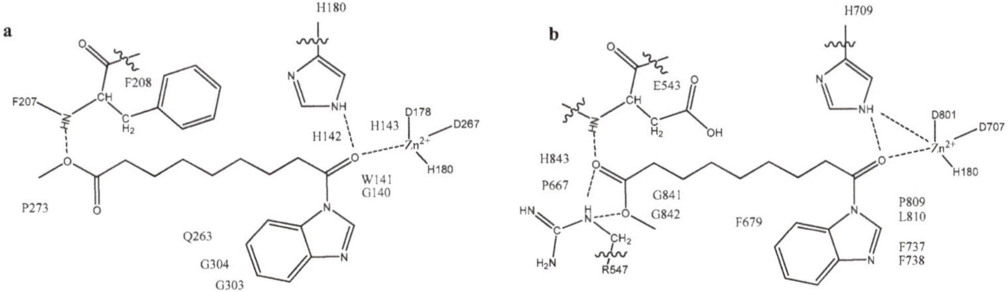

Figure 7. Hydrogen bond and chelation interactions in the active site of (**a**) complex **8a**–HDAC8-4QA3, and (**b**) complex **8a**–HDAC7-3C0Z. Amino-acid residues within 5 Å are indicated without a dashed line.

As can be seen, the coordination sphere of the zinc ion involved, in both reported cases in Figure 7, interactions with the oxygen atom of amide group of compound **8a** and two histidine residues (one of them also involved in a hydrogen bond with the oxygen atom of the amide group, and two aspartic acid residues).

Tables S1–S3 (Supplementary Materials) report the "virtual" interactions of all compounds docked with amino-acid residues and the zinc ion in the active site of HDAC8-4QA3, HDAC7-3C0Z, and HDAC2-4LXZ.

3. Materials and Methods

3.1. Chemical Syntheses

The reagents used, unless stated otherwise, were purchased from Sigma-Aldrich (Milan, Italy). CH_2Cl_2 was anhydrified by distillation over P_2O_5. Chromatographic purifications (FC) were carried out on glass columns packed with silica gel (Merck grade 9385, 230–400 mesh particle size, 60 Å pore size) at medium pressure. Thin-layer chromatography (TLC) was performed on silica gel 60 F254-coated aluminum foils (Fluka, Buchs, Switzerland). The spots related to 9-methoxy-9-oxononanoic acid were revealed using a bromocresol green solution (6% in ethanol).

The nuclear magnetic resonance spectra were recorded at 25 °C on Varian spectrometers Mercury 400 or Inova 600 (Varian, Palo Alto, CA, USA) operating at 400 or 600 MHz (for ^{1}H-NMR) and 100.56 or 150.80 MHz (for ^{13}C-NMR), respectively. Signal multiplicities were established by DEPT-135 experiments. Chemical shifts were measured in δ (ppm) with reference to the solvent (δ= 7.26 ppm and 77.00 ppm for $CDCl_3$, for ^{1}H- and ^{13}C-NMR, respectively). *J*-values are given in Hz. Electrospray ionization (ESI)-MS and ESI high-resolution (HR)MS spectra were recorded using a Waters ZQ 4000 and Xevo instrument, respectively. Melting points (m.p.) were measured on a Büchi 535 apparatus (Flawil, Zwitzerland) and are uncorrected.

3.1.1. Synthesis of Methyl 9-Chloro-9-oxononanoate (**1**)

9-Methoxy-9-oxononanoic acid (1.12 mL, 5.87 mmol) was introduced in a dried three-necked round-bottom flask (equipped with a dropping funnel and kept under nitrogen atmosphere) and dissolved in 45 mL of anhydrous CH_2Cl_2. Oxalyl chloride (0.53 mL, 6.26 mmol), diluted in 5 mL of anhydrous CH_2Cl_2, was introduced in the funnel and added dropwise to the magnetically stirred solution over 10 min. The reaction was monitored through ^{1}H-NMR spectroscopy. When the signals of the starting acid disappeared, the solvent was removed under reduced pressure, with care to avoid contact with moisture, and the residue was dissolved in anhydrous CH_2Cl_2, in an amount calculated in order to obtain a 0.5 M solution of acyl chloride. The chemical–physical data of compound **1** agreed with those reported in the literature [18].

3.1.2. General Procedure for the Synthesis of Compounds **4a–c**, **5a–c**, **6**, and **8a,b**

In a dried apparatus and under nitrogen atmosphere, 1.0 mL of a 0.5 M solution of **1** in CH_2Cl_2 (0.0005 mol) was added to a magnetically stirred solution containing 0.001 mol of the selected heterocyclic compound **2a–c** (or **3a–c**, or **7a,b**) dissolved in 5 mL of anhydrous CH_2Cl_2. The reaction course was monitored by ^{1}H-NMR spectroscopy until the acyl chloride signals disappeared. Then, water (10 mL) was added, and the mixture was extracted with dichloromethane (3 × 10 mL). The organic layer was dried over anhydrous $MgSO_4$ and filtered. After removal of the solvent under reduced pressure, the products were purified by column chromatography on silica gel or by bulb-to-bulb distillation.

Methyl 9-oxo-9-(pyridin-2-ylamino)nonanoate (**4a**), purified by FC (ethyl acetate/dichloromethane 40/60). White solid; 0.054 g (39%); m.p. 59–60 °C; ^{1}H-NMR (600 MHz, CDCl$_3$), δ (ppm): 8.43 (br. s, 1H, N$\underline{H}$), 8.25 (d, J = 5.2 Hz, 1 H, C$\underline{H}$), 8.21 (d, J = 8.3 Hz, 1 H, C$\underline{H}$), 7.69 (dt, J_1 = 7.2 Hz, J_2 = 1.7 Hz, 1H, C$\underline{H}$), 7.02 (dd, J_1 = 7.1 Hz, J_2 = 5.1 Hz, 1 H, C$\underline{H}$), 3.65 (s, 3 H, COOC$\underline{H}_3$), 2.37 (t, J = 7.6 Hz, 2 H, C$\underline{H}_2$CON), 2.28 (t, J = 7.6 Hz, 2 H, C$\underline{H}_2$COOCH$_3$), 1.71 (quint., J = 7.5 Hz, 2 H, C$\underline{H}_2$CH$_2$CON), 1.60 (quint, J = 7.1 Hz, 2 H, C$\underline{H}_2$CH$_2$COOCH$_3$), 1.39–1.27 (m, 6 H, C$\underline{H}_2$); ^{13}C-NMR (150 MHz, CDCl$_3$) δ (ppm): 174.2 (C), 171.8 (C), 151.5 (C), 147.6 (CH), 138.4 (CH), 119.6 (CH), 114.1 (CH), 51.4 (CH$_3$), 37.6 (CH$_2$), 34.0 (CH$_2$), 28.94 (CH$_2$), 28.9 (CH$_2$), 28.8 (CH$_2$), 25.2 (CH$_2$), 24.8 (CH$_2$); ESI-MS$^-$ (m/z): 313 [M + Cl]$^-$; ESI-HRMS: calculated for C$_{15}$H$_{22}$N$_2$NaO$_3^+$: 301.1523, found: 301.1528

Methyl 9-oxo-9-(pyridin-3-ylamino)nonanoate (**4b**), purified by FC (ethyl acetate/dichloromethane 40/60). White solid; 0.076 g (55%); m.p. 75.5–77.7 °C; ^{1}H-NMR (600 MHz, CDCl$_3$) δ (ppm) 8.80 (br.s, 1 H, N$\underline{H}$); 8.58 (d, J = 2.3 Hz, 1 H, C$\underline{H}$); 8.26 (d, J = 4.5 Hz, 1 H, C$\underline{H}$); 8.20 (d, J = 8.1 Hz, 1 H, C$\underline{H}$); 7.23 (dd, J_1 = 8.2 Hz, J_2 = 4.6 Hz, 1 H, C$\underline{H}$); 3.63 (s, 3 H, COOC$\underline{H}_3$); 2.35 (t, J = 7.8 Hz, 2 H, C$\underline{H}_2$CON); 2.26 (t, J = 7.8 Hz, 2 H, C$\underline{H}_2$COOCH$_3$); 1.67 (quint., J = 7.3 Hz, 2 H, C$\underline{H}_2$CH$_2$CON); 1.56 (quint., J = 7.3 Hz, 2 H, C$\underline{H}_2$CH$_2$COOCH$_3$); 1.34–1.23 (m, 6 H, C$\underline{H}_2$); ^{13}C-NMR (150 MHz, CDCl$_3$) δ (ppm): 174.4 (C), 172.4 (C), 144.3 (CH), 140.7 (CH), 135.5 (C), 127.4 (CH), 123.8 (CH), 51.4 (CH$_3$), 37.2 (CH$_2$), 33.9 (CH$_2$), 28.8 (CH$_2$), 28.76 (CH$_2$), 28.72 (CH$_2$), 25.2 (CH$_2$), 24.7 (CH$_2$); ESI-MS$^+$ (m/z): 279 [M + H]$^+$, 301 [M + Na]$^+$; ESI-HRMS: calculated for C$_{15}$H$_{22}$N$_2$NaO$_3^+$: 301.1523, found: 301.1528.

Methyl 9-oxo-9-(pyridin-4-ylamino)nonanoate (**4c**), purified by FC (methanol/ethyl acetate 5/95). White solid; 0.059 g (42%); m.p. 83–85 °C; ^{1}H-NMR (600 MHz, CDCl$_3$) δ (ppm): 8.91 (s, 1 H, N$\underline{H}$), 8.43 (d, J = 5.1 Hz, 2 H, C$\underline{H}$), 7.56 (d, J = 5.1 Hz, 2 H, C$\underline{H}$), 3.64 (s, 3 H, COOC$\underline{H}_3$), 2.35 (t, J = 7.6 Hz, 2 H, C$\underline{H}_2$CON), 2.28 (t, J = 7.6 Hz, 2 H, C$\underline{H}_2$COOCH$_3$), 1.67 (quint., J = 7.4 Hz, 2 H, C$\underline{H}_2$CH$_2$CON), 1.57 (quint., J = 7.4 Hz, 2 H, C$\underline{H}_2$CH$_2$COOCH$_3$), 1.35–1.25 (m, 6 H, C$\underline{H}_2$); ^{13}C-NMR (150 MHz, CDCl$_3$) δ (ppm): 174.4 (C), 172.7 (C), 149.9 (CH), 146.0 (C), 113.7 (CH), 51.5 (CH$_3$), 37.5 (CH$_2$), 33.9 (CH$_2$), 28.8 (CH$_2$), 28.77 (CH$_2$), 28.73 (CH$_2$), 25.1 (CH$_2$), 24.7 (CH$_2$); ESI-MS$^-$ (m/z): 277 [M − H]$^-$, 313 [M + Cl]$^-$; ESI-HRMS: calculated for C$_{15}$H$_{22}$N$_2$NaO$_3^+$: 301.1523, found: 301.1528.

Methyl 9-oxo-9-(pyrimidin-2-ylamino)nonanoate (**5a**), purified by FC (methanol/ethyl acetate 5/95). White solid; yield 0.067 g (32%); m.p. 86.2–87.3 °C; ^{1}H-NMR (400 MHz, CDCl$_3$) δ (ppm): 9.55 (br.s, 1 H, NH), 8.61 (d, J = 4.9 Hz, 2 H, C$\underline{H}$), 6.97 (7, J = 4.9 Hz, 1 H, C$\underline{H}$) 3.62 (s, 3 H, COOC$\underline{H}_3$), 2.73 (t, J = 7.4 Hz, 2 H, C$\underline{H}_2$CON), 2.26 (t, J = 7.6 Hz, 2 H, C$\underline{H}_2$COOCH$_3$), 1.70 (quint., J = 7.3 Hz, 2 H, C$\underline{H}_2$CH$_2$CON), 1.59 (quint., J = 7.0 Hz, 2 H, C$\underline{H}_2$CH$_2$COOCH$_3$), 1.45–1.2 (m, 6H, C$\underline{H}_2$); ^{13}C-NMR (100 MHz, CDCl$_3$) δ (ppm): 174.2 (C), 173.7 (C), 158.2 (CH), 157.6 (C), 115.9 (CH), 51.3 (CH$_3$), 37.3 (CH$_2$), 33.9 (CH$_2$), 28.93 (CH$_2$), 28.90 (CH$_2$), 28.8 (CH$_2$), 24.8 (CH$_2$) (one signal overlapped); ESI-MS$^+$ (m/z): 280 [M + H]$^+$, 302 [M + Na]$^+$, 318 [M + K]$^+$; ESI-HRMS: calculated for C$_{14}$H$_{22}$N$_3$NaO$_3^+$: 302.1475, found: 302.1481.

Methyl 9-oxo-9-(pyrimidin-4-ylamino)nonanoate (**5b**), purified by FC (methanol/ethyl acetate 5/95). White solid; 0.056 g (40%); m.p. 92.0–94.2 °C; ^{1}H-NMR (600 MHz, CDCl$_3$) δ (ppm): 8.84 (s, 1 H, C$\underline{H}$), 8.62 (d, J = 5.7 Hz, 1 H, C$\underline{H}$), 8.18 (dd, J_1 = 5.9 Hz, J_2 = 1.2 Hz, 1 H, C$\underline{H}$), 8.13 (br.s, 1 H, N$\underline{H}$), 3.66 (s, 3 H, COOC$\underline{H}_3$), 2.42 (t, J = 7.5 Hz, 2 H, C$\underline{H}_2$CON), 2.30 (t, J = 7.2 Hz, 2 H, C$\underline{H}_2$COOCH$_3$), 1.72 (quint., J = 7.2 Hz, 2 H, C$\underline{H}_2$CH$_2$CON), 1.62 (quint., J = 7.1 Hz, 2 H, C$\underline{H}_2$CH$_2$COOCH$_3$), 1.42–1.24 (m, 6 H, C$\underline{H}_2$); ^{13}C-NMR (150 MHz, CDCl$_3$) δ (ppm): 174.2 (C), 172.5 (C), 158.4 (CH), 158.2 (CH), 156.9 (C), 110.2 (CH), 51.5 (CH$_3$), 37.7 (CH$_2$), 34.0 (CH$_2$), 28.9 (CH$_2$), 28.8 (CH$_2$), 24.9 (CH$_2$), 24.8 (CH$_2$) (one signal overlapped); ESI-MS$^+$ (m/z): 280 [M + H]$^+$, 302 [M + Na]$^+$, 318 [M + K]$^+$; ESI-HRMS: calculated for C$_{14}$H$_{22}$N$_3$NaO$_3^+$: 302.1475, found: 302.1481.

Methyl 9-oxo-9-(pyrimidin-5-ylamino)nonanoate (**5c**), purified by FC (methanol/ethyl acetate 5/95). Orange solid; 0.036 g (26%); m.p. 78.6–80.2 °C; ^{1}H-NMR (300 MHz, CDCl$_3$) δ (ppm) 9.29 (s, 2 H, C$\underline{H}$), 8.98 (s, 1 H, C$\underline{H}$); 8.63 (br. s, 1 H, N$\underline{H}$), 3.66 (s, 3 H, COOC$\underline{H}_3$), 2.47 (t, J = 7.6 Hz, 2 H, C$\underline{H}_2$CON), 2.31 (t, J = 7.3 Hz, 2 H, C$\underline{H}_2$COOCH$_3$), 1.74 (quint., J = 7.4 Hz, 2 H, C$\underline{H}_2$CH$_2$CON), 1.62 (quint., J = 7.4 Hz, 2H, C$\underline{H}_2$CH$_2$COOCH$_3$), 1.42–1.28 (m, 6 H, C$\underline{H}_2$); ^{13}C-NMR (150 MHz, CDCl$_3$) δ (ppm): 174.4 (C), 172.5 (C), 150.7 (CH), 147.2 (CH), 134.8 (C), 51.5 (CH$_3$), 37.1 (CH$_2$), 34.0 (CH$_2$), 28.7 (CH$_2$), 25.0 (CH$_2$),

24.7 (CH_2) (two signals overlapped); ESI-MS$^+$ (*m/z*): 280 [M + H$^+$]$^+$, 302 [M + Na$^+$]$^+$, 318 [M + K$^+$]$^+$; ESI-HRMS: calculated for $C_{14}H_{22}N_3NaO_3{}^+$: 302.1475, found: 302.1481.

Dimethyl 9,9′-(pyrimidin-2-ylazanediyl)bis(9-oxononanoate) (**6**), purified by FC (methanol/ethyl acetate 5/95). Colorless liquid; 0.016 g (14%); ^{1}H-NMR (600 MHz, CDCl$_3$) δ (ppm): 8.81 (d, *J* = 4.7 Hz, 2 H, C<u>H</u>), 7.33 (t, *J* = 4.8 Hz, 1 H, C<u>H</u>), 3.59 (s, 6 H, COOC<u>H</u>$_3$), 2.47 (t, *J* = 7.4 Hz, 4 H, C<u>H</u>$_2$CON), 2.22 (t, *J* = 7.6 Hz, 4 H, C<u>H</u>$_2$COOCH$_3$), 1.60–1.48 (m, 8 H, C<u>H</u>$_2$CH$_2$COOCH$_3$), 1.29–1.18 (m, 12 H, C<u>H</u>$_2$), ^{13}C-NMR (150 MHz, CDCl$_3$) δ (ppm): 175.0 (C), 174.0 (C), 159.5 (C), 159.3 (CH), 120.3 (CH), 51.2 (CH$_3$), 37.9 (CH$_2$), 33.8 (CH$_2$), 28.75 (CH$_2$), 28.70 (CH$_2$), 28.66 (CH$_2$), 28.60 (CH$_2$), 24.6 (CH$_2$), 24.0 (CH$_2$); ESI-MS$^-$ (*m/z*): 498 [M + Cl]$^-$; ESI-HRMS: calculated for $C_{24}H_{37}N_3NaO_6{}^+$: 486.2575, found: 486.2580.

Methyl 9-(1H-benzo[d]imidazol-1-yl)-9-oxononanoate (**8a**), purified by bulb-to-bulb distillation at 150 °C and 0.1 mmHg. White solid; 0.075 g (50%); m.p. 83.0–84.3 °C; ^{1}H-NMR (600 MHz, CDCl$_3$) δ (ppm): 8.40 (s, 1 H, C<u>H</u>), 8.25 (d, *J* = 7.9 Hz, 1 H, C<u>H</u>), 7.80 (d, *J* = 7.0 Hz, 1 H, C<u>H</u>), 7.43 (dt, *J*$_1$ = 8.5 Hz, *J*$_2$ = 1.2 Hz, 1 H, C<u>H</u>), 7.40 (dt, *J*$_1$ = 9.3 Hz, *J*$_2$ = 1.6 Hz, 1 H, C<u>H</u>), 3.67 (s, 3 H, COOC<u>H</u>$_3$), 3.00 (t, *J* = 7.4 Hz, 2 H, C<u>H</u>$_2$CON), 2.32 (t, *J* = 7.3 Hz, 2H, C<u>H</u>$_2$COOCH$_3$), 1.88 (quint., *J* = 7.8 Hz, 2 H, C<u>H</u>$_2$CH$_2$CON), 1.64 (quint., *J* = 7.8 Hz, 2 H, C<u>H</u>$_2$CH$_2$COOCH$_3$), 1.47 (quint., *J* = 7.7 Hz, 2H, C<u>H</u>$_2$), 1.43–1.32 (m, 4 H, C<u>H</u>$_2$); ^{13}C-NMR (150 MHz, CDCl$_3$) δ (ppm): 174.2 (C), 170.3 (C), 143.9 (CH), 140.9 (C), 131.5 (C), 125.9 (CH), 125.0 (CH), 120.5 (CH), 115.6 (CH), 51.5 (CH$_3$), 35.9 (CH$_2$), 34.0 (CH$_2$), 28.9 (CH$_2$), 28.86 (CH$_2$), 28.84 (CH$_2$), 24.8 (CH$_2$), 24.2 (CH$_2$); ESI-MS$^+$ (*m/z*): 303 [M + H$^+$]$^+$, 325 [M + Na$^+$]$^+$; ESI-HRMS: calculated for $C_{17}H_{22}N_2NaO_3{}^+$: 325.1523, Found: 325.1528

Methyl 9-(1H-benzo[d][1,2,3]triazol-1-yl)-9-oxononanoate (**8b**), purified by bulb-to-bulb distillation at 150 °C and 0.1 mmHg. White solid; 0.032 g (20%); m.p: 55.5–57.2 °C; ^{1}H-NMR (600 MHz, CDCl$_3$) δ (ppm): 8.28 (d, *J* = 8.3 Hz, 1 H, C<u>H</u>), 8.10 (d, *J* = 8.0 Hz, 1 H, C<u>H</u>), 7.64 (t, *J* = 7.8 Hz, 1 H, C<u>H</u>), 7.49 (t, *J* = 7.8 Hz, 1 H, C<u>H</u>), 3.65 (s, 3 H, COOC<u>H</u>$_3$), 3.40 (t, *J* = 7.4 Hz, 2 H, C<u>H</u>$_2$CON), 2.30 (t, *J* = 7.6 Hz, 2 H, C<u>H</u>$_2$COOCH$_3$), 1.89 (quint., *J* = 7.8 Hz, 2 H, C<u>H</u>$_2$CH$_2$CON), 1.63 (quint., *J* = 7.5 Hz, 2 H, C<u>H</u>$_2$CH$_2$COOCH$_3$), 1.48 (quint., *J* = 7.7 Hz, 2 H, C<u>H</u>$_2$), 1.42–1.32 (m, 4 H, C<u>H</u>$_2$); ^{13}C-NMR (150 MHz, CDCl$_3$) δ (ppm) 174.1 (C), 172.5 (C), 146.1 (C), 131.1 (C), 130.3 (CH), 126.0 (CH), 120.1 (CH), 114.4 (CH), 51.4 (CH$_3$), 35.4 (CH$_2$), 34.0 (CH$_2$), 28.9 (CH$_2$), 28.8 (CH$_2$), 24.8 (CH$_2$), 24.3 (CH$_2$) (one signal overlapped); ESI-MS$^+$ (*m/z*): 304 [M + H$^+$]$^+$, 326 [M + Na$^+$]$^+$.

3.2. Cell Culture and Treatments

3.2.1. Cell Culture

The human prostate cancer (PC3), human colon cancer (HT29), human bone osteosarcoma (U2OS), and normal human adult fibroblast (HDFa) cell lines were purchased from American Type Culture Collection (ATCC, Manassas, VA, USA), while the human ovarian cancer cell line (IGROV1) was kindly provided by Istituto Nazionale Tumori (IRCCS, Milano, Italy). Cells were cultured in Roswell Park Memorial Institute (RPMI)-1640 medium (Labtek Eurobio, Milan, Italy), supplemented with 10% fetal calf serum (FCS; Euroclone, Milano, Italy) and 2mM L-glutamine (Sigma-Aldrich, Milano, Italy), at 37 °C and 5% CO$_2$ atmosphere. The compounds were dissolved in dimethyl sulfoxide (DMSO) in a 60 mM stock solution. In cell treatments, the final DMSO concentration never exceeded 0.1%.

3.2.2. MTT (3-(4,5-dimethylthiazol-2-yl)-2,5-diphenyltetrazolium bromide) Assay

U2OS were seeded at 1.5×10^4 cells/well in a 96-well culture plastic plate (Orange Scientific, Braine-l'Alleud, Belgium) and, after 24 h of growth, cells were exposed for additional 48 h to increasing concentrations of compounds (0.1 µM and 500 µM) solubilized in RPMI-1640 medium. On the day of measurement, the culture medium was replaced with 0.1 mL of 3-(4,5-dimethylthiazolyl-2)-2,5-diphenyltetrazolium bromide (MTT, Sigma-Aldrich) dissolved in phosphate-buffered saline (PBS) at the concentration of 0.2 mg/mL, and samples were incubated for 2 h at 37 °C. To dissolve the blue-violet formazan salt crystals formed, 0.1 mL of isopropyl alcohol was added to each well and incubated for 20 min. The absorbance at 570 nm was measured using a multiwell plate reader (Wallac Victor2, PerkinElmer, Milano, Italia), and viability was compared with that of untreated cells, used as controls.

3.2.3. Cell Cycles

For the cell-cycle assay, U2OS cells were treated for 24 h with 35 µM of compound **6**, 50 µM of **5a**, or 50 µM of **8a**, detached with 0.11% trypsin (Sigma-Aldrich, S.Louis, MO, USA)/0.02% ethylenediaminetetraacetic acid (EDTA) (Sigma-Aldrich, S.Louis, MO, USA), washed in PBS, and centrifuged. The pellet was resuspended in 0.01% Nonidet P-40 (Sigma-Aldrich, S.Louis, MO, USA), 10 µg/mL RNase (Sigma-Aldrich, S.Louis, MO), 0.1% sodium citrate, and 50 µg/mL propidium iodide (PI) (Sigma-Aldrich, S.Louis, MO, USA), for 30 min at room temperature in the dark. PI fluorescence and forward scatter (FS) signals were analyzed using a Beckman Coulter Epics XL-MCL flow cytometer. DNA distribution in the cell cycle was analyzed by the MODFIT 5.0 software. As the signal collected by a multichannel analyzer, the mean of the FS distribution is expressed as mean channel.

3.2.4. Histone Extraction, SDS-PAGE, and Western Blot

U2OS were cultured with **5a**, **6**, or **8a**, for 6 h, and the histone fraction was immediately extracted. Cells were harvested using 0.11% trypsin and 0.02% EDTA, washed twice with 10 mM sodium butyrate (NaBu) in PBS, and nuclei were isolated according to Amellem et al. [46]. The nuclear pellet was suspended in 0.1 mL of ice-cold water using a Vortex mixer, and concentrated H_2SO_4 was added to the suspension to give a final concentration of 0.4 N. After incubation at 4 °C for 1 h, the suspension was centrifuged for 5 min at 14,000× *g*, and the supernatant was taken and mixed with 1 mL of acetone. After overnight incubation at −20 °C, the coagulate material was collected by microcentrifugation and air-dried, and proteins were quantified using a protein assay kit (Bio-Rad, Hercules, CA, USA). Histones were analyzed as previously described [24]. Briefly, histones were resolved by 15% SDS-PAGE and immunoblotted with anti-acetylated lysine antibody (Cell Signaling Technology, Beverly, MA, USA); the detection of immunoreactive bands was performed with a secondary antibody conjugated with horseradish peroxidase and developed with an enhanced chemiluminescence (ECL) system, developed with the enhanced chemiluminescence system Clarity Western (Bio-Rad, Hercules, CA, USA), and quantification was done by Fluor-S Max MultiImager (Bio-Rad). Histone acetylation signals were quantified by densitometry and normalized on histone H1.

3.2.5. Hoechst 33,342 Staining

Nuclear morphology was assayed using a specific dye Hoechst 33,342. U2OS treated cells were washed, fixed, and then stained with 1 µM Hoechst 33,342 (Sigma-Aldrich, St Louis, MO, USA) for 15 min. Samples were embedded in Mowiol and analyzed using a Nikon C1s confocal laser-scanning microscope, equipped with a Nikon PlanApo 40×, 1.4 numerical aperture (NA) oil immersion lens. Images were quantified by ImageJ software (IMAJ 1.x).

3.3. Docking

Crystal structures of target HDACs, HDAC1 (PDB: 5ICN), HDAC2 (PDB: 4LXZ), HDAC3 (PDB: 4A69, HDAC8 (PDB: 4QA3), HDAC4 (PDB: 2VQM), and HDAC7 (PDB: 3C0Z), used for the docking study, were obtained from the Protein Data Bank (http://www.rcsb.org/pdb/). The target proteins were prepared using Maestro 9.1 software (Schrödinger, LLC, New York, NY, USA 2010) deleting waters, optimizing H-Bond assignments, and deleting original ligands. Before starting the docking calculations, ligand geometry optimization was done using Gaussian 3 [47] by semi-empirical AM1 [48] in order to obtain the minimum-energy conformation. The computational analysis was executed using AutoDock 4.2 [45] and Autogrid 4.0 on a Dual-Xeon T7400 Dell workstation. Grids (one grid for each atom type in the ligand, plus an electrostatic and a desolvation map), were centered on the binding site and were chosen to be large enough (60 × 60 × 60 Å) to allow the ligand to rotate freely, even in its most fully extended conformation. Docking was performed using the AutoDock empirical free energy function and the Lamarckian genetic algorithm with local search. Lamarckian genetic algorithms can handle ligands with more degrees of freedom than the simulated annealing method. In total, 150 docking runs with 2,500,000 energy evaluations for each run were performed for each molecule and all the evaluated HDACs. Cluster analysis (root-mean-square (RMS) tolerance equal to 0.5 Å) was then carried out on the docked results. Inhibitors were compared according to the cluster with lowest docked energy found. The inhibition constants (Ki) were calculated from the docked energy.

The same protocol was used as in previous studies [49].

4. Conclusions

A series of novel compounds bearing an aza-heterocyclic moiety bound to the azelayl scaffold by an amide bond was synthesized to evaluate the biological effects induced on a panel of both normal and tumor cell lines. Noteworthy, none of the compounds induced cytotoxicity in the normal fibroblast cell line, while only osteosarcoma (U2OS) among the tumor cell lines appeared to be sensitive to compounds such as **5a**, **6**, and **8a**. The treatment with the compounds **5a** and **8a** induced an accumulation in the S phase, while compound **6** caused an increase in the G0/G1 fraction with concomitant decrease in the G2/M phase, and these changes in the cell cycle were associated with a post-transcriptional modification of both H2/H3 and H4 histones. The above cited three active molecules underwent an in silico study using histone deacetylases as the molecular target, revealing that they were able to interact with HDAC 7. These findings are in line with the recent studies, which disclosed an unexpected function for HDAC7 in osteoclasts, which is distinct from the function of HDAC3. Interestingly, compound **6** dramatically affected the FS signal of U2OS nuclei, indicating a change of the nuclei in both shape and size. Overall, compounds **5a**, **6**, and **8a** could be selective HDACi, achieving a greater clinical utility by elimination or reduction of serious side effects, associated with the current non-selective first-generation HDACi.

Supplementary Materials: The following are available online: Figures S1–S27: ^{1}H- and ^{13}C-NMR spectra of compounds **4a–c**, **5a–c**, **6**, **8a**, **8b**; Figure S28: docking pose for **8a** in HDAC2; Figure S29: Docking pose for molecule **5a** in binding site of HDAC8, HADC7, and HDAC2; Figure S30: Docking pose for molecule **6** in binding site of HDAC8, HADC7, and HDAC2; Figure S31: Docking pose for all the evaluated compounds in the HDAC8 and HDAC7 active binding sites.

Author Contributions: Conceptualization, C.B. and N.C.; investigation, G.M., D.T., V.P., N.C., G.F. (Giovanna Farruggia), G.F. (Giulia Frisco), and E.S.; data curation, G.M., G.F. (Giovanna Farruggia), N.C., and E.S.; writing—original draft preparation, G.M., C.B., N.C., G.F. (Giovanna Farruggia), E.S., and S.B.; writing—review and editing, S.B.; funding acquisition, C.B., N.C., G.F. (Giovanna Farruggia), and S.B. All authors have read and agreed to the published version of the manuscript.

Funding: This research received no external funding

Acknowledgments: This work was supported by Alma Mater Studiorum, Università di Bologna (RFO funds). The authors thank Daniel Pecorari and Luca Zuppiroli for running the mass spectra.

Conflicts of Interest: The authors declare no conflicts of interest.

References

1. Anderson, T. Ueber die Producte der trocknen Destillation thierischer Materien. *Justus Liebigs Ann. Chem.* **1851**, *80*, 44–55. [CrossRef]

2. Moen, M.D.; McKeage, K.; Plosker, G.L.; Siddiqui, M.A. Imatinib: A review of its use in chronic myeloid leukaemia. *Drugs* **2007**, *67*, 299–320. [CrossRef] [PubMed]

3. Siddiqui, M.A.; Scott, L.J. Imatinib: A review of its use in the management of gastrointestinal stromal tumours. *Drugs* **2007**, *67*, 805–820. [CrossRef] [PubMed]

4. Zwergel, C.; Stazi, G.; Valente, S.; Mai, A. Histone Deacetylase Inhibitors: Updated Studies in Various Epigenetic-Related Diseases. *J. Clin. Epigen.* **2016**, *2*, 1–15. [CrossRef]

5. Dolezal, M.; Zitko, J. Pyrazine derivatives: A patent review (June 2012 – present). *Expert Opin. Ther. Patents* **2015**, *25*, 33–47. [CrossRef] [PubMed]

6. Miniyar, P.B.; Murumkar, P.R.; Patil, P.S.; Barmade, M.A.; Bothara, K.G. Unequivocal Role of Pyrazine Ring in Medicinally Important Compounds: A Review. *Mini-Rev. Med. Chem.* **2013**, *13*. [CrossRef]

7. Xi, N.; Huang, Q.; Liu, L. Imidazoles. In *Comprehensive Heterocyclic Chemistry III*; Katritzky, A.R., Ramsden, C.A., Scriven, E.F.V., Taylor, R.J.K., Eds.; Volume 4 (Five-membered Rings with Two Heteroatoms, each with their Fused Carbocyclic Derivatives); Elsevier: Amsterdam, The Netherlands, 2008; pp. 143–364.

8. Chen, B.; Heal, H. Thiazoles. In *Comprehensive Heterocyclic Chemistry III*; Katritzky, A.R., Ramsden, C.A., Scriven, E.F.V., Taylor, R.J.K., Eds.; Volume 4 (Five-membered Rings with Two Heteroatoms, each with their Fused Carbocyclic Derivatives); Elsevier: Amsterdam, The Netherlands, 2008; pp. 635–754.

9. Forlani, L.; Lugli, A.; Boga, C.; Bonamartini Corradi, A.; Sgarabotto, P. Mechanism of the formation of 1,2,4-Thiadiazoles by Condensation of Aromatic Thioamides and of N-Substituted Thioureas. *J. Heterocycl. Chem.* **2000**, *37*, 63–69. [CrossRef]

10. Boga, C.; Stengel, R.; Abdayem, R.; Del Vecchio, E.; Forlani, L.; Todesco, P.E. Regioselectivity in the Addition of Vinylmagnesium Bromide to Heteroarylic Ketones: *C-* versus *O*-Alkylation. *J. Org. Chem.* **2004**, *69*, 8903–8909. [CrossRef]

11. Boga, C.; Micheletti, G. Regioselectivity in the Addition of Grignard Reagents to Bis(2-benzothiazolyl) ketone. *C-* versus *O*-alkylation Using Aryl Grignard Reagents. *Eur. J. Org. Chem* **2010**, 5659–5665.

12. Boga, C.; Del Vecchio, E.; Forlani, L.; Goumont, R.; Terrier, F.; Tozzi, S. Evidence for the Intermediacy of Wheland–Meisenheimer Complexes in S_EAr Reactions of Aminothiazoles with 4,6-Dinitrobenzofuroxan. *Chem. Eur. J.* **2007**, *13*, 9600–9607. [CrossRef]

13. Forlani, L.; Boga, C.; Mazzanti, A.; Zanna, N. Trapping and Analysing Wheland–Meisenheimer σ Complexes, Usually Labile and Escaping Intermediates. *Eur. J. Org. Chem.* **2012**, 1123–1129.

14. Boga, C.; Cino, S.; Micheletti, G.; Padovan, D.; Prati, L.; Mazzanti, A.; Zanna, N. New azo-decorated N-pyrrolidinylthiazoles: Synthesis, properties and an unexpected remote substituent effect transmission. *Org. Biomol. Chem.* **2016**, *14*, 7061–7068. [CrossRef] [PubMed]

15. Boga, C.; Bordoni, S.; Casarin, L.; Micheletti, G.; Monari, M. Regioselectivity in Reactions between Bis(2-benzothiazolyl)ketone and Vinyl Grignard Reagents: *C-* versus *O*-alkylation—Part III. *Molecules* **2018**, *23*, 171. [CrossRef] [PubMed]

16. Chugunova, E.; Boga, C.; Sazykin, I.; Cino, S.; Micheletti, G.; Mazzanti, A.; Sazykina, M.; Burilov, A.; Khmelevtsova, L.; Kostina, N. Synthesis and antimicrobial activity of novel structural hybrids of benzofuroxan and benzothiazole derivatives. *Eur. J. Med. Chem.* **2015**, *93*, 349–359. [CrossRef] [PubMed]

17. Boga, C.; Micheletti, G.; Orlando, I.; Strocchi, E.; Vitali, B.; Verardi, L.; Sartor, G.; Calonghi, N. New Hybrids with 2-aminobenzothiazole and Azelayl Scaffolds: Synthesis, Molecular Docking and Biological Evaluation. *Curr. Org. Chem.* **2018**, *22*, 1649–1660. [CrossRef]

18. Bertucci, C.; Hudaib, M.; Boga, C.; Calonghi, N.; Cappadone, C.; Masotti, L. Gas chromatography/mass spectrometry assay of endogenous cellular lipid peroxidation products: Quantitative analysis of 9- and 10-hydroxystearic acids. *Rapid Commun. Mass Spectrom.* **2002**, *16*, 859–864. [CrossRef]

19. Calonghi, N.; Cappadone, C.; Pagnotta, E.; Farruggia, G.; Buontempo, F.; Boga, C.; Brusa, G.L.; Santucci, M.A.; Masotti, L. 9-Hydroxystearic acid upregulates p21WAF1 in HT29 cancer cells. *Biochem. Biophys. Res. Commun.* **2004**, *314*, 138–142. [CrossRef]

20. Calonghi, N.; Pagnotta, E.; Parolin, C.; Tognoli, C.; Boga, C.; Masotti, L. 9-Hydroxystearic acid interferes with EGF signalling in a human colon adenocarcinoma. *Biochem. Biophys. Res. Commun.* **2006**, *342*, 585–588. [CrossRef]

21. Busi, A.; Aluigi, A.; Guerrini, A.; Boga, C.; Sartor, G.; Calonghi, N.; Sotgiu, G.; Posati, T.; Corticelli, F.; Fiori, J.; et al. Unprecedented behavior of (9R)-9-hydroxystearic acid loaded keratin nanoparticles on cancer cell cycle. *Mol. Pharm.* **2019**, *16*, 931–942. [CrossRef]

22. Calonghi, N.; Pagnotta, E.; Parolin, C.; Molinari, C.; Boga, C.; Dal Piaz, F.; Brusa, G.L.; Santucci, M.A.; Masotti, L. Modulation of apoptotic signalling by 9-hydroxystearic acid in osteosarcoma cells. *Biochim. Biophys. Acta, Mol. Cell. Biol. Lipids* **2007**, *1771*, 139–146. [CrossRef]

23. Boanini, E.; Torricelli, P.; Boga, C.; Micheletti, G.; Cassani, M.C.; Fini, M.; Bigi, A. (9R)-9-Hydroxystearate-Functionalized Hydroxyapatite as Anti-Proliferative and Cytotoxic Agent towards Osteosarcoma Cells. *Langmuir* **2016**, *32*, 188–194. [CrossRef]

24. Calonghi, N.; Boga, C.; Telese, D.; Bordoni, S.; Sartor, G.; Torsello, C.; Micheletti, G. Synthesis of 9-Hydroxystearic Acid Derivatives and Their Antiproliferative Activity on HT 29 Cancer Cells. *Molecules* **2019**, *24*, 3714. [CrossRef] [PubMed]

25. Calonghi, N.; Cappadone, C.; Pagnotta, E.; Boga, C.; Bertucci, C.; Fiori, J.; Tasco, G.; Casadio, R.; Masotti, L. Histone deacetylase 1: A target of 9-hydroxystearic acid in the inhibition of cell growth in human colon cancer. *J. Lipid Res.* **2005**, *46*, 1596–1603. [CrossRef] [PubMed]

26. Parolin, C.; Calonghi, N.; Presta, E.; Boga, C.; Caruana, P.; Naldi, M.; Andrisano, V.; Masotti, L.; Sartor, G. Mechanism and stereoselectivity of HDAC I inhibition by (*R*)-9-hydroxystearic acid in colon cancer. *Biochim. Biophys. Acta* **2012**, *1821*, 1334–1340.

27. Albadri, S.; Naso, F.; Gauron, C.; Parolin, C.; Duroure, K.; Fiori, J.; Boga, C.; Vriz, S.; Calonghi, N.; Del Bene, F. Redox signaling via lipid peroxidation regulates retinal progenitor cell differentiation. *Dev. Cell* **2019**, *50*, 73–89. [CrossRef] [PubMed]

28. Richon, V.M. Cancer biology: Mechanism of antitumour action of vorinostat (suberoylanilide hydroxamic acid), a novel histone deacetylase inhibitor. *Br. J. Cancer.* **2006**, *95* (Suppl. 1), S2–S6. [CrossRef]

29. Harris, M.G.; Stewart, R. Amino group acidity in aminopyridines and aminopyrimidines. *Can. J. Chem.* **1977**, *55*, 3800–3806. [CrossRef]

30. Jiang, K.; Zhang, J.; Ji, M.; Gai, P.; Lv, Q. Inhibitory effect of 5-Fluorouracil on the proliferation of human osteosarcoma cells in vitro. *J. BUON* **2019**, *24*, 1706–1711.

31. Shin, S.H.; Choi, Y.J.; Lee, H.; Han-Soo Kim, H.-S.; Seo, S.W. Oxidative stress induced by low-dose doxorubicin promotes the invasiveness of osteosarcoma cell line U2OS in vitro. *Tumor Biol.* **2016**, *37*, 1591–1598. [CrossRef]

32. Wu, Z.; Ma, C.; Shan, Z.; Ju, Y.; Li, S.; Zhao, Q. Histone deacetylase inhibitors suppress the growth of human osteosarcomas in vitro and in vivo. *J. BUON.* **2013**, *18*, 1032–1037.

33. Bai, Y.; Chen, Y.; Chen, X.; Jiang, J.; Wang, X.; Wang, L.; Wang, J.; Zhang, J.; Gao, L. Trichostatin A activates FOXO1 and induces autophagy in osteosarcoma. *Arch. Med. Sci.* **2019**, *15*, 204–213. [CrossRef]

34. Roberts, W.E.; Mozsary, P.G.; Klinger, E. Nuclear size as a cell-kinetic marker for osteoblast differentiation. *Am. J. Anat.* **1982**, *165*, 373–384. [CrossRef] [PubMed]

35. Jevtic', P.; Edens, L.J.; Vukovic', L.D.; Levy, D.L. Sizing and shaping the nucleus: Mechanisms and significance. *Curr. Opin. Cell Bio.l* **2014**, *28*, 16–27. [CrossRef] [PubMed]

36. Chow, K.-H.; Factor, R.E.; Ullman, K.S. The nuclear envelope environment and its cancer connections. *Nat. Rev. Cancer* **2012**, *12*, 196–209. [CrossRef] [PubMed]

37. Yoon, K.B.; Park, K.R.; Kim, S.Y.; Han, S.Y. Induction of nuclear enlargement and senescence by sirtuin inhibitors in glioblastoma cells. *Immune Netw.* **2016**, *16*, 183–188. [CrossRef]

38. Bang, M.; Kim, D.G.; Gonzales, E.L.; Kwon, K.J.; Shin, C.Y. Etoposide Induces Mitochondrial Dysfunction and Cellular Senescence in Primary Cultured Rat Astrocytes. *Biomol. Ther.* **2019**, *27*, 530–539. [CrossRef]

39. Kong, J.Y.; Rabkin, S.W. Palmitate induces structural alterations in nuclei of cardiomyocytes. *Tissue Cell* **1999**, *31*, 473–479.

40. Gordon, G.B. Saturated free fatty acid toxicity II. Lipid accumulation, ultrastructural alterations and toxicity in mammalian cells in culture. *Exp. Molec. Path.* **1977**, *27*, 262–276. [CrossRef]

41. Pham, L.; Kaiser, B.; Romsa, A.; Schwarz, T.; Gopalakrishnan, R.; Jensen, E.D.; Mansky, K.C. HDAC3 and HDAC7 Have Opposite Effects on Osteoclast Differentiation. *J. Biol. Chem.* **2011**, *286*, 12056–12065. [CrossRef]

42. Benedetti, R.; Conte, M.; Altucci, L. Targeting Histone Deacetylases in Diseases: Where Are We? *Antioxid. Redox Signal.* **2015**, *23*, 99–126. [CrossRef]

43. Bottomley, M.J.; Lo Surdo, P.; Di Giovine, P.; Cirillo, A.; Scarpelli, R.; Ferrigno, F.; Jones, P.; Neddermann, P.; De Francesco, R.; Steinkühler, C.; et al. Structural and Functional Analysis of the Human HDAC4 Catalytic Domain Reveals a Regulatory Structural Zinc-Binding Domain. *J. Biol. Chem.* **2008**, *283*, 26694–26704. [CrossRef]

44. Miller, T.A.; Witter, D.J.; Belvedere, S. Histone Deacetylase Inhibitors. *J. Med. Chem.* **2003**, *46*, 5097–5116. [CrossRef] [PubMed]

45. Morris, G.M.; Goodsell, D.S.; Halliday, R.S.; Huey, R.; Hart, W.E.; Belew, R.K.; Olson, A. Automated docking using a Lamarckian genetic algorithm and an empirical binding free energy function. *J. Comput. Chem.* **1998**, *19*, 1639–1662. [CrossRef]

46. Amellem, O.; Stokke, T.; Sandvik, J.A.; Pettersen, E.O. The retinoblastoma gene product is reversibly dephosphorylated and bound in the nucleus in S and G2 phases during hypoxic stress. *Exp. Cell Res.* **1996**, *227*, 106–115. [CrossRef] [PubMed]

47. Frisch, M.J.; Trucks, G.W.; Schlegel, H.B.; Scuseria, G.E.; Robb, M.A.; Cheeseman, J.R.; Montgomery, J.A., Jr.; Vreven, T.; Kudin, K.N.; Burant, J.C.; et al. *Gaussian 03*; Gaussian, Inc.: Wallingford, CT, USA, 2004.

48. Dewar, M.J.S.; Zoebisch, E.G.; Healy, E.F.; Stewart, J.J.P. Development and use of quantum mechanical molecular models. 76. AM1: A new general purpose quantum mechanical molecular model. *J. Am. Chem. Soc.* **1985**, *107*, 3902–3909. [CrossRef]

49. Strocchi, E.; Fornari, F.; Minguzzi, M.; Gramantieri, L.; Milazzo, M.; Rebuttini, V.; Breviglieri, S.; Camaggi, C.M.; Locatelli, E.; Bolondi, L.; et al. Design, synthesis and biological evaluation of pyrazole derivatives as potential multi-kinase inhibitors in hepatocellular carcinoma. *Eur. J. Med. Chem.* **2012**, *48*, 391–401. [CrossRef] [PubMed]

Sample Availability: Samples of the all synthesized compounds are available from the authors.

 *molecules*

Article

Synthesis and Cytotoxic Evaluation of 3-Methylidenechroman-4-ones

Jacek Kędzia [1], Tomasz Bartosik [1], Joanna Drogosz [2], Anna Janecka [2], Urszula Krajewska [3] and Tomasz Janecki [1,*]

[1] Institute of Organic Chemistry, Lodz University of Technology, Żeromskiego 116, 90-924 Łódź, Poland; jacek.kedzia@p.lodz.pl (J.K.); tomasz.bartosik92@gmail.com (T.B.)

[2] Department of Biomolecular Chemistry, Medical University of Łódź, Mazowiecka 6/8, 92-215 Łódź, Poland; joanna.drogosz@studumed.lodz.pl (J.D.); anna.janecka@umed.lodz.pl (A.J.)

[3] Department of Pharmaceutical Biochemistry and Molecular Diagnostics, Faculty of Pharmacy, Medical University of Łódź, Muszyńskiego 1, 90-151 Łódź, Poland; ukrajewska@tlen.pl

* Correspondence: tomasz.janecki@p.lodz.pl; Tel.: +48-426313220

Academic Editors: Carla Boga and Gabriele Micheletti
Received: 26 April 2019; Accepted: 11 May 2019; Published: 15 May 2019

Abstract: In the search for new anticancer agents, a library of variously substituted 3-methylidenechroman-4-ones was synthesized using Horner–Wadsworth–Emmons methodology. Acylation of diethyl methylphosphonate with selected ethyl salicylates furnished 3-diethoxyphosphorylchromen-4-ones which were next used as Michael acceptors in the reaction with various Grignard reagents. The adducts were obtained as the mixtures of *trans* and *cis* diastereoisomers along with a small amount of enol forms. Their relative configuration and preferred conformation were established by NMR analysis. The adducts turned up to be effective Horner–Wadsworth–Emmons reagents giving 2-substituted 3-methylidenechroman-4-ones, which were then tested for their possible cytotoxic activity against two leukemia cell lines, HL-60 and NALM-6, and against MCF-7 breast cancer cell line. All new compounds (**14a–o**) were highly cytotoxic for the leukemic cells and showed a moderate or weak effect on MCF-7 cells. Analog **14d** exhibited the highest growth inhibitory activity and was more potent than carboplatin against HL-60 (IC_{50} = 1.46 ± 0.16 µM) and NALM-6 (IC_{50} = 0.50 ± 0.05 µM) cells. Further tests showed that **14d** induced apoptosis in NALM-6 cells, which was mediated mostly through the extrinsic pathway.

Keywords: 3-methylidenechroman-4-ones; Michael addition; Horner–Wadsworth–Emmons olefination; cancer cell lines; apoptosis

---

1. Introduction

Chroman-4-one skeleton **1** is a core structure for a large group of plant metabolites called flavonoids, which possess many desirable biological activities including anticancer, antibacterial, and antioxidant properties [1,2]. One relatively small subgroup of flavonoids are homoisoflavonoids **2–4** which are characterized by the presence of arylmethyl or arylidene group in position 3 (Figure 1). A special place within this group belongs to 3-arylidenechroman-4-ones **4** which were found in many plants. For example, Bonducellin **5** was isolated from *Caesalpina bonducella* [3] and Eucomin **6** from *Eucomis bicolor* BAK (Liliaceae) [4].

Both natural and synthetic 3-arylidenechroman-4-ones **4** display valuable biological activities. They are potent and selective monoamine oxidase-B (MAO-B) inhibitors [5,6] and possess anti-cholinesterase activity [7,8], what makes them good candidates for the treatment of various neurological diseases such as Alzheimer's or Parkinson's disease. Furthermore, they show significant cytotoxicity for several cancer cell lines [9,10] and are cytochrome P450 aromatase inhibitors [11] being used for the treatment of advanced breast cancer. Also, their antifungal [12,13], antioxidant [14,15], and anti-inflammatory [16] activity was reported.

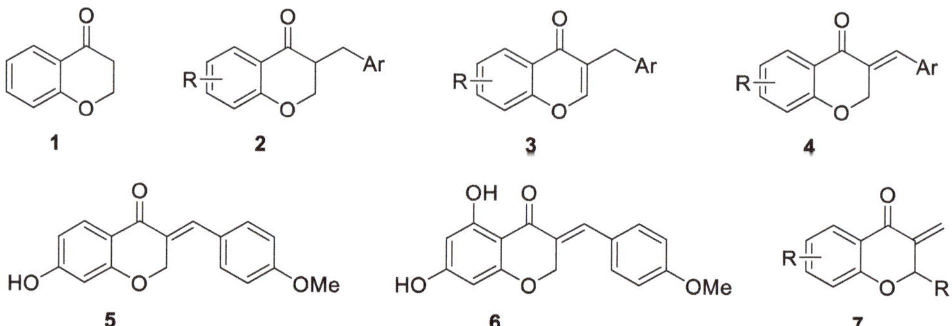

Figure 1. Structure and representative examples of homoisoflavonoids.

On the other hand, 3-methylidenechroman-4-ones **7**, which are structurally closely related to 3-arylidenechroman-4-ones **4**, have not been found in nature. Nevertheless, several syntheses of these compounds were reported. 2-Aryl-3-methylidenechroman-4-ones were obtained by Mannich reaction with 2-arylchroman-4-ones [17,18] and 2-alkyl(aryl)-3-methylidenechroman-4-ones were prepared by palladium-catalyzed tandem carbonylation-allene insertion reaction [19,20]. 3-Methylidenechroman-4-one was prepared by direct α-methylidenation mediated by diisopropylammonium trifluoroacetate [21] or by dehydration of 3-hydroxymethylchromen-4-one in the presence of methanesulfonylchloride [22]. Finally, 2-alkoxycarbonylmethyl-3-methylidenechroman-4-ones were obtained from enol silyl ethers of the corresponding chroman-4-ones [23,24]. Unfortunately, most of these methods have a very limited scope and/or are inefficient. For example, yields of the Mannich reactions were very low (3–22%) [17] or were not given because the obtained, crude 3-methylidenechroman-4-ones were used in further transformations [18]. On the other hand, palladium-catalyzed insertion reactions were more effective (23–90% yield) but required not readily available allenes as substrates. In turn the scope of direct α-methylidenation, dehydration of 3-hydroxymethylchromen-4-one or use the enol silyl ethers was limited to a single example in each case.

In contrast to 3-arylidenechroman-4-ones **4**, biological activity of 3-methylidenechroman-4-ones **7** is poorly recognized. There is only one report describing their significant bacteriostatic activity against Gram-positive microorganisms [17]. However, an *exo*-cyclic methylidene bond conjugated with a carbonyl group present in 3-methylidenechroman-4-ones **7** is a structural motif found also in a large number of natural products, such as α-methylidene-γ- and δ-lactones [25,26] which react by the Michael-type addition with various bionucleophiles, disrupting key biological processes and are considered promising anticancer agents [27,28]. Consequently, we reasoned that also 3-methylidenechroman-4-ones **7** might have considerable cytotoxic activity.

In this report, we present a new, general synthetic method for obtaining variously substituted 3-methylidenechroman-4-ones **14**, based on the (well-recognized in our laboratory) Horner–Wadsworth–Emmons approach for the construction of *exo*-methylidene bond [29–31]. All obtained 3-methylidenechroman-4-ones **14** were evaluated in terms of their cytotoxic activity against three human cancer cell lines: promyelocytic leukemia HL-60, NALM-6, and breast adenocarcinoma cell line MCF-7. The most cytotoxic compound, **14d** was selected for further experiments and its effect on the induction of apoptosis was investigated.

2. Results and Discussion

2.1. Chemistry

The first step was the synthesis of 3-diethoxyphosphorylchromen-4-ones **12a–c**, which are crucial intermediates in our methodology. Literature search revealed that there is no efficient method for obtaining 2-unsubstituted 3-phosphorylchromen-4-ones. In the only report we have found, 3-diethoxyphosphorylchromen-4-one was formed in 5% yield, as a side product in free radical phosphorylation of chromen-4-one [32]. Therefore, we worked out a two-step procedure, which starts with the reaction of commercially available ethyl salicylates **8a–c** with diethyl methylphosphonate **9** in the presence of three equivalents of LDA (Scheme 1). Using this stoichiometry, protection of the hydroxyl group is avoided and we believe this is the major improvement over the reported acylation of dimethyl methylphosphonate using benzyl protected methyl salicylate [33]. The standard work-up and column chromatography purification gave 2-(2-hydroxyphenyl)-2-oxoethylphoshonates **10a–c** in good yields (Table 1). In the next step, reaction between phophonates **10a–c** and dimethylformamide dimethyl acetal **11** gave, after purification by column chromatography, 3-diethoxyphosphorylchromen-4-ones **12a–c** in high yields (Table 1 and Supplementary Materials).

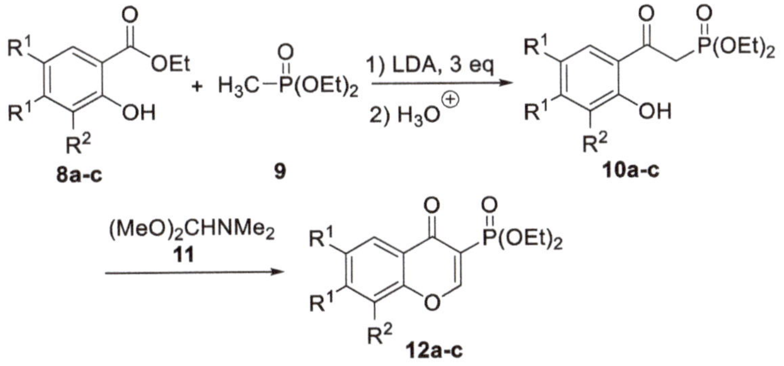

Scheme 1. Synthesis of 3-diethoxyphosphorylchromen-4-ones **12a–c**.

Table 1. Yields of phosphonates **10a-c** and 3-diethoxyphosphorylchromen-4-ones **12a–c**.

Compound	R^1,R^1	R^2	10 Yield [%] [1]	12 Yield [%] [1]
a	H,H	H	80	92
b	H,H	Me	75	89
c	CH=CH-CH=CH	H	80	78

[1] Yield of pure, isolated product, based on **8** or **10**, respectively.

With 3-diethoxyphosphorylchromen-4-ones **12a–c** in hand, we performed their reactions with various Grignard reagents (Scheme 2). In all cases, after standard work-up, we received adducts **13a–o**, which were purified by column chromatography with yields given in Table 2. Interestingly, examination of the ^{1}H, ^{13}C and ^{31}P NMR spectra revealed that all adducts **13** were formed as mixtures of *trans* and *cis* diastereoisomers (*trans-* or *cis*-**13a–o**), along with small amount of enol form (enol-**13a–o**), with *trans* diastereoisomers strongly predominating. Ratios, determined from the ^{31}P NMR spectra of the crude reaction mixtures, are given in Table 2.

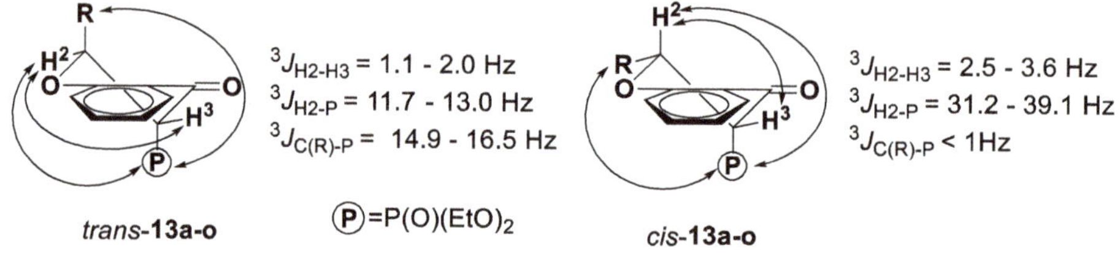

Scheme 2. Synthesis of Michael adducts **13a–o**.

Table 2. Adducts **13a–o** and methylidenechroman-4-ones **14a–o** obtained.

Compound	R^1,R^1	R^2	R^3	13		14 Yield [%] [2]
				trans/cis/enol [1]	Yield [%] [2]	
a	H,H	H	Me	67/30/3	84	68
b	H,H	H	Et	71/25/4	68	71
c	H,H	H	*n*-Bu	76/20/4	85	59
d	H,H	H	iPr	71/26/3	66	53
e	H,H	H	Ph	81/10/9	64	84
f	H,H	Me	Me	73/24/3	74	64
g	H,H	Me	Et	75/22/3	76	59
h	H,H	Me	*n*-Bu	72/25/3	83	52
i	H,H	Me	iPr	72/26/2	70	53
j	H,H	Me	Ph	83/11/6	84	59
k	CH=CH-CH=CH	H	Me	79/14/7	81	66
l	CH=CH-CH=CH	H	Et	89/4/7	64	67
m	CH=CH-CH=CH	H	*n*-Bu	73/16/11	53	60
n	CH=CH-CH=CH	H	iPr	81/15/4	66	48
o	CH=CH-CH=CH	H	Ph	65/6/29	76	70

[1] Ratios taken from ^{31}P NMR spectra of the crude mixtures. [2] Yield of pure, isolated product, based on **12** or **13**, respectively.

Careful analysis of the NMR spectra showed also that both *trans-* and *cis-***13a–o** exist in the half-chair conformation and diethoxyphosphoryl group occupies the axial position (Figure 2). Simple application of Karplus correlation between corresponding dihedral angles and coupling constants $^3J_{\text{H2-H3}}$, $^3J_{\text{H2-P}}$ and $^3J_{\text{C(R3)-P}}$, determined from the ^{1}H and ^{13}C NMR spectra of *trans-* and *cis-***13a–o** shows full agreement with the proposed configurations and conformations. Corresponding coupling constants are given in Figure 2. Similar half-chair conformation with diethoxyphosphoryl group in axial position was reported for 3-diethoxyphosphorylchroman-2-ones [34,35].

$^3J_{\text{H2-H3}}$ = 1.1 - 2.0 Hz $^3J_{\text{H2-H3}}$ = 2.5 - 3.6 Hz

$^3J_{\text{H2-P}}$ = 11.7 - 13.0 Hz $^3J_{\text{H2-P}}$ = 31.2 - 39.1 Hz

$^3J_{\text{C(R)-P}}$ = 14.9 - 16.5 Hz $^3J_{\text{C(R)-P}}$ < 1Hz

*trans-***13a–o**   (P)=P(O)(EtO)$_2$ *cis-***13a–o**

Figure 2. Half-chair conformation of *trans-* and *cis-***13a–o** and characteristic $^3J_{\text{H2-H3}}$, $^3J_{\text{H2-P}}$ and $^3J_{\text{C(R3)-P}}$ coupling constants.

On the other hand, ^{1}H NMR spectra of enols-**13a–o** revealed very characteristic doublets with coupling constant $^4J_{\text{P-H}}$ ~ 1 Hz and chemical shift in the range of 11–12 ppm, which can be assigned to the proton of the hydroxyl group. Coupling between this proton and phosphorus indicates the presence of resonance-assisted hydrogen bond (RAHB) [36] and can be visualized by resonance structures shown in Figure 3. Recently, we have reported on the existence of RAHB in

3-(dimenthoxyphosphoryl)-2-phenyl-1,2-dihydroquinolin-4-ol [37], what was the first example of this phenomenon in organophosphorus compounds. Now, we can confirm the existence of RAHB also in 3-diethoxyphosphoryl-2*H*-chromen-4-oles.

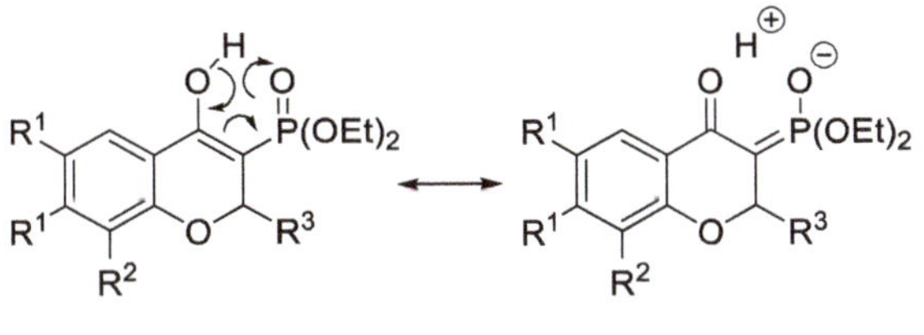

enol-13a–o

Figure 3. Two main resonance structures involved in the resonance-assisted hydrogen bond (RAHB) in 3-diethoxyphosphoryl-2*H*-chromen-4-oles.

Finally, all adducts **13a–o** were transformed into 3-methylidenechroman-4-ones **14a–o** performing Horner–Wadsworth–Emmons olefination of formaldehyde. The best results were obtained with K_2CO_3 used as a base and formalin as a source of formaldehyde (Scheme 3). Standard work-up and purification by column chromatography furnished 3-methylidenechroman-4-ones **14a–o** in moderate to good yields (Table 2).

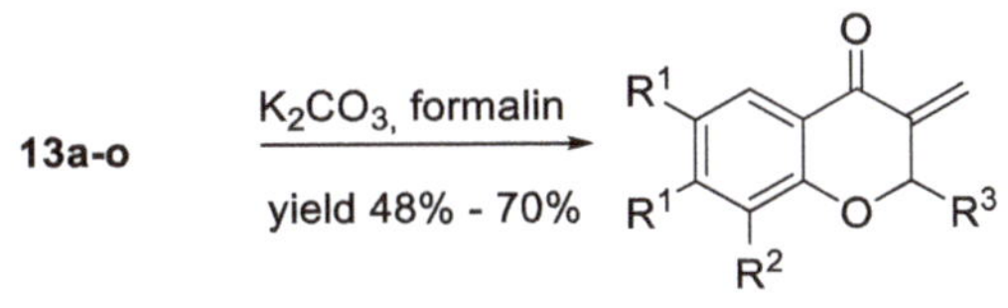

$$\textbf{13a–o} \quad \xrightarrow[\text{yield 48\% - 70\%}]{K_2CO_3,\ \text{formalin}} \quad \textbf{14a–o}$$

Scheme 3. Synthesis of 3-methylidenechroman-4-ones **14a–o**.

2.2. Biology

2.2.1. In Vitro Cytotoxicity of New Analogs Against Three Cancer Cell Lines

All obtained 3-methylidenechroman-4-ones **14a–o** were evaluated for their possible cytotoxic activity against three human cancer cell lines: leukemia HL-60 and NALM-6 and breast adenocarcinoma MCF-7 using the MTT assay (after 48 h incubation) (Table 3). Carboplatin served as a reference compound.

Analysis of the structure–activity relationship revealed that chromanones **14a–j** were, in general, more potent than benzochromanones **14k–o**, containing additional benzene ring *ortho*-fused with a chromanone skeleton. All compounds (**14a–o**) were much more cytotoxic for leukemia cells than for the solid tumor MCF-7 cells. In both series of chromanones, **14a–e** (R^2 = H) and **14f–j** (R^2 = Me), the most potent compounds against leukemic HL-60 and NALM-6 cells were these containing an *i*-propyl substituent in position 2, i.e., **14d** and **14i**, respectively. For HL-60 cells only **14d** was more cytotoxic than the reference carboplatin. The highest cytotoxicity was observed against NALM-6 cells, with three analogs **14b, 14d,** and **14i** exhibiting lower half maximal inhibitory concentration values (IC_{50}) than carboplatin. Analog **14d** was the most cytotoxic chromanone for NALM-6 cells (IC_{50} of 0.5 ± 0.05 µM) and was selected for further investigation of its potential antineoplastic properties.

Table 3. Tumor cell growth inhibitory activity of **14a–o** on three cancer cell lines.

14

14	R^1, R^1	R^2	R^3	IC_{50} [µM] [1]		
				HL-60	**NALM-6**	**MCF-7**
a	H, H	H	Me	5.91 ± 0.32	2.13 ± 0.04	9.70 ± 0.80
b	H, H	H	Et	5.24 ± 0.52	0.60 ± 0.02	12.50 ± 0.71
c	H, H	H	n-Bu	8.79 ± 0.81	4.23 ± 0.46	16.00 ± 0.20
d	H, H	H	iPr	1.46 ± 0.16	0.50 ± 0.05	19.40 ± 0.80
e	H, H	H	Ph	23.86 ± 2.30	5,76 ± 0.23	17,10 ± 0.30
f	H, H	Me	Me	7.52 ± 0.81	3.28 ± 0.35	10.50 ± 0.34
g	H, H	Me	Et	20.87 ± 1.73	4.83 ± 0.45	11.60 ± 0.07
h	H, H	Me	n-Bu	32.16 ± 3.17	6.36 ± 0.36	13.50 ± 0.60
i	H, H	Me	iPr	3.45 ± 0.34	0.58 ± 0.05	8.48 ± 0.93
j	H, H	Me	Ph	30.51 ± 3.62	6.09 ± 0.61	30.00 ± 1.90
k	CH=CH-CH=CH	H	Me	6.96 ± 0.42	4.31 ± 0.36	13.50 ± 1.00
l	CH=CH-CH=CH	H	Et	47.00 ± 4.11	10.91 ± 2.8	34.60 ± 4.50
m	CH=CH-CH=CH	H	n-Bu	8.64 ± 0.53	6.00 ± 0.23	17.40 ± 2.50
n	CH=CH-CH=CH	H	iPr	10.47 ± 2.04	5.52 ± 0.25	15.70 ± 0.70
o	CH=CH-CH=CH	H	Ph	49.96 ± 3.89	6.59 ± 0.41	71.0 ± 7.50
	Carboplatin			2.9 ± 0.1	0.7 ± 0.3	3.8 ± 0.45

[1] Compound concentration required to inhibit metabolic activity by 50%. The cells were incubated with the analogs for 48 h. Values are expressed as mean ± SEM from the concentration-response curves of at least three experiments using a nonlinear estimation (quasi-Newton algorithm) method.

2.2.2. Apoptotic Cell Death Determination

It is now well documented that most anticancer drugs induce apoptosis. One of the main characteristics of apoptosis, phosphatidylserine (PS) translocation to the outer surface of the cellular membrane, was investigated by double-staining with Annexin V and propidium iodide (PI). Annexin V is a protein exerting high affinity to PS exposed on the outer surface of the plasma membrane, enabling detection of even an early stage apoptosis. The late stage of apoptosis is characterized by loss of membrane integrity allowing permeation of a dye such as PI into the cells [38,39]. Treatment of NALM-6 cells for 24 h (Figure 4A) with the selected analog **14d** at 1.25 µM (IC_{50}) and 2.5 µM (2 IC_{50}) concentrations led to the increase of Annexin V and PI-positive cells from 2.2% to 29.4% and 91.4%, respectively (Figure 4B,C), showing that **14d** induced the late stage of apoptosis in NALM-6 cells.

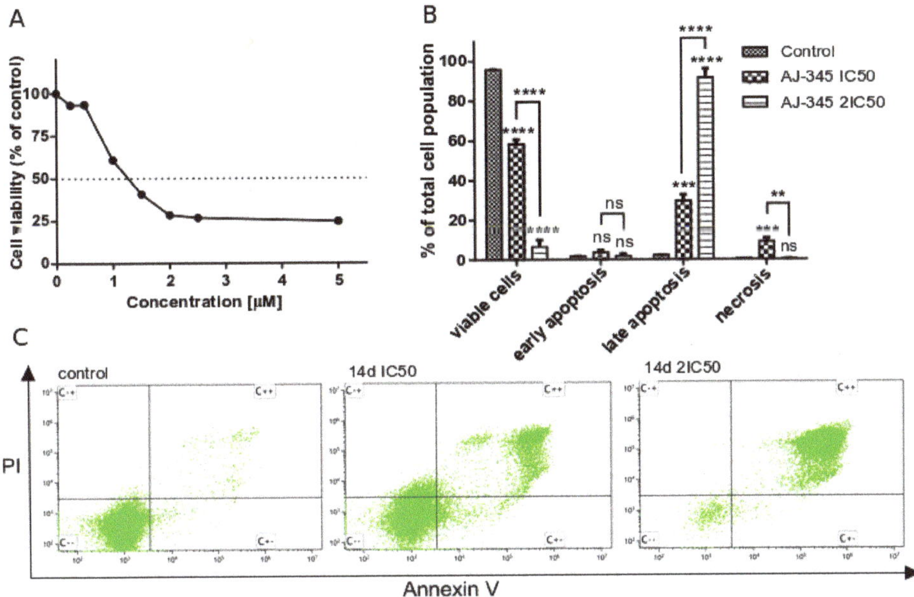

Figure 4. Effect of **14d** on induction of apoptosis in NALM-6 cells. (**A**) The cytotoxic activity of **14d** on NALM-6 cells after 24 h incubation; (**B**) Quantitative analysis of apoptosis by flow cytometry. The data are presented as mean ± SEM of three independent experiments. Statistical significance was determined using one-way ANOVA and a post-hoc multiple comparison Student–Newman–Keuls test. **** $p < 0.0001$; *** $p < 0.001$; ** $p < 0.01$; ns—not statistically significant. (**C**) Representative results of cell apoptosis obtained by Annexin V and PI staining using flow cytometry in NALM-6 cells untreated (control) or treated with **14d** at IC_{50} and 2 IC_{50} concentrations for 24 h.

Apoptosis occurs when caspases, which have proteolytic activity, cleave specific substrates, causing cell death. Depending on the initiator caspase involved in this process, the apoptosis may be mediated by the intrinsic (caspase 9) or extrinsic (caspase 8) pathway [40]. To investigate which caspases were involved in the apoptosis inducted by analog **14d** in NALM-6 cells, the cells were treated with **14d** at 1.25 µM and 2.5 µM concentrations for 6 h. Then, the activity of executioner caspase 3 and initiator caspases 8 and 9 was quantified using fluorogenic indicators. Results presented in Figure 5 indicate that the levels of caspase 3, 8 and 9 were significantly increased: 2.7- and 3.7-fold for caspase 3, 4.9- and 11.3-fold for caspase 8, and 1.25- and 5.7-fold for caspase 9 after treatment of the cells with 1.25 µM and 2.5 µM concentration of **14d**, respectively. Activation of the extrinsic pathway was more prominent than the intrinsic pathway.

Presented results indicate that compound **14d** is a potent cytotoxic agent that significantly inhibits metabolic activity of NALM-6 cells with IC_{50} value as low as 0.5 µM. Analog **14d** also promotes apoptosis in the investigated cell line, which is mediated by the extrinsic and in a much lesser extend intrinsic pathway.

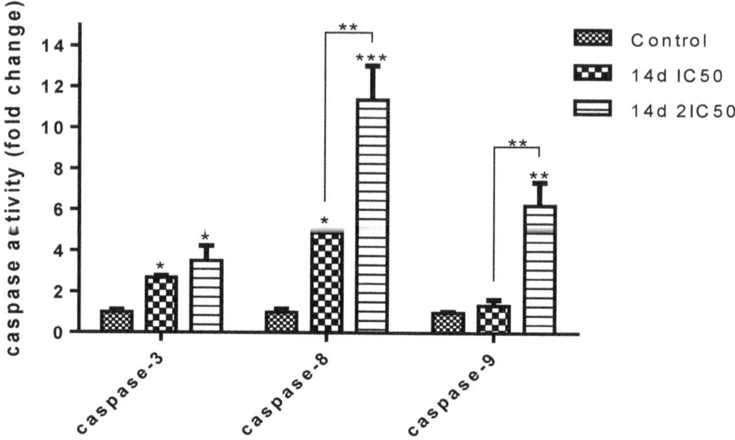

Figure 5. Activity of caspase 3, 8 and 9 in HL-60 cells after 6 h treatment with analog **14d** at 1.25 μM (IC$_{50}$) and 2.5 μM (2IC$_{50}$) concentrations. Results are expressed as mean ± SEM of triplicate experiment. Statistical significance was assessed using one-way ANOVA and a post-hoc multiple comparison Student–Newman–Keuls test; *** $p < 0.001$; ** $p < 0.01$; * $p < 0.05$.

3. Materials and Methods

3.1. Chemistry

3.1.1. General Information

NMR spectra were recorded on a Bruker DPX 250 or Bruker Avance II instrument at 250.13 MHz or 700 MHz for ^{1}H, 62.9 MHz or 176 MHz for ^{13}C, and 101.3 MHz for ^{31}P NMR with tetramethylsilane used as an internal and 85% H_3PO_4 as an external standard. ^{31}P NMR spectra were recorded using broadband proton decoupling. IR spectra were recorded on a Bruker Alpha ATR spectrophotometer. Melting points were determined in open capillaries and are uncorrected. Optical rotations were measured on a Perkin-Elmer 241 polarimeter. The $[\alpha]_D$ values are given in deg·cm2·g^{-1} and concentration c in g·(100 mL)$^{-1}$. Column chromatography was performed on silica gel 60 (230–400 mesh) (Aldrich, Steinheim, Germany). Thin-layer chromatography was performed on the pre-coated TLC sheets of silica gel 60 F254 (Aldrich, Steinheim, Germany). The purity of the synthesized compounds was confirmed by the combustion elemental analyses (CHN, elemental analyzer EuroVector 3018, Elementar Analysensysteme GmbH (Langenselbold, Germany). MS spectra were recorded on Waters 2695-Waters ZQ 2000 LC/MS apparatus (Waters Corporation, Milford, MA, USA). All reagents and starting materials were purchased from commercial vendors and used without further purification. Organic solvents were dried and distilled prior to use. Standard syringe techniques were used for transferring dry solvents.

3.1.2. General Procedure for the Synthesis of 2-Substituted 3-Metylidenechroman-4-ones **14a–o**.

To the vigorously stirred solution of 2-substituted 3-diethoxychroman-4-on **13a–o** (0.15 mmol) in THF (1.5 mL), formaldehyde (36–38% solution in water, 0.125 mL, ca. 1.50 mmol) was added at 0 °C, followed by addition of K_2CO_3 (41 mg, 0.30 mmol) in water (0.4 mL). The resulting mixture was stirred vigorously at 0 °C for 3 h. Next Et$_2$O (5 mL) was added and layers were separated. The water fraction was washed with Et$_2$O (5 mL). Organic fractions were combined, washed with brine (5 mL) and dried over $MgSO_4$. The solvents were evaporated under reduced pressure and the resulting crude product was then purified by column chromatography (eluent CH_2Cl_2).

2-Methyl-3-methylidenechroman-4-one (**14a**) (17.7 mg, 68%). Colourless oil. ^{1}H NMR (700 MHz, Chloroform-*d*) δ 1.63 (d, *J* = 6.4 Hz, 3H), 5.03–5.14 (m, 1H), 5.55 (dd, *J* = 2.0, 0.8 Hz, 1H), 6.32 (dd, *J* = 1.6, 0.8 Hz, 1H), 6.95 (dd, *J* = 8.3, 1.0 Hz, 1H), 7.03 (ddd, *J* = 8.1, 7.2, 1.1 Hz, 1H), 7.47 (ddd, *J* = 8.6, 7.1, 1.8 Hz, 1H), 7.97 (dd, *J* = 7.9, 1.8 Hz, 1H). ^{13}C NMR (176 MHz, Chloroform-*d*) δ 19.03, 76.36, 118.30, 121.44, 121.59, 121.82, 128.00, 136.14, 143.68, 161.30, 182.66. ESI-MS [M + H]$^+$ = 175.2. Anal. Calcd for $C_{11}H_{10}O_2$: C, 75.84%; H, 5.79%. Found: C, 75.61%, H, 5.71%.

2-Ethyl-3-methylidenechroman-4-one (**14b**) (20.1 mg, 71%). Colourless oil. ^{1}H NMR (700 MHz, Chloroform-*d*) δ 1.04 (t, *J* = 7.3 Hz, 3H), 1.75–1.89 (m, 1H), 1.91–2.03 (m, 1H), 4.90 (t, *J* = 7.0 Hz, 1H), 5.50 (s, 1H), 6.33 (s, 1H), 6.96 (d, *J* = 8.4 Hz, 1H), 7.02 (t, *J* = 7.5 Hz, 1H), 7.47 (s, 1H), 7.95 (d, *J* = 7.9 Hz, 1H). ^{13}C NMR (176 MHz, Chloroform-*d*) δ 9.83, 26.65, 81.89, 118.47, 121.41, 121.66, 122.33, 127.84, 136.21, 142.36, 160.77, 182.49. ESI-MS [M + H]$^+$ = 189.3. Anal. Calcd for $C_{12}H_{12}O_2$: C, 76.57%; H, 6.43%. Found: C, 76.41%, H, 6.51%.

2-Butyl-3-methylidenechroman-4-one (**14c**) (19.1 mg, 59%). Colourless oil. ^{1}H NMR (700 MHz, Chloroform-*d*) δ 0.91 (t, *J* = 7.3 Hz, 3H), 1.31–1.45 (m, 3H), 1.47–1.55 (m, 1H), 1.71–1.80 (m, 1H), 1.89–2.00 (m, 1H), 4.86–5.10 (m, 1H), 5.50 (t, *J* = 1.2 Hz, 1H), 6.32 (t, *J* = 1.0 Hz, 1H), 6.95 (dd, *J* = 8.4, 1.0 Hz, 1H), 7.02 (ddd, *J* = 8.1, 7.2, 1.1 Hz, 1H), 7.47 (ddd, *J* = 8.6, 7.1, 1.8 Hz, 1H), 7.96 (dd, *J* = 7.8, 1.8 Hz, 1H). ^{13}C NMR (176 MHz, Chloroform-*d*) δ 14.09, 22.48, 27.50, 33.18, 80.64, 118.48, 121.43, 121.66, 122.17, 127.85, 136.22, 142.65, 160.78, 182.56. ESI-MS [M + H]$^+$ = 217.4. Anal. Calcd for $C_{14}H_{16}O_2$: C, 77.75%; H, 7.46%. Found: C, 77.52%, H, 7.25%.

2-Isopropyl-3-methylidenechroman-4-one (**14d**) (16.1 mg, 53%). Colourless oil. ^{1}H NMR (700 MHz, Chloroform-*d*) δ 0.90 (d, *J* = 6.8 Hz, 3H), 1.04 (d, *J* = 6.6 Hz, 3H), 1.96–2.16 (m, 1H), 4.62 (dt, *J* = 8.5, 1.1 Hz, 1H), 5.46 (t, *J* = 1.2 Hz, 1H), 6.36 (t, *J* = 1.0 Hz, 1H), 6.95 (ddd, *J* = 8.3, 1.1, 0.5 Hz, 1H), 7.00 (ddd, *J* = 8.0, 7.1, 1.1 Hz, 1H), 7.47 (ddd, *J* = 8.3, 7.1, 1.8 Hz, 1H), 7.93 (ddd, *J* = 7.9, 1.8, 0.5 Hz, 1H). ^{13}C NMR (176 MHz, Chloroform-*d*) δ 18.51, 18.83, 30.78, 86.66, 118.41, 121.49, 121.57, 123.80, 127.66, 136.31, 141.26, 160.53, 182.44. ESI-MS [M + H]$^+$ = 203.0. Anal. Calcd for $C_{13}H_{14}O_2$: C, 77.20%; H, 6.98%. Found: C, 77.42%, H, 7.05%.

3-Methylidene-2-phenylchroman-4-one (**14e**) (29.8 mg, 84%). Yelowish oil. ^{1}H NMR (700 MHz, Chloroform-*d*) δ 5.16–5.23 (m, 1H), 6.02 (d, *J* = 1.7 Hz, 1H), 6.44 (t, *J* = 1.2 Hz, 1H), 7.35–7.39 (m, 1H), 7.39–7.45 (m, 4H), 7.50 (ddd, *J* = 8.7, 7.2, 1.8 Hz, 1H), 7.99 (dd, *J* = 7.9, 1.7 Hz, 1H). ^{13}C NMR (176 MHz, Chloroform-*d*) δ 82.49, 118.47, 121.76, 122.13, 125.09, 127.58, 128.02, 128.78, 128.87, 136.30, 137.27, 142.76, 161.14, 182.22. ESI-MS [M + H]$^+$ = 237.4. Anal. Calcd for $C_{16}H_{12}O_2$: C, 81.34%; H, 5.12%. Found: C, 81.42%, H, 5.23%.

2,8-Dimethyl-3-methylidenechroman-4-one (**14f**) (18.1 mg, 64%). Yelowish oil. ^{1}H NMR (700 MHz, Chloroform-*d*) δ 1.64 (d, *J* = 6.5 Hz, 3H), 2.23 (s, 3H), 5.10 (qt, *J* = 6.4, 1.8 Hz, 1H), 5.54 (dd, *J* = 2.0, 0.9 Hz, 1H), 6.31 (dd, *J* = 1.6, 0.9 Hz, 1H), 6.93 (t, *J* = 7.6 Hz, 1H), 7.33 (ddd, *J* = 7.2, 1.8, 1.0 Hz, 1H), 7.82 (dd, *J* = 7.9, 1.7 Hz, 1H). ^{13}C NMR (176 MHz, Chloroform-*d*) δ 15.72, 19.13, 76.19, 121.08, 121.21 × 2, 125.56, 127.50, 136.91, 143.77, 159.53, 183.04. ESI-MS [M + H]$^+$ = 189.0. Anal. Calcd for $C_{12}H_{12}O_2$: C, 76.57%; H, 6.43%. Found: C, 76.42%, H, 6.55%.

2-Ethyl-8-methyl-3-methylidenechroman-4-one (**14g**) (17.9 mg, 59%). Yelowish oil. ^{1}H NMR (700 MHz, Chloroform-*d*) δ 1.07 (t, *J* = 7.4 Hz, 3H), 1.73 – 1.86 (m, 1H), 1.90–2.03 (m, 1H), 2.25 (t, *J* = 0.7 Hz, 3H), 4.76 – 5.05 (m, 1H), 5.50 (dd, *J* = 1.5, 1.0 Hz, 1H), 6.32 (t, *J* = 1.1 Hz, 1H), 6.92 (t, *J* = 7.5 Hz, 1H), 7.34 (ddd, *J* = 7.2, 1.8, 0.9 Hz, 1H), 7.81 (ddd, *J* = 7.9, 1.8, 0.7 Hz, 1H). ^{13}C NMR (176 MHz, Chloroform-*d*) δ 10.09, 15.69, 26.69, 81.76, 121.08, 121.11, 121.88, 125.45, 127.66, 136.99, 142.60, 158.94, 182.87. ESI-MS [M + H]$^+$ = 203.0. Anal. Calcd for $C_{13}H_{14}O_2$: C, 77.20%; H, 6.98%. Found: C, 77.38%, H, 7.09%.

2-Butyl-8-methyl-3-methylidenechroman-4-one (**14h**) (20.0 mg, 52%). Yelowish oil. ^{1}H NMR (700 MHz, Chloroform-*d*) δ 0.92 (t, *J* = 7.3 Hz, 3H), 1.29–1.49 (m, 3H), 1.49–1.58 (m, 1H), 1.71–1.80 (m, 1H), 1.89–2.01 (m, 1H), 2.24 (s, 3H), 4.98 (ddt, *J* = 9.0, 5.1, 1.4 Hz, 1H), 5.50 (t, *J* = 1.2 Hz, 1H), 6.30 (d, *J* = 1.2 Hz,

1H), 6.92 (t, *J* = 7.6 Hz, 1H), 7.31–7.39 (m, 1H), 7.81 (dd, *J* = 7.8, 1.7 Hz, 1H). ^{13}C NMR (176 MHz, Chloroform-*d*) δ 14.12, 15.73, 22.44, 27.70, 33.12, 80.39, 121.12, 121.73, 125.46, 127.66, 136.99, 142.85, 158.99, 182.96. ESI-MS [M + H]$^+$ = 231.0. Anal. Calcd for C$_{15}$H$_{18}$O$_2$: C, 78.23%; H, 7.88%. Found: C, 78.44%, H, 8.05%.

2-Isopropyl-8-methyl-3-methylidenechroman-4-one (**14i**) (17.2 mg, 53%). Yelowish oil. ^{1}H NMR (700 MHz, Chloroform-*d*) δ 0.92 (d, *J* = 6.7 Hz, 3H), 1.05 (d, *J* = 6.6 Hz, 3H), 1.95–2.12 (m, 1H), 2.26 (s, 3H), 4.66 (dd, *J* = 8.3, 1.3 Hz, 1H), 5.46 (d, *J* = 1.1 Hz, 1H), 6.35 (t, *J* = 1.1 Hz, 1H), 6.91 (t, *J* = 7.6 Hz, 1H), 7.34 (ddd, *J* = 7.2, 1.8, 0.9 Hz, 1H), 7.79 (dd, *J* = 7.9, 1.7 Hz, 1H). ^{13}C NMR (176 MHz, Chloroform-*d*) δ 15.71, 18.72, 19.03, 30.58, 86.60, 120.97, 121.17, 123.42, 125.32, 127.48, 137.14, 141.40, 158.69, 182.86. ESI-MS [M + H]$^+$ = 217.0. Anal. Calcd for C$_{14}$H$_{16}$O$_2$: C, 77.75%; H, 7.46%. Found: C, 77.52%, H, 7.27%.

8-Methyl-3-methylidene-2-phenylchroman-4-one (**14j**) (22.1 mg, 59%). Yelowish oil. ^{1}H NMR (700 MHz, Chloroform-*d*) δ 2.28 (d, *J* = 0.8 Hz, 3H), 5.23 (dd, *J* = 1.8, 1.1 Hz, 1H), 6.05 (t, *J* = 1.7 Hz, 1H), 6.42 (t, *J* = 1.3 Hz, 1H), 6.95 (t, *J* = 7.6 Hz, 1H), 7.33 – 7.38 (m, 2H), 7.38–7.45 (m, 4H), 7.83 (ddd, *J* = 7.9, 1.7, 0.7 Hz, 1H). ^{13}C NMR (176 MHz, Chloroform-*d*) δ 15.86, 82.14, 121.45, 121.55, 124.67, 125.61, 127.31, 127.62, 128.71, 128.74, 137.16, 137.57, 142.76, 159.29, 182.71. ESI-MS [M + H]$^+$ = 251.0. Anal. Calcd for C$_{17}$H$_{14}$O$_2$: C, 81.58%; H, 5.64%. Found: C, 81.72%, H, 6.75%.

2-Methyl-3-methylidene-2,3-dihydro-4H-benzo[g]chromen-4-one (**14k**) (22.2 mg, 66%). Yelowish oil. ^{1}H NMR (700 MHz, Chloroform-*d*) δ 1.65 (dd, *J* = 6.5, 1.2 Hz, 4H), 5.08–5.19 (m, 1H), 5.61 (dt, *J* = 1.7, 1.1 Hz, 1H), 6.41 (dt, *J* = 1.5, 0.7 Hz, 1H), 7.33 (d, *J* = 1.5 Hz, 1H), 7.36 (ddd, *J* = 8.1, 6.8, 1.2 Hz, 1H), 7.50 (ddt, *J* = 8.2, 6.8, 1.4 Hz, 1H), 7.70 (dd, *J* = 8.4, 1.5 Hz, 1H), 7.89 (dd, *J* = 8.2, 1.6 Hz, 1H), 8.58 (s, 1H). ^{13}C NMR (176 MHz, Chloroform-*d*) δ 19.26, 76.01, 113.14, 121.83, 122.30, 124.74, 126.68, 128.80, 129.23, 129.99, 130.14, 137.88, 143.99, 156.15, 183.18. ESI-MS [M + H]$^+$ = 255.0. Anal. Calcd for C$_{15}$H$_{12}$O$_2$: C, 80.34%; H, 5.39%. Found: C, 80.22%, H, 5.27%.

2-Ethyl-3-methylidene-2,3-dihydro-4H-benzo[g]chromen-4-one (**14l**) (23.9 mg, 67%). Yelowish oil. ^{1}H NMR (700 MHz, Chloroform-*d*) δ 1.08 (t, *J* = 7.4 Hz, 3H), 1.81 (dtd, *J* = 14.6, 7.5, 1.8 Hz, 1H), 1.97 (dtd, *J* = 14.2, 7.2, 1.2 Hz, 1H), 4.93 (ddt, *J* = 8.2, 5.6, 1.2 Hz, 1H), 5.57 (t, *J* = 1.1 Hz, 1H), 6.42 (d, *J* = 1.0 Hz, 1H), 7.34 (s, 1H), 7.36 (ddd, *J* = 8.2, 6.8, 1.2 Hz, 1H), 7.51 (ddd, *J* = 8.2, 6.7, 1.2 Hz, 1H), 7.71 (d, *J* = 8.4 Hz, 1H), 7.89 (d, *J* = 8.3 Hz, 1H), 8.57 (s, 1H). ^{13}C NMR (176 MHz, Chloroform-*d*) δ 9.97, 27.15, 81.74, 113.48, 122.54, 122.75, 124.83, 126.73, 128.89, 129.35, 130.10, 130.15, 138.11, 142.84, 155.70, 183.18. ESI-MS [M + H]$^+$ = 239.2. Anal. Calcd for C$_{16}$H$_{14}$O$_2$: C, 80.65%; H, 5.92%. Found: C, 80.43%, H, 6.05%.

2-Butyl-3-methylidene-2,3-dihydro-4H-benzo[g]chromen-4-one (**14m**) (24.0 mg, 60%). Yelowish oil. ^{1}H NMR (700 MHz, Chloroform-*d*) δ 1.36 (tdd, *J* = 15.8, 8.2, 4.4 Hz, 2H), 1.45 (dddd, *J* = 13.4, 9.6, 6.6, 4.0 Hz, 1H), 1.51 – 1.59 (m, 2H), 1.75 (ddt, *J* = 14.0, 10.7, 5.6 Hz, 1H), 1.92–2.00 (m, 1H), 5.00 (dd, *J* = 8.6, 5.6 Hz, 1H), 5.56 (d, *J* = 1.3 Hz, 1H), 6.40 (s, 1H), 7.34 (s, 1H), 7.36 (ddd, *J* = 8.1, 6.8, 1.1 Hz, 1H), 7.51 (ddd, *J* = 8.2, 6.8, 1.2 Hz, 1H), 7.71 (d, *J* = 8.4 Hz, 1H), 7.90 (d, *J* = 8.3 Hz, 1H), 8.58 (s, 1H). ^{13}C NMR (176 MHz, Chloroform-*d*) δ 14.11, 22.49, 27.62, 33.69, 80.46, 113.49, 122.55, 122.58, 124.83, 126.75, 128.89, 129.35, 130.11, 130.17, 138.13, 143.11, 155.72, 183.26. ESI-MS [M + H]$^+$ = 267.2. Anal. Calcd for C$_{18}$H$_{18}$O$_2$: C, 81.17%; H, 6.81%. Found: C, 81.32%, H, 6.95%.

2-Isopropyl-3-methylidene-2,3-dihydro-4H-benzo[g]chromen-4-one (**14n**) (18.2 mg, 48%). Yelowish oil. ^{1}H NMR (700 MHz, Chloroform-*d*) δ 0.91 (d, *J* = 6.7 Hz, 3H), 1.08 (d, *J* = 6.4 Hz, 3H), 1.97–2.06 (m, 1H), 4.62 (dd, *J* = 9.0, 0.9 Hz, 1H), 5.53 (t, *J* = 1.0 Hz, 1H), 6.45 (q, *J* = 0.9 Hz, 1H), 7.33 (s, 1H), 7.36 (ddt, *J* = 7.7, 6.7, 1.0 Hz, 1H), 7.51 (ddt, *J* = 7.7, 6.8, 1.0 Hz, 1H), 7.71 (d, *J* = 8.3 Hz, 1H), 7.89 (d, *J* = 8.3 Hz, 1H), 8.56 (s, 1H). ^{13}C NMR (176 MHz, Chloroform-*d*) δ 18.64, 18.92, 31.11, 86.57, 113.41, 122.73, 124.26, 124.78, 126.70, 128.79, 129.34, 129.93, 130.16, 138.19, 141.67, 155.47, 183.19. ESI-MS [M + H]$^+$ = 253.2. Anal. Calcd for C$_{17}$H$_{16}$O$_2$: C, 80.93%; H, 6.39%. Found: C, 81.02%, H, 6.29%.

3-Methylidene-2-phenyl-2,3-dihydro-4H-benzo[g]chromen-4-one (**14o**) (30.1 mg, 70%). Yelowish oil. ^{1}H NMR (700 MHz, Chloroform-*d*) δ 5.32 (dd, *J* = 1.7, 0.9 Hz, 1H), 6.07 (d, *J* = 1.7 Hz, 1H), 6.54 (t, *J* = 1.2 Hz, 1H), 7.36 (ddd, *J* = 7.4, 3.4, 2.1 Hz, 2H), 7.38 – 7.41 (m, 4H), 7.44–7.48 (m, 2H), 7.51 (ddd, *J* = 8.2, 6.7, 1.2 Hz, 1H), 7.71 (d, *J* = 8.3 Hz, 1H), 7.89 (d, *J* = 8.3 Hz, 1H), 8.59 (s, 1H). ^{13}C NMR (176 MHz, Chloroform-*d*) δ 82.20, 113.65, 122.64, 125.00, 125.42, 126.88, 127.62, 128.82×2, 129.05, 129.44, 130.16, 130.36, 137.62, 138.05, 143.06, 156.14, 183.00. ESI-MS [M + H]$^+$ = 287.2. Anal. Calcd for $C_{20}H_{14}O_2$: C, 83.90%; H, 4.93%. Found: C, 83.72%, H, 5.05%.

3.2. Biology

3.2.1. Cell Lines

HL-60, NALM-6, MCF-7 cell lines were purchased from the European Collection of Cell Cultures (ECACC). HL-60 and NALM-6 cells were cultured in RPMI 1640 Glutamax medium (Gibco/Life Technologies, Carlsbad, CA, USA), supplemented with 10% fetal bovine serum, penicillin and streptomycin. MCF-7 cells were maintained in EMEM growth medium (Sigma-Aldrich, St. Louis, MO, USA), supplemented with 2 mM glutamine, 10% fetal bovine serum, 1% NEAA and gentamycin.

3.2.2. In Vitro Cytotoxicity Assay

The MTT (3-(4,5-dimethyldiazol-2-yl)-2,5 diphenyl tetrazolium bromide) assay was used to investigate the cytotoxicity of new analogs in HL-60, NALM-6 and MCF-7 cell lines. Briefly, cells were seeded in a 24-well plate at a concentration of 8×10^4 cells/mL. Following initial incubation (20 h; 37 °C), cells were treated with various concentrations of the analogs for 48 h. Then, cells were incubated for 1.5 h with MTT solution (100 µL/well; 5 mg/mL of PBS). The plates were centrifuged (3000 rpm, 5 min.) and the supernatant was discarded. The formazan product was dissolved by addition of DMSO (1 mL/well). The absorbance was measured using FlexStation 3 Multi-Mode Microplate Reader (Molecular Devices, LLC) at 560 nm. Assay was performed in triplicate. Cell viability rate was calculated by dividing mean sample absorbance by mean control absorbance.

3.2.3. Annexin V and Propidium Iodide Assay

Apoptotic cell death was determined using FITC Annexin V Apoptosis Detection Kit I (BD Biosciences, San Jose, CA, USA) in accordance with the manufacturer's protocol. NALM-6 cells were seeded in 6-well plates at a density of 4.0×10^5 cells/mL in 2 mL of cell culture medium. Cells were treated with 14d at IC_{50} and 2 IC_{50} concentrations (1.25 µM and 2.5 µM, respectively) for 24 h, at 37 °C. Untreated cells were used as a control. Then, cells were washed with PBS, resuspended in the binding buffer and stained with FITC Annexin V and propidium iodide (PI), followed by incubation in the dark at RT for 15 min. Finally, flow cytometry analysis of the cells was performed using CytoFLEX (Beckman Coulter, Inc., Brea, CA, USA). Data were analyzed using CytExpert Software (2.3, Beckman Coulter, Inc, Brea, CA, USA).

3.2.4. Caspase 3, Caspase 8, and Caspase 9 Activity

Activity of the key caspases involved in apoptosis was simultaneously quantified using a fluorometric Caspase 3, Caspase 8 and Caspase 9 Multiplex Activity Assay Kit (Abcam, Cambridge, UK), according to the manufacturer's protocol. Briefly, NALM-6 cells were seeded in a 96-well plate at a density of 2.0×10^5 cells/90 µl of cell culture medium per well. The plates were centrifuged at 800 rpm for 2 min. and the cells were treated with **14d** at 1.25 µM and 2.5 µM concentrations. Medium without cells and untreated cells were used as controls. The cell plate was incubated for 6 h, 37 °C, 5% CO_2. Caspase-9, -3, and -8 substrates were added to each well and the cell plate was incubated for 60 min. at RT, protected from light. The plate was centrifuged at 800 rpm for 2 min., followed by a fluorescence readout using FlexStation 3 Multi-Mode Microplate Reader (Molecular Devices, LLC) at specific wavelengths. The assay was performed in triplicate; blank readings were subtracted from all measurements and the fold change of fluorescent intensity between control and the treated cells was calculated.

4. Conclusions

A simple and efficient synthesis of 3-methylidenechroman-4-ones **14a–o**, applying Horner–Wadsworth–Emmons methodology was described. Furthermore, relative configuration of the crucial intermediates for this methodology, trans- or cis- 2-substituted 3-diethoxyphosphorylchroman-4-ones **13a–o**, was elaborated using NMR analysis. The obtained library of 3-methylidenechroman-4-ones **14a–o** was evaluated for the anti-proliferative activity against leukemia and breast cancer cell lines. Several members of this library were determined to be capable of killing leukemia cells with improved activity compared to carboplatin used as a reference compound, exhibiting IC_{50} values in the low micromolar range. The most potent 2-isopropyl-3-methylidenechroman-4-one **14d** was then evaluated for its possible apoptotic activity against NALM-6 cells. Compound **14d** promoted apoptosis mediated by caspase 3/8 and in a much lesser extend by caspase 3/9 induction, which indicated that mostly the extrinsic pathway was engaged in the programmed cell death caused by this analog. These results show that compound **14d** is a good lead in a search for new anticancer agents.

Supplementary Materials: The following are available online at http://www.mdpi.com/1420-3049/24/10/1868/s1, General procedures and characterization data for diethyl (2-(2-hydroxyarylo)-2-oxoethyl)phosphonates (**10a–c**), diethoxyphosphorylchromen-4-ones (**12a–c**) and 3-diethoxyphosphoryl-2-substituted chroman-4-ones (**13a–o**). Copies of ^{1}H, ^{13}C and ^{31}P NMR spectra of all obtained compounds.

Author Contributions: Conceptualization, T.J., J.K., and A.J.; Data curation, J.K. and J.D.; Formal analysis, J.K. and J.D.; Funding acquisition, T.J. and A.J.; Investigation, J.K., T.B., J.D., and U.K.; Methodology, T.J., J.K., J.D. and A.J.; Project administration, T.J. and A.J.; Resources, J.K. and J.D.; Supervision, T.J. and A.J.; Validation, J.K. and J.D.; Visualization, J.K. and J.D.; Writing—review and editing, T.J. and A.J.

Funding: This research was funded by the National Science Centre of Poland (project DEC-2012/07/B/ST5/02006) and by a grant from the Medical University of Łódź No. 503/1-156-02/503-11-02.

Conflicts of Interest: The authors declare no conflict of interest. The funders had no role in the design of the study, in the collection, analyses or interpretation of data, in the writing of the manuscript, or in the decision to publish the results.

References

1. Nibbs, A.E.; Scheidt, K.A. Asymmetric methods for the synthesis of flavanones, chromanones, and azaflavanones. *Eur. J. Org. Chem.* **2012**, 449–462. [CrossRef] [PubMed]
2. Huang, S.; Zhao, Y.; Zhou, X.; Wu, Y.; Wu, P.; Liu, T.; Yang, B.; Hu, Y.; Dong, X. Design, synthesis and biological evaluation of 3-benzylideneflavanone derivatives as cytotoxic agents. *Med. Chem. Res.* **2012**, *21*, 4150–4157. [CrossRef]
3. Sidwell, W.T.L.; Tamm, C. The homo-isoflavones II. Isolation and structure of 4'-O-methyl-punctatin, autumnalin and 3, 9-dihydro-autumnalin. *Tetrahedron Lett.* **1970**, 475–478. [CrossRef]
4. Böhler, P.; Tamm, C. The homo-isoflavones, a new class of natural product. Isolation and structure of eucomin and eucomol. *Tetrahedron Lett.* **1967**, *8*, 3479–3483. [CrossRef]
5. Desideri, N.; Bolasco, A.; Fioravanti, R.; Monaco, L.P.; Orallo, F.; Yañez, M.; Ortuso, F.; Alcaro, S. Homoisoflavonoids: Natural scaffolds with potent and selective monoamine oxidase-B inhibition properties. *J. Med. Chem.* **2011**, *54*, 2155–2164. [CrossRef]
6. Desideri, N.; Monaco, L.P.; Fioravanti, R.; Biava, M.; Yañez, M.; Alcaro, S.; Ortuso, F. (E)-3-Heteroarylidenechroman-4-ones as potent and selective monoamine oxidase-B inhibitors. *Eur. J. Med. Chem.* **2016**, *117*, 292–300. [CrossRef]
7. Pourshojaei, Y.; Gouranourimi, A.; Hekmat, S.; Asadipour, A.; Rahmani-Nezhad, S.; Moradi, A.; Nadri, H.; Moghadam, F.H.; Emami, S.; Foroumadi, A.; et al. Design, synthesis and anticholinesterase activity of novel benzylidenechroman-4-ones bearing cyclic amine side chain. *Eur. J. Med. Chem.* **2015**, *97*, 181–189. [CrossRef] [PubMed]

8. Li, Y.; Qiang, X.; Luo, L.; Yang, X.; Xiao, G.; Zheng, Y.; Cao, Z.; Sang, Z.; Su, F.; Deng, Y. Multitarget drug design strategy against Alzheimer's disease: Homoisoflavonoid mannich base derivatives serve as acetylcholinesterase and monoamine oxidase B dual inhibitors with multifuncional properties. *Bioorg. Med. Chem.* **2017**, *25*, 714–726. [CrossRef]

9. Perjési, P.; Das, U.; De Clercq, E.; Balzarini, J.; Kawase, M.; Sakagami, H.; Stables, J.P.; Lorand, T.; Rozmer, Z.; Dimmock, J.R. Design, synthesis and antiproliferative activity of some 3-benzylidene-2, 3-dihydro-1-benzopyran-4-ones which display selective toxicity for malignant cells. *Eur. J. Med. Chem.* **2008**, *43*, 839–845. [CrossRef]

10. Pordeli, P.; Nakhjiri, M.; Safavi, M.; Ardestani, S.K.; Foroumadi, A. Anticancer effects of synthetic hexahydrobenzo[g]chromen-4-one derivatives on human breast cancer cell lines. *Breast Cancer* **2017**, *24*, 299–311. [CrossRef]

11. Pouget, C.; Fagnere, C.; Basly, J.-P.; Habrioux, G.; Chulia, A.-J.; New aromatase inhibitors. Synthesis and inhibitory activity of pyridinyl-Substituted flavanone derivatives. *Bioorg. Med. Chem. Lett.* **2002**, *12*, 1059–1061. [CrossRef]

12. Basavaiah, D.; Bakthadoss, M.; Pandiaraju, S. A new protocol for the syntheses of (E)-3-benzylidenechroman-4-ones: A simple synthesis of the methyl ether of bonducellin. *Chem. Commun.* **1998**, *16*, 1639–1640. [CrossRef]

13. AI Nakib, T.; Bezjak, V.; Meegan, M.J.; Chandy, R. Synthesis and antifungal activity of some 3-benzylidenechroman-4-ones, 3-benzylidenethiochroman-4-ones and 2-benzylidene-1-tetralones. *Eur. J. Med. Chem.* **1990**, *25*, 455–462. [CrossRef]

14. Takao, K.; Yamashita, M.; Yashiro, A.; Sugita, Y. Synthesis and Biological Evaluation of 3-Benzylidene-4-chromanone Derivatives as Free Radical Scavengers and α-Glucosidase Inhibitors. *Chem. Pharm. Bull.* **2016**, *64*, 1203–1207. [CrossRef] [PubMed]

15. Siddaiah, V.; Rao, C.V.; Venkateswarlu, S.; Krishnarajub, A.V.; Subbaraju, G.V. Synthesis, stereochemical assignments, and biological activities of homoisoflavonoids. *Bioorg. Med. Chem.* **2006**, *14*, 2545–2551. [CrossRef] [PubMed]

16. Hung, T.M.; Cao Van, T.; Nguyen Tien, D. Homoisoflavonoid derivatives from the roots of *Ophiopogon japonicus* and their in vitro anti-inflammation activity. *Bioorg. Med. Chem. Lett.* **2010**, *20*, 2412–2416. [CrossRef]

17. Ward, F.E.; Garling, D.L.; Buckler, R.T. Antimicrobial 3-Methylene flavanones. *J. Med. Chem.* **1981**, *24*, 1073–1077. [CrossRef]

18. Wallén, E.A.A.; Dahlén, K.; Grøtli, M.; Luthman, K. Synthesis of 3-Aminomethyl-2-aryl-8-bromo-6-chlorochromones. *Org.Lett.* **2007**, *9*, 389–391.

19. Okuro, K.; Alper, H. Palladium-Catalyzed Carbonylation ofo-Iodophenols with Allenes. *J. Org. Chem.* **1997**, *62*, 1566–1567. [CrossRef]

20. Grigg, R.; Liu, A.; Shaw, D.; Suganthan, S.; Woodalla, D.E.; Yoganathan, G. Synthesis of quinol-4-ones and chroman-4-ones via a palladium-catalysed cascade carbonylation–allene insertion. *Tetrahedron Lett.* **2000**, *41*, 7125–7128. [CrossRef]

21. Bugarin, A.; Jones, K.D.; Connell, B.T. Efficient, direct α-methylenation of carbonyls mediated by diisopropylammonium trifluoroacetate. *Chem. Commun.* **2010**, *46*, 1517–1717. [CrossRef] [PubMed]

22. Vaidya, V.V.; Wankhede, K.S.; Salunkhe, M.M.; Trivedi, G.K. Synthesis of Functionalized Benzopyrans via Intramolecular 1,3-Dipolar Cycloaddition of Nitrile Oxides. *Synth. Commun.* **2008**, *38*, 2392–2403. [CrossRef]

23. Iwasaki, H.; Kume, T.; Yamamoto, Y.; Akiba, K. Reaction of 4-t-butyldimethylsiloxy-1-benzopyrylium salt with enol silyl ethers and active methylenes. *Tetrahedron Lett.* **1987**, *28*, 6355–6358. [CrossRef]

24. Lee, Y.-G.; Ishimaru, K.; Iwasaki, H.; Ohkata, K.; Akiba, K. Tandem Reactions In 4-Siloxy-1-benzopyrylium Salts: Introduction of Substituents and Cyclohexene and Cyclopentane Annulation in Chromones. *J. Org. Chem.* **1991**, *56*, 2058–2066. [CrossRef]

25. Kitson, R.R.A.; Millemaggi, A.; Taylor, R.J.K. The Renaissance of α-Methylene-γ-butyrolactones: New Synthetic Approaches. *Angew. Chem. Int. Ed.* **2009**, *48*, 9426–9452. [CrossRef] [PubMed]

26. Albrecht, A.; Albrecht, Ł.; Janecki, T. Recent Advances in the Synthesis of α-Alkylidene-Substituted δ-Lactones, γ-Lactams and δ-Lactams. *Eur. J. Org. Chem.* **2011**, *2011*, 2747–2766. [CrossRef]

27. Lagoutte, R.; Winssinger, N. Following the Lead from Nature with Covalent Inhibitors. *CHIMIA* **2017**, *71*, 703–711. [CrossRef]

28. Lagoutte, R.; Pastor, M.; Berthet, M.; Winssinger, N. Rapid and scalable synthesis of chiral bromolactones as precursors to α-exo-methylene-γ–butyrolactone-containing sesquiterpene lactones. *Tetrahedron* **2018**, *74*, 6012–6021. [CrossRef]

29. Janecki, T. Organophosphorus reagents as versatile tool in the synthesis of α-alkylidene-γ-butyrolactones and α-alkylidene-γ-butyrolactams. In *Targets in Heterocyclic Systems*; Italian Society of Chemistry: Rome, Italy, 2006; pp. 301–320.

30. Modranka, J.; Albrecht, A.; Janecki, T. A Convenient Entry to 3-Methylidenechroman-2-ones and 2-Methylidenedihydrobenzochromen-3-ones. *Synlett* **2010**, *19*, 2867–2870.

31. Pięta, M.; Kędzia, J.; Wojciechowski, J.; Janecki, T. Asymmetric synthesis of 1,4-disubstituted 3-methylidenedihydroquinolin-2(1H)-ones. *Tetrahedron Asymm.* **2017**, *28*, 567–576. [CrossRef]

32. Zhou, P.; Jiang, Y.-J.; Zou, J.-P.; Zhang, W. Manganese(III) Acetate Mediated Free-Radical Phosphonylation of Flavones and Coumarins. *Synthesis* **2012**, *44*, 1043–1050. [CrossRef]

33. Somu, R.V.; Boshoff, H.; Qiao, C.; Bennett, E.M.; Barry III, C.E.; Aldrich, C.C. Rationally Designed Nucleoside Antibiotics That Inhibit Siderophore Biosynthesis of *Mycobacterium tuberculosis*. *J. Med. Chem.* **2006**, *49*, 31–34. [CrossRef] [PubMed]

34. Janecki, T.; Wąsek, T. A novel route to substituted 3-methylidenechroman-2-ones and 3-methylchromen-2-ones. *Tetrahedron* **2004**, *60*, 1049–1050. [CrossRef]

35. Koleva, A.I.; Petkova, N.I.; Nikolova, R.D. Ultrasound-Assisted Conjugate Addition of Organometallic Reagents to 3-Diethylphosphonocoumarin. *Synlett* **2016**, *27*, 2676–2680.

36. Mahmudov, K.T.; Pombeiro, A.J.L. Resonance-Assisted Hydrogen Bonding as a Driving Force in Synthesis and a Synthon in the Design of Materials. *Chem. Eur. J.* **2016**, *22*, 16356–16398. [CrossRef] [PubMed]

37. Koszuk, J.; Bartosik, T.; Wojciechowski, J.; Wolf, W.M.; Janecka, A.; Drogosz, J.; Długosz, A.; Krajewska, U.; Mirowski, M.; Janecki, T. Synthesis of 3-Methylidene-1-tosyl-2,3-dihydroquinolin-4(1H)-ones as Potent Cytotoxic Agents. *Chem. Biodivers.* **2018**, *15*, e1800242. [CrossRef]

38. Vermes, I.; Haanen, C.; Steffens-Nakken, H.; Reutellingsperger, C. A novel assay for apoptosis flow cytometric detection of phosphatidylserine expression on early apoptotic cells using fluorescein labelled annexin V. *J. Immunol. Methods* **1995**, *184*, 39–51. [CrossRef]

39. Van Engeland, M.; Nieland, L.J.; Ramaekers, F.C.; Schutte, B.; Reutelingsperger, C.P. Annexin V-affinity assay: A review on an apoptosis detection system based on phosphatidylserine exposure. *Cytometry* **1998**, *31*, 1–9. [CrossRef]

40. Pistritto, G.; Trisciuoglio, D.; Ceci, C.; Garufi, A.; D'Orazi, G. Apoptosis as anticancer mechanism: Function and dysfunction of its modulators and targeted therapeutic strategies. *Aging* **2016**, *8*, 603–619. [CrossRef]

Sample Availability: Samples of the compounds **10a–c**,**12a–c** and **13a,c–d**, **f-h,k–l** are available from the authors.

 *molecules*

Article

Design and Synthesis of Novel Breast Cancer Therapeutic Drug Candidates Based upon the Hydrophobic Feedback Approach of Antiestrogens

Kiminori Ohta [1,*], Asako Kaise [2], Fumi Taguchi [2], Sayaka Aoto [2], Takumi Ogawa [2] and Yasuyuki Endo [2,*]

[1] School of Pharmacy, Showa University, 1-5-8, Hatanodai, Shinagawa-ku, Tokyo 142-8555, Japan
[2] Faculty of Pharmaceutical Sciences, Tohoku Medical and Pharmaceutical University, 4-4-1 Komatsushima, Aoba-ku, Sendai 981-8558, Japan; kaise@tohoku-mpu.ac.jp (A.K.); f0u1m2i9@docomo.ne.jp (F.T.); syk-at1213@docomo.ne.jp (S.A.); ogawa-takumi@pmda.go.jp (T.O.)
* Correspondence: k-ohta@pharm.showa-u.ac.jp (K.O.); yendo@tohoku-mpu.ac.jp (Y.E.)

Received: 29 September 2019; Accepted: 30 October 2019; Published: 1 November 2019

Abstract: Based upon hydrophobic feedback approaches, we designed and synthesized novel sulfur-containing ERα modulators (**4** and **5**) as breast cancer therapeutic drug candidates. The tetrahydrothiepine derivative **5a** showed the highest binding affinity toward ERα because of its high hydrophobicity, and it acted as an agonist toward MCF-7 cell proliferation. The corresponding alkylamino derivative **5d** maintained high binding affinity to ERα and potently inhibited MCF-7 cell proliferation (IC_{50}: 0.09 μM). Docking simulation studies of compound **5d** with the ERα BD revealed that the large hydrophobic moiety of compound **5d** fit well into the hydrophobic pocket of the ERα LBD and that the sulfur atom of compound **5d** formed a sulfur–π interaction with the amino acid residue His524 of the ERα LBD. These interactions play important roles for the binding affinity of compound **5d** to the ERα LBD.

Keywords: sulfur; anticancer; estrogen

1. Introduction

17β-Estradiol (E2) is an endogenous female hormone and plays an important role in the female reproductive system. Estrogens including E2 influence the growth, differentiation, and functioning of many target tissues mediated by the binding to estrogen receptor α (ERα) [1–3]. Although activation of ERα is essential for the maintenance of homeostasis of female functions, they can sometimes induce and cause progression of estrogen-dependent breast cancers [4–6]. Several antiestrogens have been developed to prevent and control hormone-responsive breast cancer [7–10]. Tamoxifen has been used worldwide for more than 40 years for the treatment of estrogen-dependent breast cancer (Figure 1) [11–13]. Interestingly, tamoxifen acts as either an agonist or an antagonist depending on tissue type; tamoxifen exhibits antiestrogenic action in breast cancer and hot flashes, and estrogenic action in bone and cholesterol metabolism, and is a selective estrogen receptor modulator (SERM) [11–13]. The dimethylaminoethyl chain of tamoxifen is attached to a benzene ring, where it plays a strategic role in the expression of antiestrogenic activity. The triphenylethylene moiety of tamoxifen plays an important role in controlling ERα binding affinity. The triphenylethylene moiety is a promising structure for the development of an ER ligand. The triphenylethylene structure exhibits geometric isomers *E* and *Z* that are caused by the key alkylamino chain, and the asymmetric hydrophobic part and the isomers easily isomerize between the *E* and *Z* forms. Therefore, the isomerization often become a big problem in the synthesis, purification, and preservation of an either isomer. Indeed, 4-hydroxytamoxifen, an active metabolite of tamoxifen, easily isomerizes from the active *Z* form to the inactive *E* form in solution and in in vitro studies [14–17]. In this regard, novel ER antagonists comprised of other skeleton structures, such as benzothiophene [18], dihydronaphthalene [19], benzopyrane [20], and steroid [21] structures, have been developed as ER antagonists.

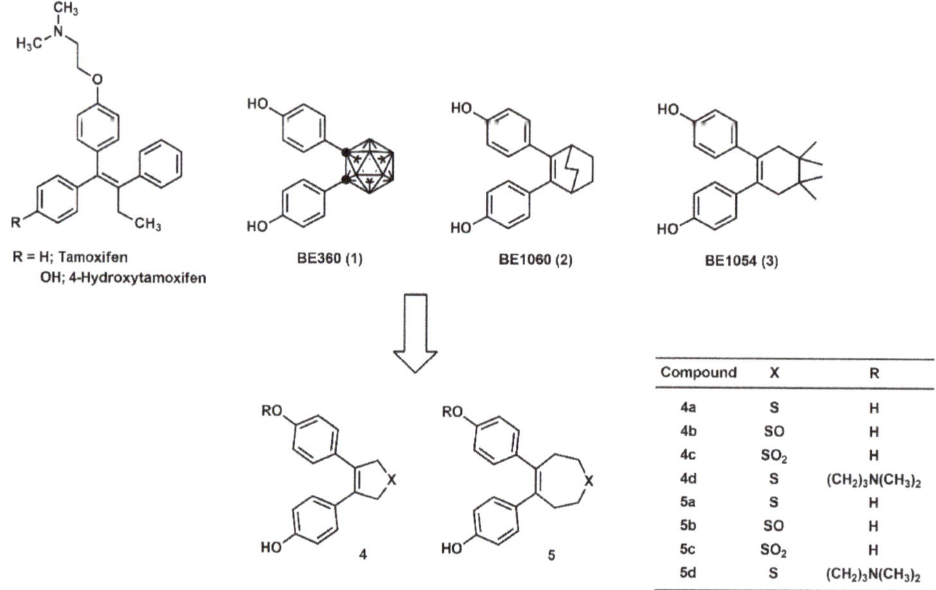

Figure 1. Structures of estrogen receptor (ER) modulators and newly designed sulfur-containing ER antagonist candidates (compounds **4a–4d** and **5a–5d**).

On the other hand, we found that a simple 1,2-bis(4-hydroxyphenyl)-*o*-carborane, BE360 (**1**) exhibited high binding affinity toward ERα and estrogenic action in bone without showing estrogenic action in the uterus of ovariectomized (OVX) and orchidectomized (ORX) mice (Figure 1) [22]. That is, compound **1** acted as an agonist in bone and an antagonist in female reproductive tissues, despite lack of an alkylamino chain. Based upon the hydrophobic feedback approaches, we focused on three-dimensional hydrocarbon units as new hydrophobic pharmacophores to design and synthesize BE1060 (**2**) and BE1054 (**3**) (Figure 1) [23]. Although the tetramethylcyclohexene-based bisphenol compound **3** acted as a partial agonist, similar to compound **1**, the bicyclo[2,2,2]octene-based bisphenol compound **2** showed potent estrogenic activity. Unlike the triphenylethylene structure, a vicinal bisphenol structure does not have a geometric isomer, even with the key alkylamino chain. We were interested in the effects of ring size and sulfur atom in the hydrophobic structure on ER activity. Furthermore, to understand the effects of sulfur, sulfone, and sulfoxide on the biological activity, we designed dihydrothiophene and tetrahydrothiepine derivatives (**4a–4d** and **5a–5d**) as novel ER antagonist candidates. Herein, we describe the facile synthesis and biological activities of the designated compounds **4a–4d** and **5a–5d** (Figure 1).

2. Results and Discussion

2.1. Chemistry

Synthesis of the dihydrothiophene derivatives (**4a–4d**) is summarized in Scheme 1. Commercially available 2-bromo-4′-methoxyacetophenone (**6**) was reacted with sodium sulfide (Na_2S) to form the sulfide compound **7** in 95% yield [24]. Intramolecular McMurry coupling of **7** resulted in formation of the dihydrothiophene derivative **8** in 60% yield, which was demethylated with BBr_3 to form the target compound **4a** in 97% yield [25]. Although the dihydrothiophene derivative **8** was treated with 1 equimolar amount of *m*-chloroperbenzoic acid (*m*-CPBA) to form the sulfoxide compound **9** in 70% yield, followed by demethylation with BBr₃ to form the decomposition product of compound **9**, it did not result in the formation of compound **4b**. Thus, we carried out oxidation of compound **4a** with *m*-CPBA to form the desired compound **4b** in 73% yield. When compound **8** was treated with two equimolar amounts of *m*-CPBA, the corresponding sulfone compound **10** was obtained and was then demethylated with BBr₃ to form compound **4c** in quantitative yield over two steps. Mitsunobu reaction of compound **4a** with 3-dimethylamino-1-propanol resulted in formation of compound **4d** with a dimethylaminopropyl chain in 37% yield [26].

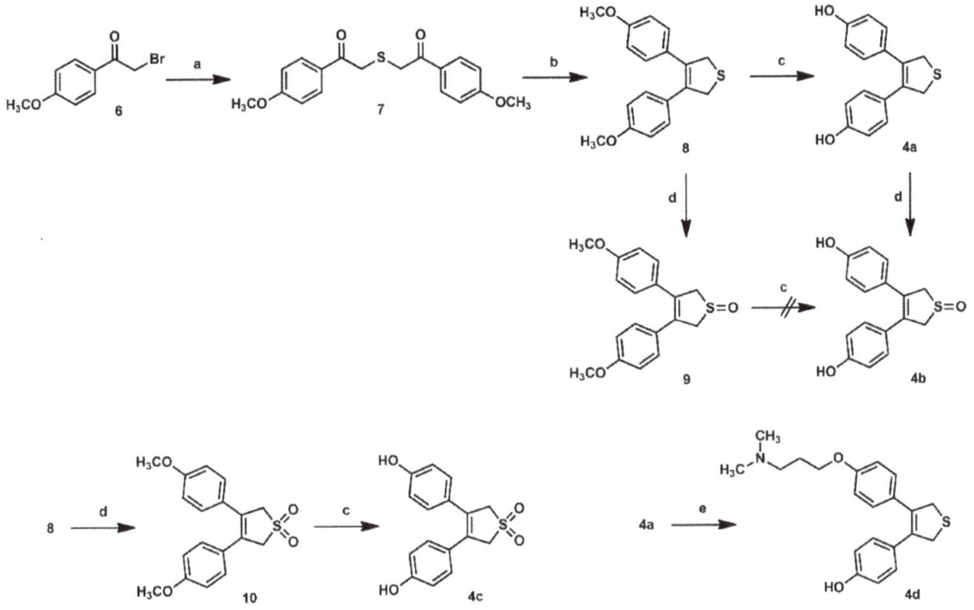

Scheme 1. Synthesis of dihydrothiophene derivatives (compounds **4a**–**4d**); Reagents and conditions: (a) Na₂S·9H₂O, acetone, H₂O, (b) Zn, TiCl₄, THF, (c) BBr₃, CH₂Cl₂, (d) *m*-CPBA, CH₂Cl₂, (e) 3-dimethylamino-1-propanol, Ph₃P, DEAD, THF.

Scheme 2 summarizes the synthesis of tetrahydrothiepine derivatives (**5a**–**5d**). A starting material (**12**) for the seven-membered ring derivative **5** was synthesized by Friedel-Craft acylation of anisole with acyl chloride **11**, followed by sulfide formation with Na₂S to form compound **13** in 84% yield over two steps [24]. The tetrahydrothiepine derivatives (**5a**–**5d**) were synthesized in the same manner as those of compounds **4a**–**4d**. Briefly, intramolecular McMurry coupling of compound **13**, followed by demethylation, resulted in formation of the sulfide compound **5a** in a 66% yield [25]. Compound **5a** was transformed into the sulfoxide compound **5b** by oxidation with one equimolar amount of *m*-CPBA in a 30% yield. Moreover, compound **5a** was reacted with 3-dimethylamino-1-propanol under Mitsunobu reaction conditions to form compound **5d** in a 28% yield [26]. Treatment of compound **14** with two equimolar amounts of *m*-CPBA followed by demethylation resulted in the formation of the sulfone compound **5c** in a 49% yield over two steps.

Scheme 2. Synthesis of tetrahydrothiepine derivatives (compounds **5a–5d**); Reagents and conditions: (a) anisol, AlCl₃, 1,2-dichloroetane, (b) Na₂S·9H₂O, acetone, H₂O, (c) Zn, TiCl₄, THF, (d) BBr₃, CH₂Cl₂, (e) *m*-CPBA, CH₂Cl₂, (f) 3-dimethylamino-1-propanol, Ph₃P, DEAD, THF.

2.2. Biological Evaluation

The binding affinities of the synthesized bisphenols (compounds **4a–4c** and **5a–5c**) toward ERα were evaluated by means of a competitive binding assay using [6,7-³H] 17β-estradiol and human recombinant ERα [27–29]. Figure 2A,B shows dose-response curves for competitive binding of the synthesized compounds **4a–4c** and **5a–5c** to ERα, respectively. Compound **4a**, containing a sulfide group, moderately bound to ERα and the binding affinity was about 100 times lower than that of E2. The sulfoxide (**4b**) and sulfone (**4c**) compounds did not bind to the ERα ligand-binding domain (LBD), even at concentrations 1000 times higher than that of E2. The binding affinity of the tetrahydrothiepine derivative **5** toward ERα showed a similar tendency to that of compound **4**, and the sulfide derivative **5a** bound to ERα more potently than compound **4a**. The tamoxifen-inspired derivatives **4d** and **5d**, both containing a dimethylaminopropyl group, showed similar binding affinities to compounds **4a** and **5a**, respectively. With the binding affinity of E2 to ERα taken as 100, the relative binding affinity (RBA) values of compounds **4d** and **5d** were 4.0 and 37.1, respectively (Table 1). It is suggested that the binding affinities of compounds **4** and **5** depend on the hydrophobicity of the ring structures and the polarity of the sulfur atoms, which can be understood from the CLogP values of compounds **4a–4d** and **5a–5d** in Table 1.

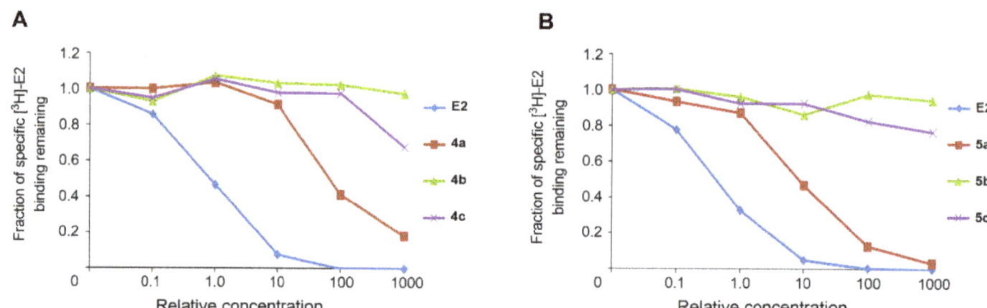

Figure 2. Dose-response curves obtained from a competitive binding assay of compounds **4a–4c** (**A**) and **5a–5c** (**B**) toward ERα. All binding assays were performed in duplicate (*n* = 2). IC₅₀ value of E2 for ERa is 4 nM.

Table 1. Biological profiles of compounds **4a–4d** and **5a–5d**.

Compound	Ring	X	R	RBA (%) [a]	EC$_{50}$ (μM) [b]	E$_{max}$ (%) [c]	IC$_{50}$ (μM) [b]	CLogP [d]
4a	5	S	H	1.2	0.39	119	–	3.1
4b		SO	H	0.003	NT [e]	NT [e]	NT [e]	0.8
4c		SO$_2$	H	0.04	0.30	130	–	0.7
4d		S	(CH$_2$)$_3$N(CH$_3$)$_2$	4.0	–	–	0.41	4.2
5a	7	S	H	4.9	0.048	99	–	3.9
5b		SO	H	0.001	NT [f]	NT [f]	NT [f]	1.6
5c		SO$_2$	H	0.01	4.7	89	–	1.5
5d		S	(CH$_2$)$_3$N(CH$_3$)$_2$	37.1	–	–	0.090	4.9

[a] Relative binding affinity (RBA) values of each compound were estimated from the sigmoidal dose-response curves in Figure 2A,B. The binding affinity of E2 was taken as 100. [b] EC$_{50}$ and IC$_{50}$ values of the test compounds were estimated from the sigmoidal dose-response curves using GraphPad Prism 7 software. [c] E$_{max}$ indicates the maximal efficacy of each compound relative to that of E2, which was taken as 100. [d] CLogP values of each compound were estimated using ChemBioDraw Ultra 12.0. [e] "NT" indicates not tested.

Next, estrogenic and antiestrogenic activities of compounds **4a–4d** and **5a–5d**, except for **4b** and **5b** (poor binding affinity toward ERa), were evaluated by means of a cell proliferation assay using the human breast cancer cell line MCF-7, which shows ER-dependent growth [27–29]. Table 1 summarizes the EC$_{50}$, IC$_{50}$, and E$_{max}$ values estimated from the MCF-7 cell proliferation assays. E$_{max}$ represents the maximal efficacy of each compound toward cell proliferation and is an index of ER partial agonistic activity. E$_{max}$ was estimated relative to that of E2, which was taken as 100. Although the binding affinity of compound **4a** was ~10 times higher than that of compound **4c**, both compounds showed estrogenic activity with similar EC$_{50}$ values. Based on the E$_{max}$ values of compounds **4a** and **4c**, we suggest that the complex formed between compound **4c** and the ERα LBD is more active for transcription than that formed between compound **4a** and the ERα LBD. Compound **5a**, which exhibited the most potent binding affinity to ERα LBD, also showed the most potent MCF-7 cell proliferation activity, with an EC$_{50}$ value of 0.048 μM. As expected from the weak binding affinity to ERα LBD, compound **5c** showed weak estrogenic activity. On the other hand, the tamoxifen-inspired derivatives **4d** and **5d**, containing alkylamino chains, potently inhibited cell proliferation of MCF-7 cells induced by 0.1 nM of E2, with IC$_{50}$ values of 0.41 and 0.09 μM, respectively.

2.3. Docking Study

To understand the docking modes of the sulfur-containing ligands **4d** and **5d** with the ERα LBD, docking simulation studies of both compounds toward the human ERα LBD (PDB: 1ERR) were carried out using an automatic docking program (Discovery Studio 2018/CDOCKER) [30]. The docking mode of compound **4d** was superimposed onto that of compound **5d** at the ERα LBD (Figure 3A,B). The docking mode of compound **4d** toward the ERα LBD was similar to that of compound **5d**. Each phenolic hydroxy group of compounds **4d** and **5d** formed hydrogen bonds with the amino acid residues Glu353 and Arg394 (Figure 3C), and each dimethylaminopropyl chain of compounds **4d** and **5d** was found to be located within the same space of the ERα LBD (Figure 3A). The hydrophobic moiety of compound **5d** was spread widely across the hydrophobic pocket of the ERα LBD and the sulfur atom of the tetrahydrothiepine ring formed a sulfur–π interaction with the amino acid residue His524 (Figure 3B,C). Docking scores (CDOCKER interaction energy) of compounds **4d** and **5d** toward the ERα LBD were 51.1 and 54.0, respectively, and the order of docking score was correlated with the results of the binding assay. Therefore, we attributed the potent binding affinity of compound **5d** toward the ERα LBD to both hydrophobic interactions and sulfur–π interactions.

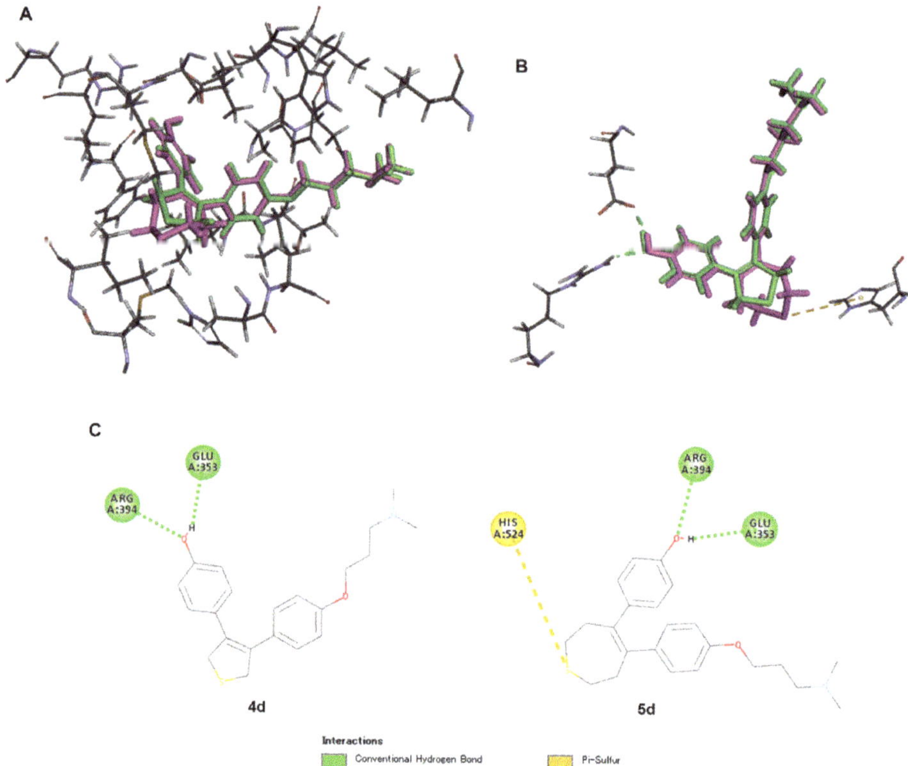

Figure 3. Docking modes of compounds **4d** and **5d** at the ERα ligand-binding domain (LBD). (**A**) Superimposed models of compounds **4d** and **5d**; (**B**) Interaction modes of compounds **4d** and **5d** with the amino acid residues of the ERα LBD; (**C**) A sulfur–π interaction of **5d** with the amino acid residue His524 of the ERα LBD.

3. Materials and Methods

3.1. General Consideration

Melting points were determined with an MP-J3 micro melting point apparatus (Yanaco, Kyoto, Japan) and were not corrected. ^{1}H NMR and ^{13}C NMR spectra were recorded with JNM-LA-400 spectrometer (JEOL, Tokyo, Japan). Chemical shifts for ^{1}H NMR spectra were referenced to tetramethylsilane (0.0 ppm) as an internal standard. Chemical shifts for ^{13}C NMR spectra were referenced to residual ^{13}C present in deuterated solvents. The splitting patterns are designated as follows: s (singlet), d (doublet), t (triplet), q (quartet) and m (multiplet). Mass spectra were recorded on a JEOL JMS-DX-303 spectrometer (JEOL, Tokyo, Japan). Elemental analyses were performed with a Perkin Elmer 2400 CHN spectrometer (PerkinElmer, Winter Street Waltham, MA, USA). Column chromatography was carried out using Merck silica gel 60 (0.063–0.200 μm) and TLC was performed on Merck silica gel F254.

3.2. Synthesis

1-(4-Methoxyphenyl)-2-[2-(4-methoxyphenyl)-2-oxo-ethylsulfanyl] ethanone (**7**). To a stirred solution of **6** (2.80 g, 12.2 mmol) in acetone (40 mL) was added a solution of Na$_2$S 9H$_2$O (1.50 g, 6.25 mmol) in H$_2$O (12 mL) at 0 °C. After completion of the addition, the mixture was warmed to room temperature and stirred for 2.5 h. The solvents were removed, and the residue was recrystallized from MeOH to give **7** (1.91 g, 5.78 mmol, 95 %) as a colorless solid; ^{1}H NMR (400 MHz, CDCl$_3$) δ (ppm) 3.87 (6H, s), 3.93 (4H, s), 6.93 (4H, d, J = 8.8 Hz), 7.95 (4H, d, J = 8.8 Hz); ^{13}C NMR (100 MHz, CDCl$_3$) δ (ppm) 37.4, 55.5, 113.9, 128.4, 131.0, 163.8, 193.0; MS (EI) m/z = 330 (M$^+$), 135 (100%).

3,4-Bis-(4-methoxyphenyl)-2,5-dihydrothiophene (**8**). To a suspension of Zn powder (1.57 g, 24.0 mmol) in dry THF (60 mL) was added TiCl$_4$ (1.3 mL, 12.0 mmol) at −30 °C, and the mixture was refluxed for 2.5 h. A solution of **7** (1.0 g, 3.0 mmol) in dry THF (45mL) was slowly added to the reaction mixture at 0°C. The reaction mixture was refluxed for 2 h and then quenched with 10% K$_2$CO$_3$ aqueous solution. The insoluble materials were filtered through Celite, and the filtrate was extracted with AcOEt. The organic layer was washed with brine, dried over Na$_2$SO$_4$, and concentrated. Purification by silica gel column chromatography (eluent: AcOEt/n-hexane = 1/30) gave **8** as a colorless solid (0.55 g, 1.84 mmol, 61 %); ^{1}H NMR (400 MHz, CDCl$_3$) δ (ppm) 3.77 (6H, s), 4.22 (4H, s), 6.75 (4H, d, J = 8.8 Hz), 7.05 (4H, d, J = 8.8 Hz); ^{13}C NMR (100 MHz, CDCl$_3$) δ (ppm) 43.8, 55.1, 113.6, 129.4, 129.7, 134.3, 158.6; MS (EI) m/z = 298 (M$^+$, 100%); Anal. Calcd for C$_{18}$H$_{18}$O$_2$S: C, 72.45; H, 6.08, Found: C, 72.13; H, 6.20.

3,4-Bis-(4-hydroxyphenyl)-2,5-dihydrothiophene (**4a**). To a solution of **8** (407 mg, 1.36 mmol) in dry CH$_2$Cl$_2$ (9 mL) was added 1M of a solution of BBr$_3$ in CH$_2$Cl$_2$ (3.4 mL, 3.4 mmol) at −80 °C, and the mixture was stirred at room temperature for 24 h. The mixture was poured onto ice and extracted with AcOEt. The organic layer was washed with brine, dried over MgSO$_4$, and concentrated. Purification by silica gel column chromatography (eluent: AcOEt/n-hexane = 1/3) gave **4a** as a colorless solid (358 mg, 1.33 mmol, 97 %); colorless needles (MeOH/CH$_2$Cl$_2$); mp 168.5–170.0 °C; ^{1}H NMR (400 MHz, CD$_3$OD) δ (ppm) 4.15 (4H, s), 6.62 (4H, d, J = 8.8 Hz), 6.95 (4H, d, J = 8.8 Hz); ^{13}C NMR (100 MHz, CD$_3$OD) δ(ppm) 44.3, 116.0, 129.8, 130.8, 135.5, 157.7; MS (EI) m/z = 270 (M$^+$, 100%); HRMS Calcd for C$_{16}$H$_{14}$O$_2$S: 270.0715, Found: 270.0707; Anal. Calcd for C$_{16}$H$_{14}$O$_2$S 0.4 H$_2$O: C, 69.23; H, 5.38, Found: C, 69.04; H, 5.66.

3,4-Bis-(4-methoxyphenyl)-2,5-dihydrothiophene-1-oxide (**9**). To a solution of **8** (406 mg, 1.36 mmol) in dry CH$_2$Cl$_2$ (15 mL) was portionwise added m-CPBA (65%, 349 mg, 1.41 mmol) at 0 °C. The reaction mixture was stirred at room temperature for 25 h and then quenched with saturated Na$_2$SO$_3$ aqueous solution. The mixture was extracted with AcOEt, washed with saturated NaHCO$_3$ aqueous solution and brine, dried over MgSO$_4$, and concentrated. Purification by silica gel column chromatography (eluent: AcOEt/n-hexane = 1/1) gave **9** as a colorless solid (302 mg, 0.96 mmol, 70%); colorless needles (CHCl$_3$/n-hexane); mp 181.0–182.0 °C; ^{1}H NMR (400 MHz, CDCl$_3$) δ (ppm) 3.78 (6H, s), 4.00 (2H, d, J = 16.6 Hz), 4.34 (2H, d, J = 16.6 Hz), 6.77 (4H, d, J = 8.8 Hz), 7.10 (4H, d, J = 8.8 Hz); ^{13}C NMR (100 MHz, CDCl$_3$) δ (ppm) 55.2, 64.3, 113.9, 128.1, 129.9, 130.1, 159.1; MS (EI) m/z = 296 (M$^+$, 100%); HRMS Calcd for C$_{18}$H$_{18}$O$_3$S: 314.0977, Found: 314.0974; Anal. Calcd for C$_{18}$H$_{18}$O$_3$S: C, 68.76; H, 5.77, Found: C, 68.42; H, 5.88.

3,4-Bis-(4-hydroxyphenyl)-2,5-dihydrothiophene-1-oxide (**4b**). To a solution of **4a** (402 mg, 1.49 mmol) in dry CH$_2$Cl$_2$ (60 mL) was portion-wise added m-CPBA (65%, 368 mg, 1.49 mmol) at 0 °C. The reaction mixture was stirred at room temperature for 20 h and then quenched with saturated Na$_2$SO$_3$ aqueous solution. The mixture was extracted with AcOEt, washed with saturated NaHCO$_3$ aqueous solution and brine, dried over MgSO$_4$, and concentrated. Purification by silica gel column chromatography (eluent: AcOEt/n-hexane = 1/1) afforded **4b** as a colorless solid (311 mg, 1.09 mmol, 73 %); colorless needles (MeOH/CH$_2$Cl$_2$); mp 219.0–220.0 °C; ^{1}H NMR (400 MHz, CD$_3$OD) δ (ppm) 3.94 (2H, d, J = 17.1 Hz), 4.49 (2H, d, J = 17.1 Hz), 6.67 (4H, d, J = 8.8 Hz), 7.05 (4H, d, J = 8.8 Hz); ^{13}C NMR (100 MHz, CD$_3$OD) δ (ppm) 64.9, 116.3, 128.4, 131.0, 131.1, 158.3; MS (EI) m/z = 286 (M$^+$), 268 (100%); HRMS Calcd for C$_{16}$H$_{14}$O$_3$S: 286.0664, Found: 286.0665; Anal. Calcd for C$_{16}$H$_{14}$O$_3$S 0.1 H$_2$O: C, 66.69; H, 4.97, Found: C, 66.47; H, 4.72.

3,4-Bis-(4-methoxyphenyl)-2,5-dihydrothiophene-1,1-dioxide (**10**). To a solution of **8** (601 mg, 2.02 mmol) in dry CH$_2$Cl$_2$ (30 mL) was portionwise added m-CPBA (65%, 1.11 g, 4.51 mmol) at 0 °C. The reaction mixture was stirred at room temperature for 3.5 h and then quenched with saturated Na$_2$SO$_3$ aqueous solution. The mixture was extracted with AcOEt, washed with saturated NaHCO$_3$ aqueous solution and brine, dried over MgSO$_4$, and concentrated. The residue was purified by recrystallization with CHCl$_3$/n-hexane to give **10** (377 mg, 1.14 mmol, 57 %) as a colorless needles; mp 141.0–141.5 °C; ^{1}H

NMR (400 MHz, CDCl$_3$) δ(ppm) 3.78 (6H, s), 4.26 (4H, s), 6.77 (4H, d, *J* = 9.3 Hz), 7.03 (4H, d, *J* = 9.3 Hz); ^{13}C NMR (100 MHz, CDCl$_3$) δ(ppm) 55.2, 61.0, 114.0, 127.2, 129.2, 129.7, 159.4; MS (EI) *m/z* = 330 (M$^+$), 266 (100%); Anal. Calcd for C$_{18}$H$_{18}$O$_4$S: C, 65.43; H, 5.49, Found: C, 65.15; H, 5.55.

3,4-Bis-(4-hydroxyphenyl)-2,5-dihydrothiophene-1,1-dioxide (**4c**). To a solution of **10** (203 mg, 0.61 mmol) in dry CH$_2$Cl$_2$ (10 mL) was added 1 M of a solution of BBr$_3$ in CH$_2$Cl$_2$ (1.4 mL, 1.4 mmol) at −80 °C. The reaction mixture was stirred at room temperature for 24 h, poured onto ice, and then extracted with AcOEt. The organic layer was washed with brine, dried over MgSO$_4$, and concentrated. Purification by silica gel column chromatography (eluent: AcOEt/*n*-hexane = 1/2) afforded **4c** as a colorless solid (185 mg, 0.61 mmol, quant); colorless needles (AcOEt/*n*-hexane); mp 180.5–181.0°C; ^{1}H NMR (400 MHz, CD$_3$OD) δ (ppm) 4.27 (4H, s), 6.64 (4H, d, *J* = 8.8 Hz), 6.99 (4H, d, *J* = 8.8 Hz); ^{13}C NMR (100 MHz, CD$_3$OD) δ (ppm) 61.7, 116.2, 127.9, 130.5, 131.0, 158.6; MS (EI) *m/z* = 302 (M$^+$), 238 (100%); HRMS Calcd for C$_{16}$H$_{14}$O$_4$S: 302.0613, Found: 302.0618; Anal. Calcd for C$_{16}$H$_{14}$O$_4$S: C, 63.56; H, 4.67, Found: C, 63.29; H, 4.75.

4-{4-[4-(3-dimethylaminopropoxy)phenyl]-2,5-dihydrothiophen-3-yl}phenol (**4d**). To a solution of **4a** (100 mg, 0.37 mmol) in dry THF (5 mL), 3-dimethylamino-1-propanol (48 μL, 0.41 mmol), triphenylphosphine (97 mg, 0.37 mmol) was added a solution of diethyl azodicarboxylate in toluene (40%, 161 μL, 0.37 mmol) at room temperature, and the mixture was stirred for 22 h. The mixture was concentrated and purified by silica gel column chromatography (eluent: MeOH/CHCl$_3$ = 1/5) gave **4d** as a colorless solid (49 mg, 37%); colorless needles (AcOEt); mp 177.0–178.5°C; ^{1}H NMR (400 MHz, DMSO-*d$_6$*) δ (ppm) 1.79 (2H, quint, *J* = 6.5 Hz), 2.11 (6H, s), 2.30 (2H, t, *J* = 7.2 Hz), 3.92 (2H, t, *J* = 6.0 Hz), 4.14 (4H, s), 6.60 (2H, d, *J* = 8.2 Hz), 6.76 (2H, d, *J* = 8.7 Hz), 6.93 (2H, d, *J* = 7.7 Hz), 7.03 (2H, d, *J* = 8.2 Hz), 9.45 (1H, s); ^{13}C NMR (100 MHz, DMSO-*d$_6$*) δ (ppm) 26.9, 43.0, 45.2, 55.6, 65.6, 114.1, 115.1, 126.8, 128.2, 128.9, 129.5, 133.1, 134.1, 157.2, 157.6; MS (EI) *m/z* 355 (M$^+$), 58 (100%); HRMS Calcd for C$_{21}$H$_{25}$NO$_2$S: 355.1606, Found: 355.1607.

3-Chloro-1-(4-methoxyphenyl)-propan-1-one (**12**). To a soltion of anisol (8.5 mL, 78.2 mmol) and AlCl$_3$ (11 g, 82.5 mmol) in dry 1,2-dichloroethane (30 mL) was added 3-chloropropionyl chloride **11** (5.5 mL, 57.61 mmol) at 0 °C, and the mixture was stirred at room temperature for 2 h. The mixture was poured onto ice and extracted with AcOEt. The organic layer was washed with brine, dried over MgSO$_4$, and concentrated. The residue was purified by recrystallization from CH$_2$Cl$_2$/*n*-hexane to give **12** (9.3 g, 46.7 mmol, 81%) as a colorless solid; ^{1}H NMR (400 MHz, CDCl$_3$) δ (ppm) 3.40 (2H, t, *J* = 6.8 Hz), 3.87 (3H, s), 3.91 (2H, t, *J* = 6.8 Hz), 6.95 (2H, d, *J* = 8.8 Hz), 7.94 (2H, d, *J* = 8.8 Hz); ^{13}C NMR (100 MHz, CDCl$_3$) δ (ppm) 38.9, 40.9, 55.5, 113.9, 129.5, 130.3, 163.8, 195.2; MS (EI) *m/z* = 198 (M$^+$), 135 (100%).

7-Methoxy-1-[3-(4-methoxy-phenyl)-3-oxo-propylsulfanyl]-4-methyl-hepta-4,6-dien-3-one (**13**). To a stirred solution of **12** (7.01 g, 35.4 mmol) in acetone (100 mL) was added a solution of Na$_2$S 9H$_2$O (4.12 g, 17.2 mmol) in H$_2$O (12 mL) at 0 °C. After completion of the addition, the mixture was warmed to room temperature and stirred for 21 h. The solvents were removed, and the residue was recrystallized from MeOH to give **13** (6.79 g, 17.4 mmol, 98 %) as a colorless solid; ^{1}H NMR (400 MHz, CDCl$_3$) δ (ppm) 3.0 (4H, t, *J* = 7.3 Hz), 3.2 (4H, t, *J* = 7.3 Hz), 3.9 (6H, s), 6.9 (4H, d, *J* = 8.8 Hz), 7.9 (4H, d, *J* = 8.8 Hz); ^{13}C NMR (100 MHz, CDCl$_3$) δ (ppm) 26.7, 38.5, 55.5, 113.8, 129.7, 130.3, 163.6, 196.8. MS (EI) *m/z* = 358 (M$^+$), 135 (100%).

4,5-Bis (4-methoxyphenyl)-2,3,6,7-tetrahydrothiepine (**14**). To a suspension of Zn powder (4.39 g, 67.1 mmol) in THF (90 mL) was added TiCl$_4$ (3.8 mL, 34.6 mmol) at −30 °C, and the mixture was refluxed for 2.5 h. A solution of **13** (3.01 g, 8.4 mmol) in THF (120 mL) was slowly added to the reaction mixture at 0 °C. The reaction mixture was refluxed for 18 h and then quenched with 10% K$_2$CO$_3$ aqueous solution. The insoluble materials were filtered through Celite, and the filtrate was extracted with AcOEt. The organic layer was washed with brine, dried over MgSO$_4$, and concentrated. The residue was purified by column chromatography on silica gel with *n*-hexane to give **14** (2.72 g, 8.34 mmol, 99%) as a colorless solid; ^{1}H NMR (400 MHz, CDCl$_3$) δ (ppm) 2.8 (4H, t, *J* = 5.3 Hz), 3.13 (4H, t, *J* = 5.3 Hz), 3.72 (6H, s), 6.65 (4H, d, *J* = 8.8 Hz), 6.88 (4H, d, *J* = 8.8 Hz); ^{13}C NMR (100 MHz, CDCl$_3$) δ (ppm) 27.3, 40.0, 55.1, 113.2, 130.0, 136.8, 139.2, 157.5; MS (EI) *m/z* = 326 (M$^+$, 100%).

3,4-Bis (4-hydroxyphenyl)-2,3,6,7-tetrahydrothiepine (**5a**). To a solution of **14** (1.01 g, 3.08 mmol) in dry CH_2Cl_2 (15 mL) was added 1M of a solution of BBr_3 in CH_2Cl_2 (7.6 mL, 7.6 mmol) at −80 °C, and the mixture was stirred at room temperature for 16 h. The mixture was poured onto ice and extracted with AcOEt. The organic layer was washed with brine, dried over $MgSO_4$, and concentrated. The residue was purified by column chromatography on silica gel with 1:2 AcOEt:*n*-hexane to give **5a** (413 mg, 2.02 mmol, 66 %) as a colorless solid; colorless needles (MeOH/CH_2Cl_2); mp 165.5–166.0 °C; ^{1}H NMR (400 MHz, CD_3OD) δ (ppm) 2.73 (4H, t, *J* = 5.4 Hz), 3.09 (4H, t, *J* = 5.4 Hz), 6.51 (4H, d, *J* = 8.8 Hz), 6.78 (4H, d, *J* = 8.8 Hz); ^{13}C NMR (100 MHz, CD_3OD) δ (ppm) 28.1, 41.0, 115.5, 131.2, 137.4, 140.5, 156.3; MS (EI) *m/z* = 298 (M$^+$, 100%); HRMS Calcd for $C_{18}H_{18}O_2S$: 298.1028, Found: 298.1034; Anal. Calcd for $C_{18}H_{18}O_2S$: C, 72.45; H, 6.08, Found: C, 72.27; H, 6.15.

4,5-Bis (4-hydroxyphenyl)-2,3,6,7-tetrahydrothiepine-1-oxide (**5b**). To a solution of **5a** (402 mg, 1.35 mmol) in CH_2Cl_2 (60 mL) was portionwise added *m*-CPBA (65%, 399 mg, 1.62 mmol) at 0 °C. The reaction mixture was stirred at room temperature for 9 h and then quenched with saturated Na_2SO_3 aqueous solution. The mixture was extracted with AcOEt, washed with saturated $NaHCO_3$ aqueous solution and brine, dried over $MgSO_4$, and concentrated. The residue was purified by column chromatography on silica gel with 3:1 AcOEt:*n*-hexane to give **5b** (127 mg, 0.40 mmol, 30 %) as a colorless solid; colorless needles (MeOH/CH_2Cl_2); mp 190–191.5 °C; ^{1}H NMR (400 MHz, CD_3OD) δ (ppm) 2.51 (2H, dd, *J* = 15.8 Hz), 3.00 (2H, dd, *J* = 14.2 Hz), 3.10 (2H, t, *J* = 14.2 Hz), 3.65 (2H, dd, *J* = 15.6 Hz), 6.54 (4H, d, *J* = 8.3 Hz), 6.83 (4H, d, *J* = 8.3 Hz); ^{13}C NMR (100 MHz, CD_3OD) δ (ppm) 26.0, 46.8, 115.9, 131.4, 136.2, 140.4, 157.1; MS (EI) *m/z* = 314 (M$^+$), 251 (100%); HRMS Calcd for $C_{18}H_{18}O_3S$: 314.0977, Found: 314.0976.

4,5-Bis (4-methoxyphenyl)-2,3,6,7-tetrahydrothiepine-1,1-dioxide (**15**). To a solution of **14** (1.00 mg, 3.07 mmol) in dry CH_2Cl_2 (20 mL) was portion-wise added *m*-CPBA (65%, 1.75 g, 7.11 mmol) at 0 °C. The reaction mixture was stirred at room temperature for 22 h and then quenched with saturated Na_2SO_3 aqueous solution. The mixture was extracted with AcOEt, washed with saturated $NaHCO_3$ aqueous solution and brine, dried over $MgSO_4$, and concentrated. The residue was purified by column chromatography on silica gel with 1:3 AcOEt:*n*-hexane to give **15** (1.05 g, 2.93 mmol, 96 %) as colorless solid; ^{1}H NMR (400 MHz, $CDCl_3$) δ (ppm) 3.02 (4H, t, *J* = 5.1 Hz), 3.17 (4H, t, *J* = 5.1 Hz), 3.74 (6H, s), 6.68 (4H, d, *J* = 8.8 Hz), 6.88 (4H, d, *J* = 8.8 Hz); ^{13}C NMR (100 MHz, $CDCl_3$) δ (ppm) 30.4, 54.0, 55.3, 113.7, 130.1, 135.1, 138.6, 158.3. MS (EI) *m/z* = 358 (M$^+$), 292 (100%).

4,5-Bis (4-hydroxyphenyl)-2,3,6,7-tetrahydrothiepine-1,1-dioxide (**5c**). To a solution of **15** (403 mg, 1.12 mmol) in dry CH_2Cl_2 (7 mL) was added 1M of a solution of BBr_3 in CH_2Cl_2 (2.8 mL, 2.8 mmol) at −80 °C. The reaction mixture was stirred at room temperature for 31 h, poured onto ice, and then extracted with AcOEt. The organic layer was washed with brine, dried over $MgSO_4$, and concentrated. The residue was purified by column chromatography on silica gel with 1 : 2 AcOEt : *n*-hexane to give **5c** (195 mg, 0.20 mmol, 53%) as a colorless solid; colorless needles (AcOEt/*n*-hexane); mp 128.5–130.0 °C; ^{1}H NMR (400 MHz, CD_3OD) δ (ppm) 2.97 (4H, t, *J* = 5.3 Hz), 3.21 (4H, t, *J* = 5.3 Hz), 6.54 (4H, d, *J* = 8.8 Hz), 6.85 (4H, d, *J* = 8.8 Hz); ^{13}C NMR (100 MHz, CD_3OD) δ (ppm) 31.2, 54.5, 115.7, 131.4, 135.8, 139.8, 157.0; MS (EI) *m/z* = 330 (M$^+$), 264 (100%); HRMS Calcd for $C_{18}H_{18}O_4S$: 330.0926, Found: 330.0920; Anal. Calcd for $C_{18}H_{18}O_4S$: C, 65.43; H, 5.49, Found: C, 65.40; H, 5.59.

4-{5-[4-(3-Dimethylaminopropoxy)phenyl]-2,3,6,7-tetrahydrothiepin-4-yl}phenol (**5d**). To a solution of **5a** (200 mg, 0.67 mmol) in dry THF (10 mL), 3-dimethylamino-1-propanol (87 μL, 0.74 mmol), triphenylphosphine (176 mg, 0.67 mmol) was added a solution of diethyl azodicarboxylate in toluene (40%, 292 μL, 0.67 mmol) at room temperature, and the mixture was stirred for 22 h. After the solvent was removed, the residue was purified by column chromatography on silica gel with 1:5 MeOH:$CHCl_3$ to give **5d** (71 mg, 28%) as a colorless solid; Colorless needles (CH_2Cl_2/MeOH); mp 176.0–177.0 °C; ^{1}H NMR (400 MHz, $CDCl_3$) δ (ppm) 1.88 (2H, quint, *J* = 6.9 Hz), 2.25 (6H, s), 2.43 (2H, t, *J* = 7.7 Hz), 2.78 (4H, t, *J* = 4.8 Hz), 3.12 (4H, t, *J* = 5.1 Hz), 3.91 (2H, t, *J* = 6.3 Hz), 6.53 (2H, d, *J* = 8.7 Hz), 6.61 (2H, d, *J* = 8.7 Hz), 6.79 (2H, d, *J* = 8.7 Hz), 6.83 (2H, d, *J* = 8.7 Hz); ^{13}C NMR (100 MHz, $CDCl_3$) δ (ppm) 27.1, 27.3,

39.9, 45.2, 56.3, 66.0, 113.8, 115.0, 130.1, 130.2, 136.3, 137.0, 139.1, 139.5, 154.4, 156.7; MS (EI) *m/z* 383 (M$^+$), 58 (100%); HRMS Calcd for $C_{23}H_{29}NO_2S$: 383.1919, Found: 383.1931.

3.3. Competitive Binding Assay Using Human ER

The ligand binding activity of human estrogen receptor α (ER α) was determined by a nitrocellulose filter binding assay method. ER α was diluted with a binding assay buffer (20 mM Tris-HCl pH 8.0, 0.3 M NaCl, 1 mM EDTA pH 8.0, 10 mM 2-mercaptoethanol, 0.2 mM phenylmethylsulfonyl floride) and incubated with 4 nM [6, 7-^{3}H]-17β-estradiol in the presence or absence of an unlabeled competitor at 4 ^{6}C for 18 h. The incubation mixture was absorbed by suction onto a nitrocellulose membrane that had been soaked in binding assay buffer. The membrane was washed two times with buffer (20 mM Tris-HCl pH 8.0, 0.15 M NaCl) and with 25% ethanol in distilled water. Radioactivity that remained in the membrane was measured in Atomlight by using a liquid scintillation counter.

3.4. MCF-7 Cell Proliferation Assay

The human breast adenocarcinoma cell line MCF-7 was routinely cultivated in DMEM supplemented with 10% FBS and 100 IU/mL penicillin and 100 mg/mL streptomycin at 37 °C in a 5% CO_2 humidified incubator. Before an assay, MCF-7 cells were switched to DMEM (low glucose phenol red-free supplemented with 5% FBS, and 100 IU/mL penicillin and 100 mg/mL streptomycin). Cells were trypsinized from the maintenance dish with phenol red-free trypsin-EDTA and seeded in a 96-well plate at a density of 2000 cells per final volume of 100 µL DMEM supplemented with 5% stripped FBS and 100 IU/mL penicillin and 100 mg/mL streptomycin. After 24 h, the medium was removed and 90 µL of fresh medium and 10 µL of drug solution, supplemented with serial dilutions of test compounds or DMSO as dilute control in the presence or absence of 0.1 nM E2, were added to triplicate microcultures. Cells were incubated for four days. At the end of the incubation time, number of cells was counted by using the WST-8, which was added to microcultures 10 µL each, and they were incubated for 2–4 h. The absorbance at 450 nm was measured. This parameter relates to and number of living cells in the culture.

3.5. Docking Simulation Study

Three-dimensional (3D) structures of protein-ligand complexes were predicted using the Discovery Studio 2018/CDOCKER software (BIOVIA) with default settings. The 3D structures of ER used in this study were retrieved from the RCSB Protein Data Bank (PDB ID: 1ERR). Missing hydrogen atoms in the crystal structure were computationally added and the center of the active site was defined as the center of raloxifen in 1ERR. The conformations of **4d** and **5d** were optimized using the CHARMm force field, and the docking simulations of **4d** and **5d** with 1ERR were performed using the CDOCKER protocol. The docking poses for **4d** and **5d** with the highest CDOCKER ENERGY were selected for the discussions of 10 docking modes, respectively.

4. Conclusions

Novel sulfur-containing ERα modulators (compounds **4** and **5**) as potential breast cancer therapeutic drug candidates were designed and synthesized based upon the hydrophobic feedback approach for the simple bisphenols **1–3** developed in our previous studies. Tetrahydrothiepine derivatives (compounds **5a–5d**) showed higher binding affinity toward ERα than the corresponding dihydrothiophene derivatives (compounds **4a–4d**) because of the hydrophobicity of the sulfur-containing ring structure. Although the bisphenol derivatives **4a**, **4c**, **5a**, and **5c** showed estrogenic activity toward the MCF-7 breast cancer cell line, the corresponding alkylamino derivatives (compounds **4d** and **5d**) acted as ER antagonists. In particular, compound **5d** showed the most potent antiestrogenic activity among the tested compounds with an IC_{50} value of 0.09 µM. A sulfur-containing structure might be a promising scaffold for antiestrogen discovery, owing to ease of synthesis, binding modes toward the ERα LBD, and availability of various substituted derivatives.

Author Contributions: Conceptualization, K.O. and Y.E.; Data curation, A.K., F.T., S.A. and T.O.; Formal analysis, K.O., A.K. and T.O.; Funding acquisition, K.O. and Y.E.; Investigation, K.O., A.K., F.T. and S.A.; Methodology, K.O. and A.K.; Project administration, K.O.; Resources, K.O.; Software, A.K., F.T., S.A. and T.O.; Supervision, K.O. and Y.E.; Validation, A.K.; Visualization, K.O. and A.K.; Writing—original draft, K.O.; Writing—review & editing, K.O.

Funding: This research was supported by a grant-in-aid from the Strategic Research Program for Private Universities (2015–2019) and a grant-in-aid for Scientific Research (C) (26460151 and 15K08029) from the Japanese Ministry of Education, Culture, Sports, Science, and Technology (MEXT).

Conflicts of Interest: The authors declare no conflict of interest.

References

1. Umetani, M.; Domoto, H.; Gormley, A.K.; Yuhanna, I.S.; Cummins, C.L.; Javitt, N.B.; Korach, K.S.; Shaul, P.W.; Mangelsdorf, D.J. 27-Hydroxycholesterol is an endogenous SERM that inhibits the cardiovascular effects of estrogen. *Nat. Med.* **2007**, *13*, 1185. [CrossRef] [PubMed]

2. Takano-Yamamoto, T.; Rodan, G.A. Direct effects of 17 beta-estradiol on trabecular bone in ovariectomized rats. *Proc. Natl. Acad. Sci. USA* **1990**, *87*, 2172. [CrossRef] [PubMed]

3. Ramirez, V.D.; Kipp, J.; Joe, I. Estradiol, in the CNS, targets several physiologically relevant membrane-associated proteins. *Brain Res. Rev.* **2001**, *37*, 141. [CrossRef]

4. Green, S.; Walter, P.; Kumar, V.; Krust, A.; Bornert, J.M.; Argos, P.; Chambon, P. Human oestrogen receptor cDNA: Sequence, expression and homology to v-erb-A. *Nature* **1986**, *320*, 134. [CrossRef] [PubMed]

5. Leng, X.H.; Bray, P.F. Hormone therapy and platelet function. *Drug Discov. Today Disease Mechanisms* **2005**, *2*, 85. [CrossRef]

6. Motivala, A.; Pitt, B. Drospirenone for Oral Contraception and Hormone Replacement Therapy. *Drugs* **2007**, *67*, 647. [CrossRef]

7. Labrie, F.; Labrie, C.; Belanger, A.; Simard, J. *Selective Estrogen Receptor Modulators*; Manni, A., Verderame, M., Eds.; Humana Press: Totowa, NJ, USA, 2002.

8. Jameera, B.A.; Jubie, S.; Nanjan, M.J. Estrogen receptor agonists/antagonists in breast cancer therapy: A critical review. *Bioorg. Chem.* **2017**, *71*, 257. [CrossRef]

9. Patel, H.K.; Bihani, T. Selective estrogen receptor modulators (SERMs) and selective estrogen receptor degraders (SERDs) in cancer treatment. *Pharmacol. Ther.* **2018**, *186*, 1. [CrossRef]

10. Valera, M.C.; Fontaine, C.; Dupuis, M.; Noirrit-Esclassan, E.; Vinel, A.; Guillaume, M.; Gourdy, P.; Lenfant, F.; Arnal, J.F. Towards optimization of estrogen receptor modulation in medicine. *Pharmacol. Ther.* **2018**, *189*, 123. [CrossRef]

11. Shiau, A.K.; Barstad, D.; Loria, P.M.; Cheng, L.; Kushner, P.J.; Agard, D.A.; Greene, G.L. The structural basis of estrogen receptor/coactivator recognition and the antagonism of this interaction by tamoxifen. *Cell* **1998**, *95*, 927. [CrossRef]

12. Jordan, V.C. Tamoxifen: A most unlikely pioneering medicine. *Nat. Rev. Drug Discov.* **2003**, *2*, 205. [CrossRef] [PubMed]

13. Shagufta; Ahmad, I. Tamoxifen a pioneering drug: An update on the therapeutic potential of tamoxifen derivatives. *Eur. J. Med. Chem.* **2018**, *143*, 515. [CrossRef] [PubMed]

14. McCague, R.; Jarman, M.; Leung, O.T.; Foster, A.B.; Leclercq, G.; Stoessel, S.J. Inhibitors of steroid hormone biosynthesis and action. *Steroid Biochem.* **1988**, *31*, 545. [CrossRef]

15. Katzenellenbogen, B.S.; Norman, M.J.; Eckert, R.L.; Peltz, S.W.; Mangel, W.F. Bioactivities, estrogen receptor interactions, and plasminogen activator-inducing activities of tamoxifen and hydroxy-tamoxifen isomers in MCF-7 human breast cancer cells. *Cancer Res.* **1984**, *44*, 112.

16. Gauthier, S.; Mailhot, J.; Labrie, F. New Highly Stereoselective Synthesis of (Z)-4-Hydroxytamoxifen and (Z)-4-Hydroxytoremifene via McMurry Reaction. *J. Org. Chem.* **1996**, *61*, 3890. [CrossRef] [PubMed]

17. Yu, D.D.; Forman, B.M. Simple and efficient production of (Z)-4-hydroxytamoxifen, a potent estrogen receptor modulator. *J. Org. Chem.* **2003**, *68*, 9489. [CrossRef]

18. Mitlak, B.H.; Cohen, F.J. Selective estrogen receptor modulators: A look ahead. *Drugs* **1999**, *57*, 653. [CrossRef]

19. Jones, C.D.; Blaszczak, L.C.; Goettel, M.E.; Suarez, T.; Crowell, T.A.; Mabry, T.E.; Ruenitz, P.C.; Srivatsan, V. Antiestrogens. 3. Estrogen receptor affinities and antiproliferative effects in MCF-7 cells of phenolic analogs of trioxifene, [3,4-dihydro-2-(4-methoxyphenyl)-1-naphthalenyl][4-[2-(1-pyrrolidinyl)ethoxy]phenyl] methanone. *J. Med. Chem.* **1992**, *35*, 931. [CrossRef]

20. Gara, R.K.; Sundram, V.; Chauhan, S.C.; Jaggi, M. Anti-cancer potential of a novel SERM ormeloxifene. *Curr. Med. Chem.* **2013**, *20*, 4177. [CrossRef]

21. Robertson, J.F.R.; Lindemann, J.; Garnett, S.; Anderson, E.; Nicholson, R.I.; Kuter, I.; Gee, J.M.W. A Good Drug Made Better: The Fulvestrant Dose-Response Story. *Clin. Breast Cancer* **2014**, *14*, 381. [CrossRef]

22. Endo, Y.; Yoshimi, T.; Miyaura, C. Boron clusters for medicinal drug design: Selective estrogen receptor modulators bearing carborane. *Pure Appl. Chem.* **2003**, *75*, 1197. [CrossRef]

23. Endo, Y.; Yoshimi, T.; Ohta, K.; Suzuki, T.; Ohta, S. Potent Estrogen Receptor Ligands Based on Bisphenols with a Globular Hydrophobic Core. *J. Med. Chem.* **2005**, *48*, 3941. [CrossRef] [PubMed]

24. Gharpure, C.J.; Anuradha, D.; Prasad, J.V.K.; Srinivasu, R.P. Stereoselective Synthesis of cis-2,6-Disubstituted Morpholines and 1,4-Oxathianes by Intramolecular Reductive Etherification of 1,5-Diketones. *Eur. J. Org. Chem.* **2015**, *86*. [CrossRef]

25. Schnapperelle, I.; Bach, T. Modular Synthesis of Phenanthro[9,10-c]thiophenes by a Sequence of C-H Activation, Suzuki Cross-Coupling and Photocyclization Reactions. *Chem. Eur. J.* **2014**, *20*, 9725. [CrossRef]

26. Ohta, K.; Chiba, Y.; Kaise, A.; Endo, Y. Novel retinoid X receptor (RXR) antagonists having a dicarba-closo-dodecaborane as a hydrophobic moiety. *Bioorg. Med. Chem.* **2015**, *23*, 861. [CrossRef]

27. Ohta, K.; Ogawa, T.; Kaise, A.; Endo, Y. Enhanced estrogen receptor beta (ERβ) selectivity of fluorinated carborane-containing ER modulators. *Bioorg. Med. Chem. Lett.* **2013**, *23*, 6555. [CrossRef]

28. Ohta, K.; Ogawa, T.; Oda, A.; Kaise, A.; Endo, Y. Aliphatic Substitution of o-Carboranyl Phenols Enhances Estrogen Receptor Beta Selectivity. *Chem. Pharm. Bull.* **2014**, *62*, 386. [CrossRef]

29. Ohta, K.; Ogawa, T.; Oda, A.; Kaise, A.; Endo, Y. Design and synthesis of carborane-containing estrogen receptor-beta (ERb)-selective ligands. *Bioorg. Med. Chem. Lett.* **2015**, *25*, 4174. [CrossRef]

30. Jones, G.; Willett, P.; Glen, R.C. Molecular recognition of receptor sites using a genetic algorithm with a description of desolvation. *J. Mol. Biol.* **1995**, *245*, 43. [CrossRef]

Sample Availability: Samples of all the compounds are available from the authors.

Article

3-Aryl-4-nitrobenzothiochromans *S,S*-dioxide: From Calcium-Channel Modulators Properties to Multidrug-Resistance Reverting Activity

Matteo Micucci [1], Maurizio Viale [2], Alberto Chiarini [1], Domenico Spinelli [3], Maria Frosini [4], Cinzia Tavani [5], Massimo Maccagno [5], Lara Bianchi [5], Rosaria Gangemi [2] and Roberta Budriesi [1,*]

[1] Dipartimento di Farmacia & Biotecnologia, Alma Mater Studiorum-University of Bologna, Via Belmeloro 6, 40126 Bologna, Italy; matteo.micucci2@unibo.it (M.M.); alberto.chiarini@unibo.it (A.C.)

[2] IRCCS Ospedale Policlinico San Martino, U.O. Bioterapie, L.go R. Benzi 10, 16132 Genova, Italy; maurizio.viale@hsanmartino.it (M.V.); rosaria.gangemi@hsanmartino.it (R.G.)

[3] Dipartimento di Chimica "G. Ciamician", Alma Mater Studiorum-University of Bologna, via F. Selmi 2, 40126 Bologna, Italy; domenico.spinelli@unibo.it

[4] Dipartimento di Scienze della Vita, Università degli Studi di Siena, Via A. Moro 2, 53100 Siena, Italy; maria.frosini@unisi.it

[5] Dipartimento di Chimica e Chimica Industriale, Università degli Studi di Genova, Via Dodecaneso 31, 16146 Genova, Italy; cinzia.tavani@unige.it (C.T.); massimo.maccagno@unige.it (M.M.); lara.bianchi@unige.it (L.B.)

* Correspondence: roberta.budriesi@unibo.it; Tel.: +39-051-2099721

Academic Editors: Carla Boga and Gabriele Micheletti
Received: 8 January 2020; Accepted: 23 February 2020; Published: 27 February 2020

Abstract: Our research groups have been involved for many years in studies aimed at identifying new active organic compounds endowed with pharmacological properties. In this work, we focused our attention on the evaluation of cardiovascular and molecular drug resistance (MDR) reverting activities of some nitrosubstituted sulphur-containing heterocycles. Firstly, we have examined the effects of 4-nitro-3-(4-methylphenyl)-3,6-dihydro-2*H*-thiopyran *S,S*-dioxide **5**, and have observed no activity. Then we have extended our investigation to the 3-aryl-4-nitrobenzothiochromans *S,S*-dioxide **6** and **7**, and have observed an interesting biological profile. Cardiovascular activities were assessed for all compounds using ex vivo studies, while the MDR reverting effect was evaluated only for selected compounds using tumor cell lines. All compounds were shown to affect cardiovascular parameters. Compound **7i** exerted the most effect on negative inotropic activity, while **6d** and **6f** could be interesting molecules for the development of more active ABCB1 inhibitors. Both **6** and **7** represent structures of large possible biological interest, providing a scaffold for the identification of new ABCB1 inhibitors.

Keywords: 3-aryl-4-nitrothiochromans *S,S*-dioxide; L-Type Calcium Channels (LTCC); anticancer therapy; multidrug resistance (MDR1); in vitro experiments; cardiovascular activity

---

1. Introduction

Cancer chemotherapy, as in the case of antimicrobial chemotherapy, has had trouble with resistance to treatment. This phenomenon occurs mainly through molecular mechanisms that may be affected by novel compounds that may be associated to cancer chemotherapy [1]. Nowadays, cancer treatment based on a combination of drugs with different mechanisms of action seems to be the best strategy to avoid multi drug resistance (MDR) [2,3]. This approach has proved to be successful for some kinds of cancer such as leukemias, while for other types of high-spread cancers, including breast and lung cancer, the failure rate is still very high.

Studies dating back to the second half of the last century identified the presence of a P-glycoprotein (Pgp) that is correlated to the onset of drug resistance [4,5]. The discovery that the high concentration of Pgp in resistant cells is independent of the chemical class of drugs used enlightens a further complicating factor. The mechanisms underlying multi drug resistance (MDR) are not fully understood, and they are responsible for the decrease of intracellular drugs concentration [6]. To counteract this phenomenon, various strategies, including the search for chemical entities able to inhibit Pgp, have been investigated.

Calcium channels might also be targets for pathologies other than cardiovascular diseases. Many calcium channel blockers of different structural classes, including diltiazem-like drugs, in fact, represent modulators of MDR (resistance modifiers, chemosensitizers). Since the sensitized tumor cells do not express voltage-gated calcium channels, the chemosensitization exerted by these drugs is independent on their activity as intracellular calcium-regulators. Unfortunately, most of them affect cardiovascular parameters at the concentrations required for a successful reversal of MDR, preventing its clinical use. Thus, novel compounds endowed with low calcium channel blocking activity may be considered in the search for relatively specific MDR reverting agents.

Our research group has been involved for a long time in a research program aimed at finding new L-Type calcium channel modulators for peripheral or central application [7,8] by modifying the chemical structure to select alternative action for the cardiovascular system.

In the frame of our research, we examined the activity of a small library (21 compounds) of variously substituted 8-aryl-8-hydroxy-8*H*-[1,4]thiazino[3,4-*c*][1,2,4]oxadiazol-3-ones (**1**) as calcium-entry blockers by means of functional studies on isolated guinea-pig left and right atria and K$^+$-depolarized aortic strips [9]. This new class of compounds is structurally correlated to diltiazem (**2**) [10] and to pyrrolobenzothiazines (**3**) [11], which are well known as potent calcium antagonists with a strong selectivity for cardiac tissues over vascular ones, and thus widely used as therapeutic agents in cardiovascular diseases. Interestingly, we observed [9] that at least two of the examined compounds (**4a**: X = Cl; **4b**: X = Br) are more potent than diltiazem itself as L-type calcium channel modulators, and they selectively affect the heart contractility without effect on vascular tissues (Figure 1).

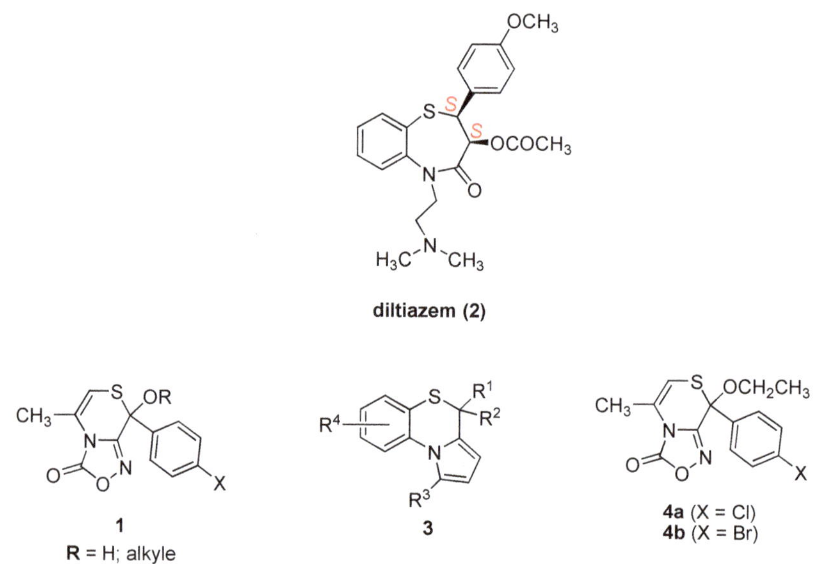

diltiazem (2)

Figure 1. Chemical structure of diltiazem and related compounds purposely designed and synthesized.

Following the identification of this new chemotype that is able to selectively bind to L-type calcium channels, we explored the possibility for this class of compounds to exhibit MDR reverting properties, such as the parent compound diltiazem [12], independent of the action on L-type calcium channels even if the resistant cells have high levels of intracellular calcium accumulation [13]. For this reason, some selected oxadiazolones have been studied on doxorubicin-resistant cell lines. Some of these compounds can increase accumulation of doxorubicin in the ovarian carcinoma multidrug resistant cell line A2780 through inhibition of MDR1 [14].

Considering the calcium-entry blocker (CEB) activity and the chemical structures of the diltiazem related compounds purposely designed and studied [7,9,15], we focused our attention on a new class of compounds, namely 4-nitrothiopyran *S,S*-dioxide (**5**) [16] and on the benzocondensed analogues (viz. the thiochromans *S,S*-dioxide **6** and **7**) [17,18] (Figure 2).

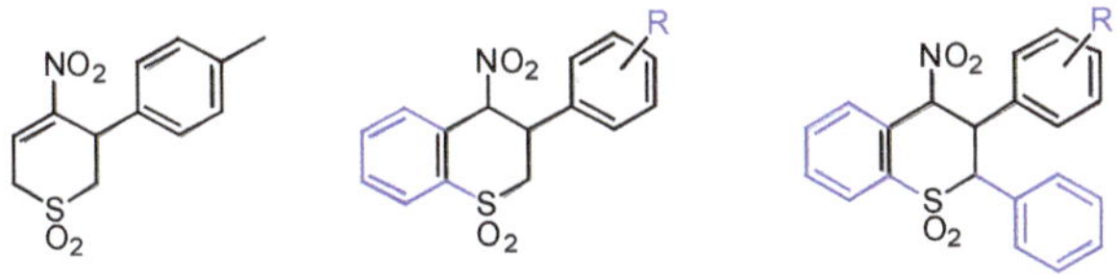

Figure 2. 4-Nitrothiopyran *S,S*-dioxide **5** and benzocondensed analogues **6** and **7**.

In the latter chemotypes, the possibility of different type/degrees of substitution at C-2 and/or C-3 of the thiopyran ring should expand the opportunity for a significant biological activity from L-type calcium channel antagonism to MDR reverting activity. Relevant results on the cardiovascular characterization of the new compounds are reported hereinafter. In addition, we also investigated the effects on MDR reverting activity of some selected compounds to identify new chemical entities with this peculiarity.

2. Results

2.1. Chemistry

The synthetic procedure for **5** (Scheme 1 [16]) and for the benzocondensed derivatives **6a–d, f–j** and **7a–e, g–l**, isolated as diastereomeric mixtures, with a generally high degree of diastereoselectivity (Scheme 2 [17,18]), features an overall ring-opening/ring-closing protocol starting from 3-nitrothiophene (**8**) and 3-nitrobenzo[*b*]thiophene (**9**), respectively. It encompasses (a) substrate ring-opening with pyrrolidine/AgNO₃, (b) trapping of the intermediate silver sulfide with a suitable halide, (c) replacement of the pyrrolidino moiety with the aryl of a Grignard reagent, (d) oxidation of sulfur, and (e) cyclization to the final thiopyran (**5**) or thiochromans (**6** or **7**) dioxide. The procedure is based on a long-standing project on the synthetic exploitation [16,18–21] of the valuable polyfunctionalized nitrobutadienic building-blocks, which originate from the non-benzenoid behavior of nitrothiophenes, viz. the ring-opening reaction that follows the nucleophilic attack of primary or secondary amines.

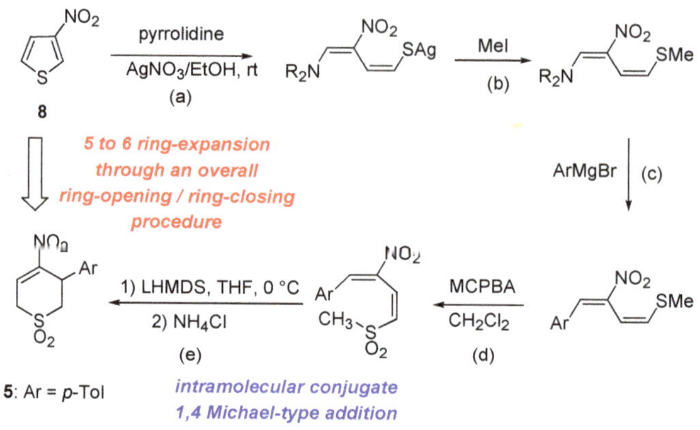

Scheme 1 reagents and labels:

- **8**
- pyrrolidine, AgNO$_3$/EtOH, rt (a)
- MeI (b)
- ArMgBr (c)
- MCPBA, CH$_2$Cl$_2$ (d)
- 1) LHMDS, THF, 0 °C; 2) NH$_4$Cl (e)
- *5 to 6 ring-expansion through an overall ring-opening / ring-closing procedure*
- *intramolecular conjugate 1,4 Michael-type addition*
- **5**: Ar = *p*-Tol

Scheme 1. Synthetic protocol for the preparation of thiopyrans *S,S*-dioxide **5** [16].

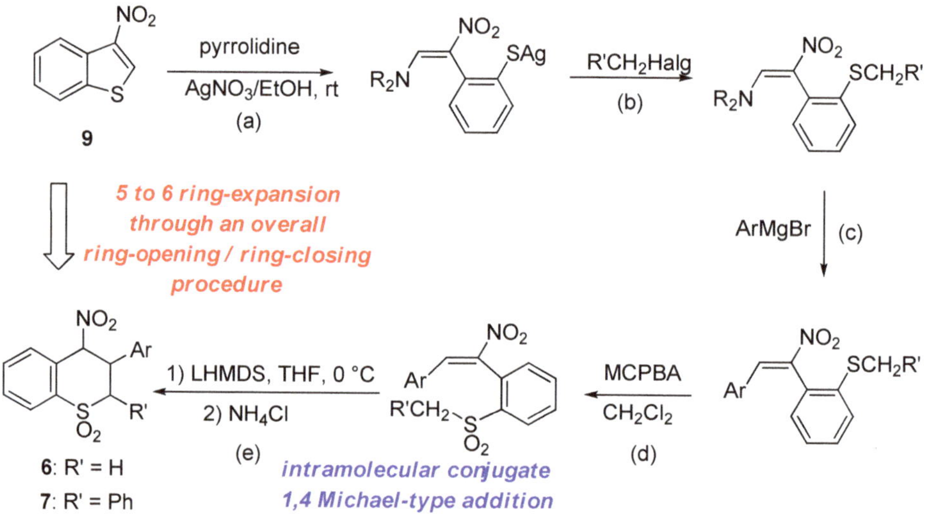

Scheme 2 reagents and labels:

- **9**
- pyrrolidine, AgNO$_3$/EtOH, rt (a)
- R'CH$_2$Halg (b)
- ArMgBr (c)
- MCPBA, CH$_2$Cl$_2$ (d)
- 1) LHMDS, THF, 0 °C; 2) NH$_4$Cl (e)
- *5 to 6 ring-expansion through an overall ring-opening / ring-closing procedure*
- *intramolecular conjugate 1,4 Michael-type addition*
- **6**: R' = H
- **7**: R' = Ph

Ar		Ar	
Ph	**6a** and **7a**	*m*-ClC$_6$H$_4$	**6f**
o-MeC$_6$H$_4$	**6b** and **7b**	*p*-ClC$_6$H$_4$	**6g** and **7g**
m-MeC$_6$H$_4$	**6c** and **7c**	*p*-BrC$_6$H$_4$	**6h** and **7h**
p-MeC$_6$H$_4$	**6d** and **7d**	1-Naphthyl	**6i** and **7i**
p-MeOC$_6$H$_4$	**7e**	2-Thienyl	**7j**
		3-Thienyl	**7l**

Scheme 2. Synthetic protocol for the preparation of thiochromans *S,S*-dioxide **6** and **7** [17,18].

As a matter of fact, the 5- to 6-member ring-expansion protocol had been earlier applied to the synthesis of 4-nitro-3-(4-methylphenyl)-3,6-dihydro-2*H*-thiopyran *S,S*-dioxide **5** [16], revealing for this heterocycle a thermal instability, which possibly contributes to the disappointing preliminary results on its pharmacological activity (see below in the text). This outcome has led us to turn our attention to the benzocondensed derivatives, on the grounds of both the experimental higher stability of 3-aryl-4-nitrothiocromans *S,S*-dioxide (**6**) and 3-aryl-4-nitro-2-phenylthiochromans *S,S*-dioxide (**7**) vs. **5** and the results of some virtual-screening investigations.

This fact would most likely lead to an improved activity on shifting from **5** to **6** or **7**. Furthermore, the encouraging results on the activity of **6** (to be discussed below) have fostered our interest in a deeper evaluation of this class of sulfur heterocycles as potential cardioregulators. Thus, the thiochromans **7** have been designed as likely suitable candidates for the programmed screening as far as (a) the phenyl at C-2 should conceivably alter the geometric array of the molecule, and (b) different and/or variously-substituted aryl/heteroaryl moieties can be introduced at either C-2 (by choosing a proper trapping halide for the silver thiolate in step (b) of Scheme 2), C-3 (by choosing a proper Grignard reagent for step (c)) or at both; this "diversity" should of course increase the probability of highlighting more effective drugs.

From the stereochemical point of view, in the case of compounds **7**, the situation was expected to be definitely more complex because of the presence of the additional stereocenter at C-2, which is generated together with those at C-3 and C-4 during the cyclization step (e). As a matter of fact, interestingly enough, a significant stereoselectivity occurs, strongly favoring the *cis,trans*-racemate **7** (Figure 3) (the assignment of structure being based on ^{1}H NMR data), which is an aspect that is conceivably related to the crowding of the molecule and that has been already discussed in detail [18].

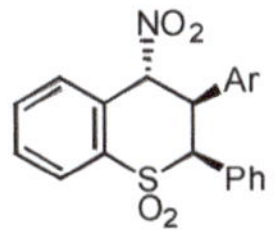

NO$_2$

Ar

S

Ph

O$_2$

cis,trans-**7** (racemate)

Figure 3. Main diastereoisomer isolated from the preparation of **7** (see Scheme 2).

2.2. Ex Vivo Functional Studies

2.2.1. Cardiac Activity

The cardiac profile of all new compounds (**5**, **6a–d, f–j**, and **7a–e, g–l**) was investigated using guinea-pig isolated left and right atria to evaluate their inotropic and/or chronotropic effects, respectively. In particular, the percent decrease of developed tension on isolated left atrium driven at 1 Hz and on the spontaneously beating right atrium (negative inotropic activity), and the percent decrease in atrial rate on spontaneously beating right atrium (negative chronotropic activity), were checked at increasing concentration.

Thiopyran **5** is the only compound lacking cardiac action; in fact, it has no negative chronotropic activity on the spontaneously beating right atrium and no negative inotropic activity either on the left atrium driven at 1 Hz or on the right atrium spontaneously beating. All the other compounds are also devoid of bradycardic effects while all have a negative inotropic action both on the left atrium and on the right atrium. Thiochromans **6a, 7e, 7g, 6h, 7i**, and **7l** are, respectively, about 1.8, 4.2, 1.6, 2.5, 4.6, and 2.5 times more potent than diltiazem. Compounds **6d, 6f, 6g**, and **6i** are less potent than diltiazem 1.7, 4.5, 2.6, and 1.8 times, respectively; while compounds **7a, 6b, 7b, 6c, 7c, 7d, 7h**, and **7j** have potency comparable to that of diltiazem (Table 1). The most potent compounds on this parameter are **7i** and **7e** (EC$_{50}$ = 0.17 µM (c.l. 0.12–0.21); EC$_{50}$ = 0.19 µM (c.l. 0.14–0.26), respectively). In the same way, all the tested compounds, with the exception of **5**, have a negative inotropic activity even on the spontaneously beating right atrium. The most potent compounds on this parameter are **6f** and **7j** (EC$_{50}$ = 0.28 µM (c.l. 0.17–0.38); EC$_{50}$ = 0.19 µM (c.l. 0.13–0.28), respectively). For diltiazem, this activity has not been evaluated because it has a bradycardiac activity, which affects the inotropic effect.

Table 1. Cardiac Activity of Tested Compounds.

| Compd | Left Atrium | | | Right Atrium | | | |
| | Negative Inotropy | | | Negative Inotropy | | | Negative Chronotropy |
	Activity [a] M ± S.E.M.	EC$_{50}$ [b] (µM)	95% c.l.	Activity [c] M ± S.E.M.	EC$_{50}$ [b] (µM)	95% c.l.	Activity [d] M ± S.E.M.
Diltiazem	78 ± 3.5 [e]	0.79	0.70–0.85				94 ± 5.6 [h,i]
5	40 ± 2.3			31 ± 2.1			3 ± 0.2
6a	77 ± 1.5 [f]	0.42	0.28–0.63	79 ± 3.3	1.88	1.13–2.12	15 ± 1.3
7a	76 ± 0.9	0.95	0.66–1.37	80 ± 1.6	1.20	0.95–1.92	6 ± 0.4
6b	90 ± 2.7	1.17	0.80–1.31	77 ± 2.8	4.17	3.79–5.23	3 ± 0.2
7b	74 ± 2.2 [f]	0.98	0.63–1.34	78 ± 2.1	1.45	1.03–2.03	6 ± 0.4
6c	83 ± 2.4	0.90	0.57–1.21	78 ± 3.5	3.34	2.66–4.01	7 ± 0.5
7c	85 ± 5.9	0.66	0.48–0.88	77 ± 1.3	1.13	0.78–1.65	6 ± 0.5
6d	95 ± 2.6	1.38	0.97–1.96	80 ± 2.2	9.52	8.73–10.38	8 ± 0.4
7d	86 ± 3.3	1.08	0.71–1.50	79 ± 4.5	4.19	3.90–5.04	1 ± 0.1
7e	88 ± 3.4 [g]	0.19	0.14–0.26	56 ± 2.4	1.62	0.98–2.06	2 ± 0.2
6f	89 ± 0.9	3.56	2.37–5.34	60 ± 0.7 [e]	0.28	0.17–0.38	14 ± 0.8
6g	94 ± 2.5	2.05	1.76–2.38	84 ± 3.6	4.21	3.87–5.03	25 ± 1.0
7g	82 ± 3.6	0.48	0.32–0.74	72 ± 2.8	2.61	2.03–3.04	6 ± 0.3
6h	74 ± 3.4	0.32	0.21–0.48	72 ± 1.9	2.24	1.27–3.95	5 ± 0.3
7h	69 ± 1.6	0.79	0.55–1.15	71 ± 1.6	7.29	4.13–10.89	25 ± 0.7
6i	78 ± 2.6	1.41	0.90–2.00	92 ± 0.8	8.12	5.76–11.44	40 ± 2.4
7i	82 ± 0.3 [e]	0.17	0.12–0.21	73 ± 1.9 [e]	0.66	0.49–0.90	2 ± 0.1 [f]
7j	80 ± 1.9 [f]	0.57	0.42–0.78	56 ± 2.2 [g]	0.19	0.13–0.28	5 ± 0.1
7l	90 ± 3.4	0.31	0.21–0.45	86 ± 2.4	2.05	1.57–2.98	29 ± 2.4

[a] Decrease in developed tension on isolated guinea-pig left atrium driven at 1 Hz at 100 µM, expressed as percent changes from the control (n = 5–6). The 100 µM concentration gives the maximum effect for most compounds. [b] Calculated from log concentration-response curves (Probit analysis by Litchfield and Wilcoxon [22] with n = 6–7). When the maximum effect was <50%, the EC$_{50}$ ino. and EC$_{50}$ chrono. Values were not calculated. [c] Decrease in developed tension on guinea-pig spontaneously beating isolated right atrium at 100 µM, expressed as percent changes from the control (n = 7–8). The 100 µM concentration gives the maximum effect for most compounds. [d] Decrease in atrial rate on guinea-pig spontaneously beating isolated right atrium at 100 µM, expressed as percent changes from the control (n = 7–8). Pre-treatment heart rate ranged from 165 to 190 beats/min. The indicated concentration gives the maximum effect for most compounds. [e] At 10 µM. [f] At 50 µM. [g] At 5 µM. [h] At 1 µM. [i] EC$_{50}$ = 0.07 µM (c.l. 0.064–0.075).

2.2.2. Smooth Muscle Spasmolytic Activity

All compounds were tested on K^+-depolarized (80 mM) guinea-pig vascular (aortic strips) smooth muscle to assess calcium antagonist activity. Some selected compounds were tested also on guinea pig ileum to test the activity on K^+-depolarized (80 mM) non-vascular smooth muscle. Compounds were checked at increasing concentration to evaluate the percent inhibition of calcium-induced contraction on K^+-depolarized vascular (aortic strips) and non-vascular (ileum) smooth muscle. Data are presented in Table 2 together with those for diltiazem.

None of the studied compounds had spasmolytic effects on K^+-depolarized vascular smooth muscle in contrast to diltiazem, which, in analogy with the other classes of calcium antagonists, has an intrinsic activity greater than 50% and µM potency (Table 2).

Some compounds, selected on the basis of their chemical structure, have been studied for their spasmolytic effects on non-vascular smooth muscle (ileum); **5**, **6b**, **7c**, **7e**, **6g**, **7g**, **6h**, **7h**, and **7l**. **5** and **7c** have no effect in agreement with what observed on vascular smooth muscle. The other tested compounds showed intrinsic activity greater than 50% and interesting potency. In particular **6b**, **7c**, **7e**, **7h**, and **7l** are 117, 71, 121, 7, and 25 times less potent than diltiazem, respectively. Surprisingly, **6g** has a potency comparable to that of diltiazem, while **6h** is 611 times more potent than diltiazem (IC$_{50}$ = 0.00018 µM (c.l. 0.00014–0.00024); IC$_{50}$ = 0.11 µM (c.l. 0.085–0.13) respectively).

Table 2. Relaxant activity of tested compounds on K^+-depolarized guinea pig aortic strips and ileum longitudinal smooth muscle.

Compd	Aorta			Ileum		
	Activity [a] M ± S.E.M.	EC$_{50}$ [b] (µM)	95% c.l.	Activity [a] M ± S.E.M.	EC$_{50}$ [b] (µM)	95% c.l.
Diltiazem	88 ± 2.3	2.6	2.2–3.1	98 ± 1.5 [c]	0.11	0.085–0.13
5	15 ± 0.9			26 ± 2.1		
6a	3 ± 0.2					
7a	18 ± 0.9					
6b	5 ± 0.3			88 ± 2.7	12.89	10.06–15.66
7b	2 ± 0.1					
6c	4 ± 0.2					
7c	10 ± 0.1			96 ± 3.5	7.81	6.73–9.06
6d	14 ± 0.3					
7d	28 ± 1.4					
7e	17 ± 0.9			59 ± 3.9	13.38	10.12–17.68
6f	3 ± 0.2					
6g	2 ± 0.1			79 ± 1.6 [d]	0.17	0.10–0.21
7g	14 ± 0.6			38 ± 1.7		
6h	2 ± 0.1			84 ± 3.1 [e]	0.00018	0.00014–0.00024
7h	2 ± 0.1			58 ± 2.3 [d]	0.82	0.53–1.27
6i	2 ± 0.1					
7i	2 ± 0.1					
7j	1 ± 0.09					
7l	4 ± 0.2			83 ± 3.3	2.79	2.03–3.82

[a] Percent inhibition of calcium-induced contraction on K^+-depolarized (80 mM) guinea pig aortic strips (at 100 µM) and longitudinal smooth muscle (at 50 µM). The 100 µM and the 50 µM concentration gave the maximum effect for most compounds, respectively. [b] Calculated from log concentration-response curves (Probit analysis by Litchfield and Wilcoxon [22] with n = 6–7). When the maximum effect was <50%, the IC$_{50}$ values were not calculated. [c] At the 1 µM. [d] At the 10 µM. [e] At the 0.5 nM.

2.3. Antiproliferative Activity of Compounds 6 and 7

None of the compounds **6** and **7** showed a significant pharmacological antiproliferative activity against any cell targets (Table 3). In any combination of compounds and cell lines, the IC$_{50}$ were mostly much higher than 30 µM, an arbitrary limit chosen to define a significant pharmacological activity in our experimental conditions. Most importantly, these values differ by 3 orders of magnitude than the mean IC$_{50}$ for doxorubicin (0.01 µM).

Table 3. IC$_{50}$ and IC$_5$ (µM) of compounds **6** and **7** in SHSY5Y, MDA-MB-453, sensitive A2780, and multidrug resistant A2780/DX3 cells.

Compd	A2780 IC$_{50}$ [a]	A2780/DX3 IC$_{50}$	A2780/DX3 IC$_5$ [a]	MDA-MB-453 IC$_{50}$	SHSY5Y IC$_{50}$
6a	>30	>30	4.7	>30	>30
6d	>30	>30	12.7	>30	>30
6f	>30	>30	1.0	>30	>30
7a	>30	>30	3.6	>30	>30
7d	>30	>30	6.8	>30	>30
7h	>30	>30	6.4	>30	>30
7i	>30	>30	34.7	>30	>30
7l	>30	>30	1.0	>30	>30

[a] IC$_{50}$ and IC$_5$ represent the concentration inhibiting 50% and 5% of cell proliferation. Each value is expressed in µM and represents the mean of four data.

Based on the mean curves of human ovarian cancer cell line A2780/DX3, which was selected for doxorubicin, we also calculated the IC_5 parameters (Table 3). These values were used in the following experiments to evaluate the inhibition of doxorubicin efflux by ABCB1 in these cells.

2.4. Determination of Doxorubicin Accumulation by Flow Cytometry

All compounds were then analyzed for their ability to neutralize the efflux of doxorubicin in A2780/DX3 cells. These cells show a multidrug resistance linked to the over-expression of ABCB1 on the membrane (Figure 4).

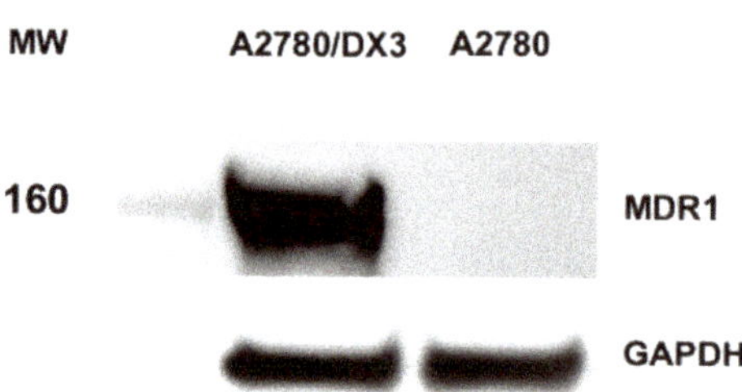

Figure 4. Differential expression of ABCB1 in sensitive A2780 and multidrug resistant A2780/DX3 cells.

The flow cytometric analysis results for the reversing activity of compounds **6** and **7**, in resistant A2780/DX3 cells, are summarized in Table 4.

Table 4. Percent doxorubicin accumulation in multidrug resistant A2780/DX3 cells after co-treatment with compounds **6** and **7**.

	Compd							
	6a	**6d**	**6f**	**7a**	**7d**	**7h**	**7i**	**7l**
% doxorubicin accumulation [a]	102 ± 12 [b]	128 ± 9	124 ± 4	104 ± 7	111 ± 5	98 ± 8	102 ± 10	103 ± 8

[a] As related to A2780/DX3 cells treated with doxorubicin alone taken as 100%. [b] Data are reported as mean ± SD.

In none of the cases did the co-treatment with our compounds cause a significant pharmacological increase of doxorubicin content in resistant cells, suggesting the absence of inhibition of ABCB1 activity.

It is important to note that in similar experimental conditions, doxorubicin accumulation in A2780 cells (not co-treated with compounds **6** and **7**) was about 4.6 ± 1.3 (460%) times higher than in equally treated resistant A2780/DX3 cells, and that in similar experimental conditions, diltiazem caused a mean increase of doxorubicin accumulation of 699 ± 75% (Standard Error) [14].

3. Discussion

Resistance to anticancer therapy is one of the main causes of therapy failure, mostly due to the activation of transport mechanisms responsible for the extrusion of the drug from the cells. Of all the transporters known at the cellular level to be responsible for MDR, Pgps are among the most important [5]. As a matter of fact, many studies have shown a direct correlation between P-glycoprotein (Pgp) increase and resistance to anticancer therapy in various human cancers [23–27]. A wide range of unrelated chemical structures are able to inhibit Pgps [28]. Among these, L-type calcium channel modulators [29,30], used in therapy for cardiovascular diseases, provided this "side effect." In particular, among the most representative chemotypes—1,4-dihydropyridines, benzothiazepines, and phenylalkylamines—the latter represents the most interesting ones [31–33]. The lead compound of the phenylalkylamine class, verapamil, was the first compound to present enormous therapeutic potential as an MDR reverting agent, but its well-known effects on the cardiovascular system prevent its clinical application. This "side effect" of calcium channel modulators cannot be related only to the ability to reduce Ca^{2+} entry into cells, but also to a direct action on Pgps [34]. In fact, verapamil directly inhibits Pgp (PM 170 kDa), while nifedipine with diltiazem binds a 65 kDa protein, thus demonstrating that calcium modulators exert the effects of MDR reverting not only by direct interaction with Pgp, but also by affecting other proteins that can indirectly alter Pgps transport/activity.

Additionally, if the calcium antagonist classes are chemically very heterogeneous, it is possible to identify some common characteristics that are necessary for MDR reverting action: aromatic ring, protonable nitrogen, high lipophilicity, and H-bond interactions. In particular, the lipophilicity is important to help stabilize bonding between drug and Pgps.

Many studies have been devoted to chemical modifications of lead compounds to increase the MDR reverting "side effect" at the expense of cardiovascular effects, therefore finding pharmacophores that are able to relegate the cardiovascular ones to the role of "side effects." In addition, we have shown that other chemotypes capable of blocking L-type calcium channels also have the same MDR reverting action and that, based on chemical modifications, it is possible to modulate their activity [30]. With this in mind and having long been involved in the study of new chemotypes as calcium channel ligands, we started from this ground to find new MDR reverting candidates [31], demonstrating that appropriate structural changes were requested to increase this action to the detriment of the cardiovascular system. In line with previous studies, we have now examined a new chemotype: 4-nitrothiopyran *S,S*-dioxide (**5**).

In compound **5**, the contemporary presence of two pharmacophores, such as the nitro (several cardiovascular drugs contain the nitrogroup: 1,4-dihydropyridines, such as nifedipine and nicardipine, are just two examples) and the sulphonyl groups, seemed a good premise for opening the way to the discovery of new leads with CEB activity and for evidencing new structure-activity relationships in this field. Biological characterization of **5** has not been able to confirm any significant cardiovascular activity. For this reason, considering that in all of the previously studied compounds [9–11,15,35] condensed homo- or hetero-cycles are present, we have turned our attention to the benzocondensed sulfur heterocycles **6** and **7**.

Regarding cardiac parameters, all thiochromans *S,S*-dioxide **6** and **7** are devoid of chronotropic effects, while all derivatives have negative inotropic activity both on the electrically stimulated left atrium and on the spontaneous beating right atrium. In general, 2 phenyl derivatives are more potent than the corresponding 2-H compounds (Figure 5). The most promising pharmacophores for negative inotropic activity are the *p*-methoxyphenyl (**7e**) and the phenyl (**7a**) substituted. None of the studied compounds has spasmolytic effects on vascular smooth muscle. On the contrary, the compounds selected for action on non-vascular smooth muscle all have intrinsic activity greater than 50% except for compound **5**. The spasmolytic activity is ca. 2 times higher for the 3-aryl-4-nitrothiocromans *S,S*-dioxide (**6**), compared to 3-aryl-4-nitro-2-phenylthiochromans *S,S*-dioxide (**7**). The action increases if the phenyl in 3 carries an electron-withdrawing group in *meta* or *para* position.

In no case were our selected myocardial L-Type calcium channel modulators **6** and **7** able to significantly inhibit the proliferation of tumor cells, and they did nothing to hamper the ABCB1 activity at low equitoxic concentration.

In fact, compounds **6d** and **6f** were able only to slightly improve doxorubicin accumulation in multidrug resistant A2780/DX3 cells (28% and 24%, respectively), as shown by our determinations by flow cytometry. Although significant on a statistical point of view ($p < 0.05$) our results are not pharmacologically so relevant, our target being an increase of doxorubicin accumulation of at least 400–500% for reaching a comparable rate of doxorubicin accumulation in A2780 cells, the sensitive parental cell line. Thus, it is worth further investigation of **6d** and **6f** in order to enhance their potentiation of intracellular doxorubicin accumulation using molecular docking analysis and drug modeling methods [36], and in order to possibly synthesize new and more active molecules starting from these lead compounds.

In conclusion, on the basis of the preliminary data here reported, we can state that derivatives **6** and **7** are able to modulate L-Type calcium channels, in particular those of non-vascular smooth muscle, while for derivatives of chemotype **6**, there is little selectivity in the MDR reverting action (Figure 6). Appropriate chemical modifications on the scaffold **6** could lead to more interesting results in MDR reverting activities.

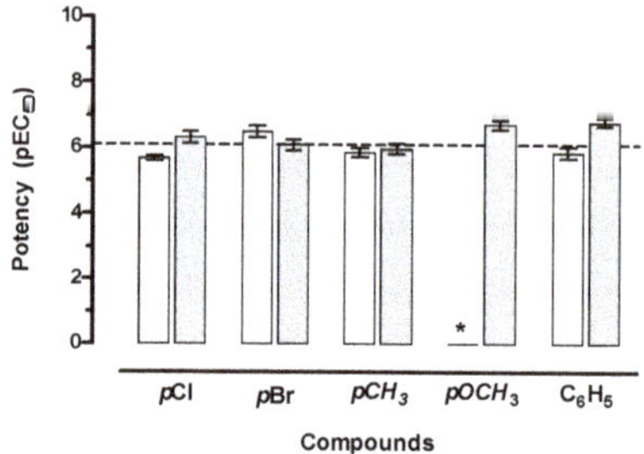

Figure 5. Potency (pEC$_{50}$) of **6** (R′ = H; white) and **7** (R′ = Ph; grey) derivative compounds on guinea pig left atria driven at 1 Hz. The dotted lines show the reference value of negative inotropic potency for diltiazem (pEC$_{50}$). * The *p*-MeO of **6** has not been synthesized.

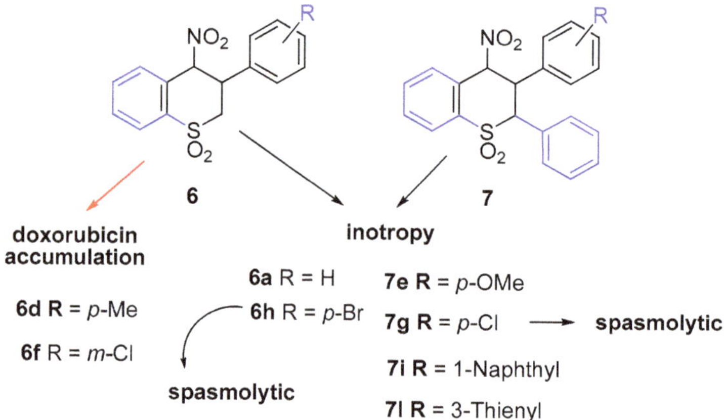

Figure 6. Highlights in structure/activity relationship.

4. Materials and Methods

4.1. Chemistry

The preparation and characterization of thiopyrans *S,S*-dioxide **5** [16], and of the thiochromans *S,S*-dioxide **6** and **7** [17,18], has been already reported.

4.2. Ex Vivo Studies

The functional profile of compounds **5**, **6**, and **7** was studied on guinea-pig isolated left and right atria to assess their inotropic and/or chronotropic effects, respectively, and on K$^+$-depolarized (80 mM) guinea-pig vascular aortic strips to assess calcium antagonist activity. Relevant details have been previously reported [37]. All compounds were checked at increasing concentration to evaluate the percent decrease of developed tension on the isolated left atrium driven at 1 Hz (negative inotropic activity), and the percent decrease in the atrial rate on the spontaneously beating right atrium (negative chronotropic activity). For compounds lacking negative chronotropic effects, the inotropy was also checked in the spontaneously beating right atria. Selected compounds were studied on the K^+-depolarized ileum smooth muscle strips to evaluate percent inhibition of calcium-induced contraction.

Data were analyzed using Student's t-test and are presented as mean ± S.E.M. [22]. Since the drugs were added in cumulative manner, the difference between the control and the experimental values at each concentration were tested for a P value < 0.05. The potency of drugs defined as EC_{50} was evaluated from log concentration-response curves (Probit analysis using Litchfield and Wilcoxon [37] or GraphPad Prism® software [38,39]) in the appropriate pharmacological preparations. When the compounds were devoid of negative chronotropic activity, the inotropic activity was checked for the spontaneously beating right atria using diltiazem as reference compounds.

Experimental values at each concentration were tested for a P value < 0.05. The potency of drugs defined as IC_{50} was evaluated from log concentration-response curves as previously described in Section 2.2.1.

4.3. In Vitro Studies

4.3.1. Chemicals

Doxorubicin in clinical formulation was obtained from Pfizer Italia (Milano, Italy) and diluted in normal saline solution to opportune concentrations. All examined compounds **6** (**a, d, f**) and **7** (**a, d, h, l**), except **7i**, were dissolved in 100% dimethylsulfoxide (DMSO) and then diluted in fetal calf serum, in order to obtain a final 1% DMSO concentration (0.1% on cells). **7i** was dissolved to a final 3% DMSO concentration (0.3% on cells).

4.3.2. Cell Lines

Tumor cell lines SHSY5Y (neuroblastoma), and A2780 (ovary) sensitive to doxorubicin and multidrug-resistant A2780/DX3 cells (provided by Dr. Y.M. Rustum and obtained with exposure to increasing concentrations of doxorubicin), were cultured in an RPMI 1640 medium plus 10% fetal calf serum, 1% glutamine, and 1% penicillin-streptomycin (complete medium). Resistant cells were maintained in a presence of doxorubicin 0.1 µM, removed 3–6 days before experiments. The MDA-MB-453 (breast) cell line was cultured in DMEM medium containing 10% fetal calf serum, 1% glutamine, 1% non-essential amino-acids, and 1% penicillin-streptomycin (complete medium).

4.3.3. MTT Assay

Cell lines were plated at opportune densities/well in 96-well microtiter plates for 6–8 h. To evaluate the concentrations of compounds **6** and **7** that were able to inhibit 5% and 50% (IC_5 and IC_{50}) cell growth, they were administered in 5 different concentrations (10-fold serial dilutions, starting from the maximum concentration of 100 µM); the maximal final volume/well was 200 µL. After three days, 50 µL of MTT (Sigma, St. Louis, MO, USA) solution (2 mg/mL in PBS) was added to each well, and microtiter plates were incubated at 37 °C for 4 h. Thereafter, the culture medium was carefully aspirated; 100 µL of 100% DMSO was added and incubated for 20 min, and a complete and homogeneous solubilization of formazan crystals was achieved by shaking the well contents.

The absorbance was measured with a Microplate Reader iMark (Bio-Rad Laboratories, Milano, Italy) at 595 nm. IC_5s (compound concentration inhibiting 5% cell proliferation) and IC_{50}s (compound concentration inhibiting 50% cell proliferation) were calculated with the analysis of single dose response curves basis. Each experiment was repeated four times.

4.3.4. Study of Doxorubicin Intracellular Accumulation

Multidrug-resistant A2780/DX3 cells were plated in 25 cm^2 flasks at 1.5×10^6 cells/flask in 10 mL. Eighteen hours later, the cells reached about 65–75% confluence and were treated with **6** and **7** compounds (final volume: 10 mL). The concentration used for each compound was five times higher than the specific IC_5. After 2 h, a small volume of doxorubicin (29 µL) was added to reach a final concentration of 10 µM, and the incubation was prolonged for other 2 h. After removing the medium, cells were washed once with cold phosphate buffered saline (PBS) in a flask, and then were harvested by trypsin at 37 °C for 5 min, and washed again quickly with cold PBS. Pelleted cells were then fixed at 4° C for 24 h with 3.7% paraformaldehyde in PBS containing 2% sucrose. Untreated (negative control) and positive control cells treated without or with doxorubicin were assayed as well.

The intensity of intracellular fluorescence was assayed with flow cytometry (FACScan, BD Biosciences, Milano, Italy) using 488 nm excitation and 575 nm bandpass filter for doxorubicin detection. Values were expressed in arbitrary units as mean fluorescence intensity (MFI), while doxorubicin accumulation was calculated as percentage of positive controls as follow (Equation (1)):

$$\text{(MFI treated} - \text{MFI negative control)/MFI positive control} - \text{MFI negative control.} \tag{1}$$

Author Contributions: Conceptualization, R.B. and D.S.; formal analysis, M.M. (Matteo Micucci), R.G. and C.T.; investigation, M.M. (Massimo Maccagno), M.M. (Matteo Micucci), R.G., M.V., C.T. and L.B.; data curation, M.M. (Matteo Micucci), R.G. and L.B.; writing—original draft preparation, R.B., L.B. and M.V.; writing—review and editing, R.B., M.F., M.V. and A.C.; visualization, M.M. (Matteo Micucci) and M.V.; supervision, D.S. and A.C.; project administration, R.B.; funding acquisition, R.B. All authors have read and agreed to the published version of the manuscript.

Funding: This research was funded by University of Bologna, grant number RFO 2018.

Acknowledgments: A special thank to Silvia Marconi for the preparation of ABCB1 western blot.

Conflicts of Interest: The authors declare no conflict of interest.

References

1. Gillet, J.P.; Gottesman, M.M. Mechanisms of multidrug resistance in cancer. *Methods Mol. Biol.* **2010**, *596*, 47–76. [PubMed]

2. Trédan, O.; Galmarini, C.M.; Patel, K.; Tannock, I.F. Drug resistance and the solid tumor microenvironment. *J. Natl. Cancer Inst.* **2007**, *99*, 1441–1454.

3. Dagogo-Jack, I.; Shaw, A.T. Tumour heterogeneity and resistance to cancer therapies. *Nat. Rev. Clin. Oncol.* **2018**, *15*, 81–94. [CrossRef] [PubMed]

4. Beck, W.T. Strategies to circumvent multidrug resistance due to P-glycoprotein or to altered DNA topoisomerase II. *Bull. Cancer* **1990**, *77*, 1131–1141. [PubMed]

5. Breier, A.; Barancík, M.; Sulová, Z.; Uhrík, B. P-glycoprotein–implications of metabolism of neoplastic cells and cancer therapy. *Curr. Cancer Drug Targets* **2005**, *6*, 457–468. [CrossRef]

6. Kessel, D.; Botterill, V.; Wodinsky, I. Uptake and retention of daunomycin by mouse leukemic cells as factors in drug response. *Cancer Res.* **1968**, *28*, 938–941.

7. Carosati, E.; Cosimelli, B.; Ioan, P.; Severi, E.; Katneni, K.; Chiu, F.C.; Saponara, S.; Fusi, F.; Frosini, M.; Matucci, R.; et al. Understanding Oxadiazolothiazinone Biological Properties: Negative Inotropic Activity versus Cytochrome P450-Mediated Metabolism. *J. Med. Chem.* **2016**, *59*, 3340–3352. [CrossRef]

8. Leoni, A.; Frosini, M.; Locatelli, A.; Micucci, M.; Carotenuto, C.; Durante, M.; Budriesi, R. 4-Imidazo[2,1-*b*] Thiazole-1,4-DHPs and Neuroprotection: Preliminary Study in Hits Searching. *Eur. J. Med. Chem.* **2019**, *169*, 89–102. [CrossRef]

9. Budriesi, R.; Cosimelli, B.; Ioan, P.; Lanza, C.Z.; Spinelli, D.; Chiarini, A. Cardiovascular characterization of [1,4]thiazino[3,4-*c*][1,2,4]oxadiazol-1-one derivatives: Selective myocardial calcium channel modulators. *J. Med. Chem.* **2002**, *45*, 3475–3481. [CrossRef]

10. Dustan, H.P. Calcium channel blockers. Potential medical benefits and side effects. *Hypertension* **1989**, *13*, I137–I140. [CrossRef]

11. Corelli, F.; Manetti, F.; Tafi, A.; Campiani, G.; Nacci, V.; Botta, M. Diltiazem-like calcium entry blockers: A hypothesis of the receptor-binding site based on a comparative molecular field analysis model. *J. Med. Chem.* **1997**, *40*, 125–131. [CrossRef] [PubMed]

12. Tsuruo, T.; Iida, H.; Tsukagoshi, S.; Sakurai, Y. Potentiation of vincristine and Adriamycin effects in human hemopoietic tumor cell lines by calcium antagonists and calmodulin inhibitors. *Cancer Res.* **1983**, *43*, 2267–2272. [PubMed]

13. Tsuruo, T.; Iida, H.; Tsukagoshi, S.; Sakurai, Y. Vincristine-resistant P388 leukemia cells contain a large amount of calcium. *GANN Jpn. J. Cancer Res.* **1983**, *74*, 619–621. [PubMed]

14. Viale, M.; Cordazzo, C.; Cosimelli, B.; de Totero, D.; Castagnola, P.; Aiello, C.; Severi, E.; Petrillo, G.; Cianfriglia, M.; Spinelli, D. Inhibition of MDR1 activity in vitro by a novel class of diltiazem analogues: Toward new candidates. *J. Med. Chem.* **2009**, *52*, 259–266. [CrossRef] [PubMed]

15. Budriesi, R.; Carosati, E.; Chiarini, A.; Cosimelli, B.; Cruciani, G.; Ioan, P.; Spinelli, D.; Spisani, R. A new class of selective myocardial calcium channel modulators. 2. Role of the acetal chain in oxadiazol-3-one derivatives. *J. Med. Chem.* **2005**, *48*, 2445–2456. [CrossRef] [PubMed]

16. Bianchi, L.; Maccagno, M.; Petrillo, G.; Rizzato, E.; Sancassan, F.; Severi, E.; Spinelli, D.; Stenta, M.; Galatini, A.; Tavani, C. A new route to thiopyran *S,S*-dioxide derivatives via an overall ring-enlargement protocol from 3-nitrothiophene. *Tetrahedron* **2009**, *65*, 336–343. [CrossRef]

17. Bianchi, L.; Dell'Erba, C.; Maccagno, M.; Morganti, S.; Novi, M.; Petrillo, G.; Rizzato, E.; Sancassan, F.; Severi, E.; Spinelli, D.; et al. Easy access to 4-nitrothiochroman *S,S*-dioxides via ring-enlargement from 3-nitrobenzo[*b*]thiophene. *Tetrahedron* **2004**, *60*, 4967–4973. [CrossRef]

18. Bianchi, L.; Maccagno, M.; Petrillo, G.; Rizzato, E.; Sancassan, F.; Spinelli, D.; Tavani, C. Access to 2,3-diaryl-4-nitrothiochroman *S,S*-dioxides from 3-nitrobenzo[*b*]thiophene. *Tetrahedron* **2011**, *67*, 8160–8169. [CrossRef]

19. Dell'Erba, C.; Spinelli, D.; Leandri, G. Ring-opening Reaction in the Thiophen Series: Reaction between 3,4-Dinitrothiophen and Secondary Amines. *Chem. Commun.* **1969**, *10*, 549–550.

20. Bianchi, L.; Maccagno, M.; Petrillo, G.; Sancassan, F.; Spinelli, D.; Tavani, C. 2,3-Dinitro-1,3-butadienes: Versatile building-blocks from the ring opening of 3,4-dinitrothiophene. In *Targets in Heterocyclic Systems: Chemistry and Properties*; Attanasi, O.A., Spinelli, D., Eds.; Società Chimica Italiana: Rome, Italy, 2006; Volume 10, pp. 1–23.

21. Bianchi, L.; Maccagno, M.; Petrillo, G.; Rizzato, E.; Sancassan, F.; Severi, E.; Spinelli, D.; Tavani, C.; Viale, M. Versatile nitrobutadienic building-blocks from the ring-opening of 2- and 3-nitrothiophenes. In *Targets in Heterocyclic Systems: Chemistry and Properties*; Attanasi, O.A., Spinelli, D., Eds.; Società Chimica Italiana: Rome, Italy, 2007; Volume 11, pp. 1–20.

22. Tallarida, R.J.; Murray, R.B. *Manual of Pharmacologic Calculations with Computer Programs*, 2nd ed.; Springer: New York, NY, USA, 1987.

23. Hirose, M.; Hosoi, E.; Hamano, S.; Jalili, A. Multidrug Resistance in Hematological Malignancy. *J. Med. Investig.* **2003**, *50*, 126–135.

24. Sonneveld, P. Multidrug Resistance in Haematological Malignancies. *J. Intern. Med.* **2000**, *247*, 521–534. [CrossRef] [PubMed]

25. Kaye, S.B. Multidrug Resistance: Clinical Relevance in Solid Tumours and Strategies for Circumvention. *Curr. Opin. Oncol.* **1998**, *10*, S15–S19. [PubMed]

26. Ren, F.; Shen, J.; Shi, H.; Hornicek, F.J.; Kan, Q.; Duan, Z. Novel mechanisms and approaches to overcome multidrug resistance in the treatment of ovarian cancer. *Biochim. Biophys. Acta* **2016**, *1866*, 266–275. [CrossRef] [PubMed]

27. Liu, T.; Li., Z.; Zhang, Q.; De Amorim Bernstein, K.; Lozano-Calderon, S.; Choy, E.; Hornicek, F.J.; Duan, Z. Targeting ABCB1 (MDR1) in multi-drug resistant osteosarcoma cells using the CRISPR-Cas9 system to reverse drug resistance. *Oncotarget* **2016**, *50*, 83502–83513. [CrossRef] [PubMed]

28. Breier, A.; Drobná, Z.; Docolomansky, P.; Barancik, M. Cytotoxic activity of several unrelated drugs on L1210 mouse leukemic cell sublines with P-glycoprotein (PGP) mediated multidrug resistance (MDR) phenotype. A QSAR study. *Neoplasma* **2000**, *2*, 100–106.

29. Triggle, D.J. Calcium channel antagonists: Clinical uses: Past, present and future. *Biochem. Pharm.* **2007**, *74*, 1–9. [CrossRef]

30. Palmeira, A.; Sousa, E.; Vasconcelos, M.T.; Pinto, M.M. Tree decades of P-gp inhibitors: Skimming through several generations and scaffolds. *Curr. Med. Chem.* **2012**, *19*, 1946–2025. [CrossRef]

31. Spinelli, D.; Budriesi, R.; Cosimelli, B.; Severi, E.; Micucci, M.; Baroni, M.; Fusi, F.; Joan, P.; Cross, S.; Frosini, M.; et al. Playing with Opening and Closing of Heterocycles: Using the Cusmano-Ruccia Reaction to Develop a Novel Class of Oxadiazolothiazinones, Active as Calcium Channel Modulators and P-Glycoprotein Inhibitors. *Molecules* **2014**, *19*, 16543–16572. [CrossRef]

32. Viale, M.; Cordazzo, C.; de Torero, D.; Budriesi, R.; Rosano, C.; Leoni, A.; Ioan, P.; Aiello, C.; Croce, M.; Andreani, A.; et al. Inhibition of MDR1 activity and induction of apoptosis by analogues of nifedipine and diltiazem: An in vitro analysis. *Investig. New Drugs* **2011**, *29*, 98–109. [CrossRef]

33. Teodori, E.; Dei, S.; Quidu, P.; Budriesi, R.; Chiarini, A.; Garnier-Suillerot, A.; Gualtieri, F.; Manetti, D.; Romanelli, M.N.; Scapecchi, S. Design, synthesis, and in vitro activity of catamphiphilic reverters of multidrug resistance: Discovery of a selective, 37highly efficacious chemosensitizer with potency in the nanomolar range. *J. Med. Chem.* **1999**, *42*, 1687–1697. [CrossRef]

34. Sulová, Z.; Seres, M.; Barancík, M.; Gibalová, L.; Uhrík, B.; Poleková, L.; Breier, A. Does any relationship exist between P-glycoprotein-mediated multidrug resistance and intracellular calcium homeostasis. *Gen. Physiol. Biophys.* **2009**, *28*, F89–F95.

35. Carosati, E.; Cruciani, G.; Chiarini, A.; Budriesi, R.; Ioan, P.; Spisani, R.; Spinelli, D.; Cosimelli, B.; Fusi, F.; Frosini, M.; et al. Calcium channel antagonists discovered by a multidisciplinary approach. *J. Med. Chem.* **2006**, *49*, 5206–5216. [CrossRef] [PubMed]

36. Rosano, C.; Viale, M.; Cosimelli, B.; Strict, E.; Gangemi, R.; Ciogli, A.; De Totero, D.; Spinelli, D. ABCB1 structural models, molecular docking, and synthesis of new oxadiazolothiazin-3-one inhibitors. *ACS Med. Chem. Lett.* **2013**, *4*, 694–698. [CrossRef] [PubMed]

37. Bisi, A.; Micucci, M.; Gobbi, S.; Belluti, F.; Budriesi, R.; Rampa, A. Cardiovascular Profile of Xanthone-Based 1,4 Dihydropyridines Bearing a Lidoflazine Pharmacophore Fragment. *Molecules* **2018**, *23*, 3088. [CrossRef] [PubMed]

38. Motulsky, H.; Christopoulos, A. *Fitting Models to Biological Data Using Linear and Non Linear Regression: A Practical Guide to Curve Fitting*; Oxford University Press: New York, NY, USA, 2003.

39. Motulsky, H.J. *Prism 5 Statistics Guide*; GraphPad Software Inc.: San Diego, CA, USA, 2007.

Sample Availability: Samples of the compounds are available from the authors.

Article

Synthesis and Evaluation of Anticancer Activities of Novel C-28 Guanidine-Functionalized Triterpene Acid Derivatives

Anna Spivak *, Rezeda Khalitova, Darya Nedopekina, Lilya Dzhemileva *, Milyausha Yunusbaeva, Victor Odinokov, Vladimir D'yakonov and Usein Dzhemilev

Institute of Petrochemistry and Catalysis, Russian Academy of Sciences, 141 Prospekt Oktyabrya, Ufa 450075, Russia; rezedamuf@yandex.ru (R.K.); dashana25@gmail.com (D.N.); milyausha_ufa@mail.ru (M.Y.); odinokov@anrb.ru (V.O.); DyakonovVA@gmail.com (V.D.); dzhemilev@anrb.ru (U.D.)
* Correspondence: spivak.ink@gmail.com (A.S.); dzhemilev@mail.ru (L.D.);
  Tel.: +7-917-421-71-06 (A.S.); +7-917-458-47-29 (L.D.)

Academic Editors: Carla Boga and Gabriele Micheletti
Received: 8 October 2018; Accepted: 13 November 2018; Published: 16 November 2018

Abstract: Triterpene acids, namely, 20,29-dihydrobetulinic acid (BA), ursolic acid (UA) and oleanolic acid (OA) were converted into C-28-amino-functionalized triterpenoids **4–7, 8a, 15, 18** and **20**. These compounds served as precursors for the synthesis of novel guanidine-functionalized triterpene acid derivatives **9b–12b, 15c, 18c** and **20c**. The influence of the guanidine group on the antitumor properties of triterpenoids was investigated. The cytotoxicity was tested on five human tumor cell lines (Jurkat, K562, U937, HEK, and Hela), and compared with the tests on normal human fibroblasts. The antitumor activities of the most tested guanidine derivatives was lower, than that of corresponding amines, but triterpenoids with the guanidine group were less toxic towards human fibroblasts. The introduction of the tris(hydroxymethyl)aminomethane moiety into the molecules of triterpene acids markedly enhanced the cytotoxic activity of the resulting conjugates **15, 15c, 18b,c** and **20b,c** irrespective of the triterpene skeleton type. The dihydrobetulinic acid amine **15**, its guanidinium derivative **15c** and guanidinium derivatives of ursolic and oleanolic acids **18c** and **20c** were selected for extended biological investigations in Jurkat cells, which demonstrated that the antitumor activity of these compounds is mediated by induction of cell cycle arrest at the S-phase and apoptosis.

Keywords: triterpenoids; betulinic acid; ursolic acid; oleanolic acid; amino group; guanidine group; cytotoxicity; apoptosis; cell cycle

---

1. Introduction

Among natural products of plant origin that are considered as abundant sources of lead structures for the discovery of new drugs, the pentacyclic lupane, ursane, and oleanane triterpenoids occupy a prominent place [1,2]. Triterpene acids (betulinic, ursolic, and oleanolic acids, Figure 1) are of interest for pharmacological research, as they exhibit a variety of biological activities including antimicrobial, antiparasitic, antitumor, and antiviral, in particular, anti-HIV, types of activity [3–6]. Among these properties of triterpenoids, of special interest is their anticancer activity and the ability to trigger the mitochondrial apoptosis pathway in various types of human cancer cells [5–9]. Thus, betulinic acid is capable of inducing apoptosis in tumor cells, such as melanoma, adenocarcinoma, neuroblastoma, medulloblastoma, glioblastoma and neuroectodermal tumors [7–9]. The *in vivo* anticancer activity of betulinic acid was identified using xenograft models [10,11]. The ursolic acid can also induce apoptosis, autophagy, and cell cycle arrest through various pathways, such as inhibition of DNA replication, stimulation of reactive oxygen species (ROS) production, and affecting the balance between proapoptotic and antiapoptotic proteins [6,12,13].

The useful pharmacological properties of triterpene acids are successfully combined with their acceptable systemic toxicity towards animals. However, the relatively low anticancer potential and high hydrophobicity of these secondary metabolites markedly hamper their advancement as anticancer drug candidates. For this reason, active search is in progress for analogues of natural triterpenoids with a higher biological potential and enhanced pharmacological characteristics (hydrophilicity, bioavailability) [8,14,15]. It has been shown [16–25] that conversion of triterpene compounds to cationic derivatives such as quaternary ammonium [16,17], pyridinium [18,19] or triphenylphosphonium salts [20–25] may serve as an efficient approach to improving bioavailability and selectivity of their biological action. Our recent study has shown that triphenylphosphonium derivatives of betulinic and ursolic acids are substantially superior over their prototypes in the antitumor activity and in the triggering mitochondria-dependent apoptosis of cancer cells [24,25].

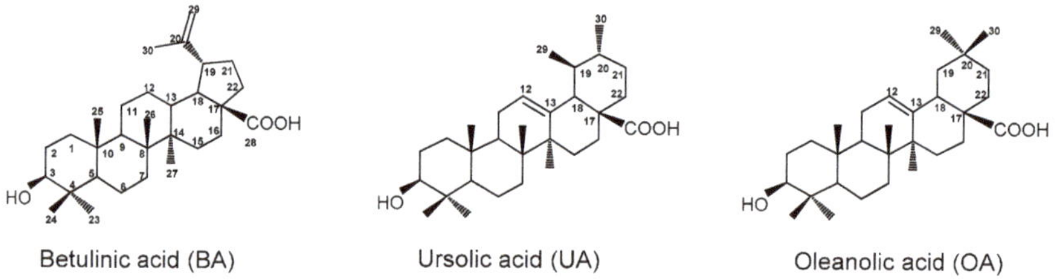

Figure 1. Betulinic, ursolic and oleanolic acids.

However, the cytotoxic activity of the phosphonium salts was comparable with their cytotoxic activity against normal peripheral blood cells. In continuation of the search for efficient and selective antitumor agents, we have investigated novel cationic derivatives of pentacyclic triterpenoids containing guanidine groups, which are readily protonated at a physiological pH level. The introduction of hydrophilic guanidine groups into hydrophobic triterpene acid molecules may enhance their transmembrane transport and physicochemical characteristics. Meanwhile, the new hybrid molecules may preserve the selectivity of cytotoxic action against normal cells inherent in the natural triterpene acids. The guanidine group is a common key unit in various natural and synthetic compounds demonstrating antimicrobial, antiviral, and antitumor activities [26]. High symmetry of the Y-shaped guanidinium group promotes the formation of two parallel hydrogen bonds with the biologically relevant counterparts. Unlike ammonium groups, in which the charge is localized on one nitrogen atom (hard cations), guanidinium groups with a delocalized charge actively interact through hydrogen bonds with soft ions such as phosphates and sulfates. This feature of the guanidinium cation induce the efficient transport of biologically active substances through liposomal and cell membranes [27–29]. Furthermore, because of high basicity (pKa 13.5), the guanidinium group is important for selective delivery of cytotoxic molecules to tumor cells. Guanidine derivatives can be accumulated in the mitochondria of tumor cells, thus destroying the mitochondrial potential and inhibiting the mitochondrial respiratory chain [29,30].

Polyamines, which are precursors of aminoalkylguanidines, are also used to develop chemotherapeutic agents, including antibacterial and antitumor compounds [31,32]. Structurally, polyamine molecules contain positively charged nitrogen atoms at physiological pH value and can serve as electrostatic bridges between negatively charged phosphates. They are able to bind to negatively charged DNA macromolecules. However, some of physiological diamines, polyamines, and their synthetic analogues have exhibited high toxicity toward normal cells. A large body of data has now been accumulated on the biological activity of polyaminosterols, among which squalamine, trodusquemine, and their synthetic analogues are best known [33–36]. The synthesis and biological properties of polyamino triterpene acids are described in several publications [6,37–40]; the effect of introduction of the guanidine group into triterpenoid molecules has not been studied so far. Here we describe the synthesis and comparative evaluation of the cytotoxic and apoptosis-inducing activities of new guanidine derivatives of pentacyclic lupane, ursane, and oleanane triterpenoids and their precursors—C-28 conjugates of triterpene acids with some linear and branched mono-, di-, and triaminoalkanes.

2. Results and Discussion

2.1. Chemistry

While synthesizing the target compounds, we found that the Boc-deprotection of guanidine derivatives of betulinic and betulonic acids in acid medium (50% TFA in CH_2Cl_2) is complicated by skeletal rearrangements of the lupane skeleton. It is known from the literature [41,42] that hydrogenation of the C-20 double bond of lupane triterpenoids does not considerably affect their cytotoxic activity and selectivity between normal and tumor cell lines; therefore, in the subsequent experiments, we used 20,29-dihydrobetulinic and 20,29-dihydrobetulonic acids **1** and **2** to prepare target compounds **9b–12b** and **14** (Scheme 1).

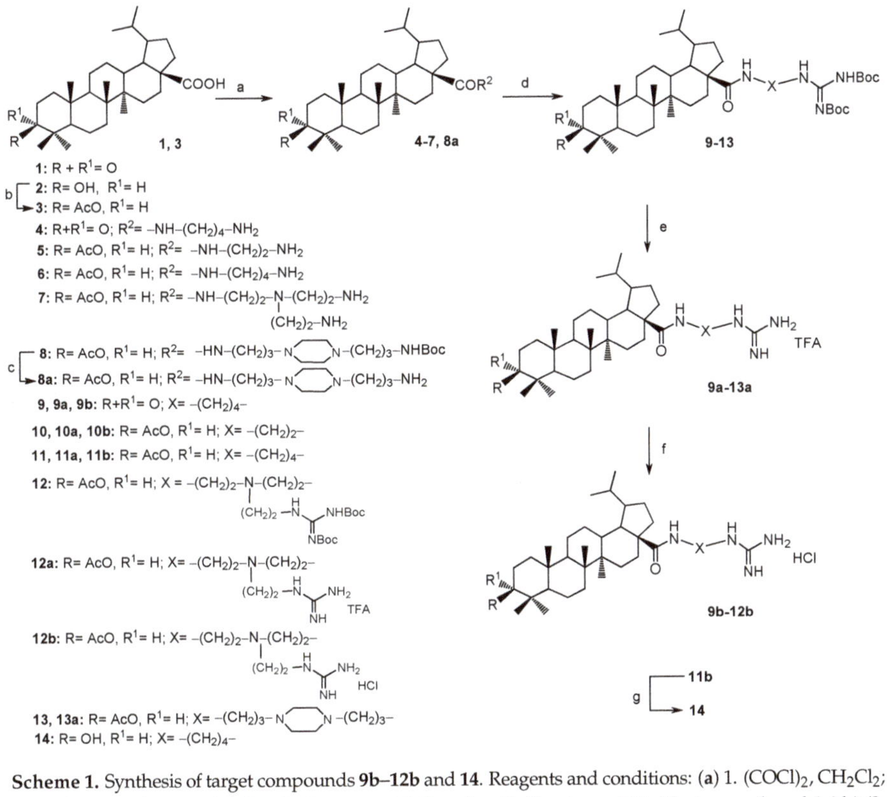

1: R + R^1= O

b 2: R= OH, R^1= H
3: R= AcO, R^1= H

4: R+R^1= O; R^2= –NH–(CH$_2$)$_4$–NH$_2$

5: R= AcO, R^1= H; R^2= –NH–(CH$_2$)$_2$–NH$_2$

6: R= AcO, R^1= H; R^2= –NH–(CH$_2$)$_4$–NH$_2$

7: R= AcO, R^1= H; R^2= –NH–(CH$_2$)$_2$–N–(CH$_2$)$_2$–NH$_2$, (CH$_2$)$_2$–NH$_2$

8: R= AcO, R^1= H; R^2= –HN–(CH$_2$)$_3$–N(piperazine)N–(CH$_2$)$_3$–NHBoc

8a: R= AcO, R^1= H; R^2= –HN–(CH$_2$)$_3$–N(piperazine)N–(CH$_2$)$_3$–NH$_2$

9, 9a, 9b: R+R^1= O; X= –(CH$_2$)$_4$–

10, 10a, 10b: R= AcO, R^1= H; X= –(CH$_2$)$_2$–

11, 11a, 11b: R= AcO, R^1= H; X= –(CH$_2$)$_4$–

12: R= AcO, R^1= H; X = –(CH$_2$)$_2$–N–(CH$_2$)$_2$–, (CH$_2$)$_2$–N(H)–C(=NBoc)–NHBoc

12a: R= AcO, R^1= H; X= –(CH$_2$)$_2$–N–(CH$_2$)$_2$–, (CH$_2$)$_2$–N(H)–C(=NH)–NH$_2$ · TFA

12b: R= AcO, R^1= H; X= –(CH$_2$)$_2$–N–(CH$_2$)$_2$–, (CH$_2$)$_2$–N(H)–C(=NH)–NH$_2$ · HCl

13, 13a: R= AcO, R^1= H; X= –(CH$_2$)$_3$–N(piperazine)N–(CH$_2$)$_3$–

14: R= OH, R^1= H; X= –(CH$_2$)$_4$–

Scheme 1. Synthesis of target compounds **9b–12b** and **14**. Reagents and conditions: (**a**) 1. (COCl)$_2$, CH$_2$Cl$_2$; 2. 1,2-diaminoethane or 1,4-diaminobutane or tris(2-aminoethyl)amine or *N-tert*-butoxy-carbonyl-1,4-bis(3-aminopropyl)piperazine, Et$_3$N, CH$_2$Cl$_2$, r.t.; (**b**) AcCl, THF, Py, DMAP, r.t.; (**c**) 10% TFA, CH$_2$Cl$_2$, 3 h, r.t.; (**d**) 1,3-di-Boc-2-(trifluoromethylsulfonyl)guanidine, Et$_3$N, CH$_2$Cl$_2$; (**e**) 50% TFA, CH$_2$Cl$_2$, 2–4 h, r.t.; (**f**) 5 M HCl, MeOH; (**g**) 4 N NaOH, MeOH, THF.

Acetate **3** obtained upon hydroxyl group protection in dihydrobetulinic acid **2**, and dihydrobetulonic acid **1** were converted to C-28 amide derivatives **4–7** and **8a** via relatively unstable acid chlorides (Scheme 1).

The reactions were carried out with 1,2-diaminoethane, 1,4-diaminobutane (putrescine), tris(2-aminoethyl)amine, and *N-tert*-butyloxycarbonyl-1,4-bis(3-aminopropyl)piperazine. The use of branched triamine resulted in two guanidine functions being introduced into the triterpenoid molecule.

Amides **4–7** were synthesized using a 3-fold molar excess of amines over triterpenes in order to avoid the formation of dimeric products. Compound **8a** was formed in a good yield only with the use of *N*-Boc bis-aminopropylpiperazine, which was synthesized as described in reference [43]. The *N*-Boc protection was then removed by treatment with 10% trifluoroacetic acid in CH_2Cl_2. Guanylation of amines **4–7** and **8a** was carried out according to the typical procedure [44] by using commercially available *N,N′*-di-Boc-*N″*-triflylguanidine. The expected compounds **9–13** were obtained in 60–88% yields after column chromatography on SiO_2. The subsequent treatment of Boc derivatives **9–13** with CF_3COOH and then with HCl/MeOH gave hydrochlorides **9b–13b**. Dihydrobetulinic acid guanidinium hydrochloride **14** was prepared by saponification of the 3-OAc function in conjugate **11b** A well studied and promising betulinic acid derivative is 2-amino-3-hydroxy-2-(hydroxymethyl)propyl 3-*O*-acetyl betulinate, known as the anticancer agent NVX-207 [45–48]. We have synthesized its C-20,29 hydrogenated derivative **15**. Ursolic and oleanolic acetates **16** and **17** were converted into NVX-207 analogues **18** and **20** with ursane and oleanane skeletons (Scheme 2).

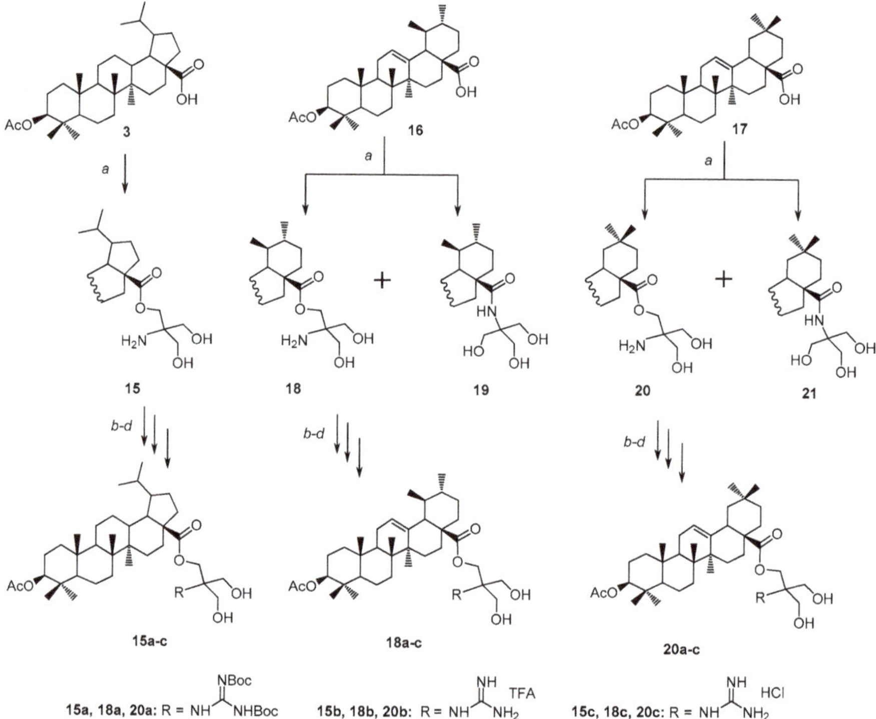

Scheme 2. Synthesis of compounds **15a–c**, **18a–c**, **20a–c**. *Reagents and conditions*: (**a**) 1. $(COCl)_2$, CH_2Cl_2; 2. tris(hydroxymethyl)aminomethane, DMAP, Py, CH_2Cl_2; (**b**) 1,3-di-Boc-2-(trifluoromethyl-sulfonyl) guanidine, Et_3N, $CHCl_3$, reflux; (**c**) 50% TFA, CH_2Cl_2, 2–4 h, r.t.; (**d**) 5M HCl, MeOH.

The reactions under reported conditions [45] afforded the desired esters **15**, **18**, and **20** containing the terminal NH_2 group in the branched C-28 chain and amides **19** and **21** in 15–23% and 24–32% yields, respectively. Guanylation of the free amino groups in the resulting amines **15**, **18**, and **20** on treatment with *N,N′*-di-Boc-*N″*-triflylguanidine gave rise to guanidine derivatives **15a**, **18a**, and **20a** in 63–79% yields. The Boc-deprotection with 50% TFA/CH_2Cl_2 and the subsequent treatment of salts **15b**, **18b**, and **20b** with 5M HCl in MeOH yielded hydrochlorides **15c**, **18c**, and **20c**.

The structures of all products were confirmed by 1D (^{1}H, ^{13}C, ^{19}F) and 2D (HSQC, HMBC, COSY) NMR experiments and MALDI TOF MS. The ^{13}C-NMR spectra of compounds **4–7**, **8a**, **15**, **18**, and **20** show signals for amide carbon atoms at 176.3–179.3 ppm, while the amide group protons resonate at 5.86–6.79 ppm in the ^{1}H-NMR spectra. In the spectra of guanidine derivatives **9–13**, **15a**, **18a**, and **20a**, the NH-C=N and NH-Boc proton signals occurred at 8.30–9.06 ppm and 11.47–11.53 ppm, respectively. The signals for the guanidine group carbon atoms were observed at 161.8–163.6 ppm.

2.2. Biological Evaluation

The *in vitro* cytotoxic activity of triterpene acids (dihydrobetulinic, ursolic and oleanolic acids), twelve guanidinium salts, and some of their precursors, primary amines **4–6**, **8a**, and **15**, was evaluated on five human tumor cell lines: Jurkat (T-lymphoblastic leukemia), K562 (chronic myeloid leukemia), U937 (histiocytic lymphoma), HEK 293 (embryonic kidney), and HeLa (cervical cancer). The possible cell toxicity was assessed against normal human fibroblasts. Most of the tested compounds showed moderate or significant activity against Jurkat, K562, and U937 cells as compared to triterpenoic acids (Table 1).

Table 1. Cytotoxicity of dihydrobetulinic-, ursolic- and oleanolic acids and compounds **4–6**, **8a**, **9b–12b**, **12a**, **13a**, **14**, **15**, **15c**, **18b,c**, and **20b,c** against Jurkat, K562, Hek293, HeLa, U937, and normal fibroblast cells.

Compound	IC_{50} (μmol/L) [a]					
	Jurkat	K562	U937	HEK293	HeLa	Fibroblasts
Dihydro BA	59 ± 0.31	44 ± 0.24	39 ± 0.38	128 ± 0.32	132 ± 0.47	517 ± 0.26
UA	23 ± 0.34	68 ± 0.11	17 ± 0.12	96 ± 0.22	88 ± 0.35	324 ± 0.16
OA	271 ± 0.19	235 ± 0.24	186 ± 0.18	247 ± 0.27	258 ± 0.31	694 ± 0.17
4	1.4 ± 0.41	1.3 ± 0.49	1.7 ± 0.36	2.4 ± 0.35	2.8 ± 0.28	3.4 ± 0.19
5	2.3 ± 0.11	2.8 ± 0.23	1.9 ± 0.34	4.8 ± 0.25	3.9 ± 0.41	5.3 ± 0.27
6	7.7 ± 0.31	1.8 ± 0.12	4.5 ± 0.43	6.7 ± 0.21	7.1 ± 0.18	8.3 ± 0.29
8a	4.5 ± 0.29	3.0 ± 0.19	8.2 ± 0.13	5.9 ± 0.14	6.1 ± 0.38	7.3 ± 0.43
9b	21 ± 0.24	57 ± 0.36	14 ± 0.16	92 ± 0.34	81 ± 0.19	179 ± 0.43
10b	67 ± 0.21	98 ± 0.35	82 ± 0.14	128 ± 0.35	152 ± 0.49	354 ± 0.46
11b	16 ± 0.18	55 ± 0.38	63 ± 0.29	98 ± 0.27	106 ± 0.51	117 ± 0.38
12a	22 ± 0.21	89 ± 0.18	28 ± 0.25	112 ± 0.19	101 ± 0.22	149 ± 0.43
12b	14 ± 0.37	52 ± 0.26	13 ± 0.06	68 ± 0.29	72 ± 0.34	106 ± 0.23
13a	24 ± 0.27	11 ± 0.33	18 ± 0.21	74 ± 0.15	79 ± 0.29	104 ± 0.26
14	17 ± 0.34	36 ± 0.42	20 ± 0.12	59 ± 0.21	62 ± 0.24	96 ± 0.17
15	3.3 ± 0.11	2.1 ± 0.13	6.6 ± 0.23	17 ± 0.25	19 ± 0.49	31 ± 0.29
15c	3.1 ± 0.41	2.3 ± 0.34	15 ± 0.23	16 ± 0.27	14 ± 0.31	27 ± 0.14
18b	6.8 ± 0.32	40 ± 0.29	2.1 ± 0.41	13 ± 0.19	11 ± 0.27	47 ± 0.15
18c	3.8 ± 0.23	11 ± 0.18	5.3 ± 0.29	14 ± 0.34	12 ± 0.29	51 ± 0.26
20b	6.7 ± 0.15	20 ± 0.14	6.1 ± 0.43	12 ± 0.23	13 ± 0.41	49 ± 0.25
20c	7.6 ± 0.33	13 ± 0.45	6.8 ± 0.11	19 ± 0.17	17 ± 0.32	54 ± 0.12

[a] IC_{50} (μM) is the half maximal inhibitory concentration for viable cells. Each IC_{50} (mean ± SE) has been derived from at least three experiments in duplicate.

Amino derivatives **4–6** and **8a** showed cytotoxic activity against all tumor cell cultures with IC_{50} values of 1.3–8 μM. However, these compounds were also cytotoxic against fibroblasts. Contrary to our expectations, guanylation of the terminal amino groups of lupane triterpenoids with C-28 linear aminoalkane chains (compounds **9b–11b**, **13a**, and **14**) or the branched tris-aminoethyl moiety (compounds **12a,b**) did not considerably enhance the cytotoxic action. Of the listed guanidinium salts, only compounds **9b**, **11b**, and **14** showed an approximately 3-fold increase in the cytotoxic activity against Jurkat cells in comparison with dihydrobetulinic acid **2** and also showed selectivity towards Jurkat cells with an SI of 8.5, 7.3, and 5.6 respectively (SI = IC_{50} fibroblasts/IC_{50} Jurkat cells). The introduction of the tris(hydroxymethyl)aminomethane moiety into the molecules of triterpene acids **3**, **16** and **17** markedly enhanced the cytotoxic activities for the resulting conjugates **15**, **15c**, **18b,c**, and **20b,c**, irrespective of the triterpene skeleton type. The C-28 esters of dihydrobetulinic, ursolic, and oleanolic acids with amino and guanidine groups in the ester side chain had from moderate to good activity against Jurkat and K562 cell lines. For example, the IC$_{50}$ values of compounds **15c** and **18c** were 3.1 and 3.8 µM for T-lymphoblastic leukemia cells and 2.3 and 11.0 µM for chronic myeloid leukemia cells. The most pronounced differences in the antitumor activity were found for oleanolic acid and its conjugates **20b,c**. Indeed, the IC$_{50}$ values of oleanolic acid, **20b**, and **20c** for Jurkat cells were 271, 6.7, and 7.6 µM, respectively. It is worth noting that amine **15**, its guanidine derivative **15c**, and guanidinium salts of ursolic and oleanolic acids **18b,c** and **20b,c** showed acceptable selectivity towards Jurkat tumor cell with an SI from 6.9 to 13.4.

The identified lead compounds with the highest cytotoxicity characteristics, **15**, **15c**, **18c**, and **20c**, were further evaluated for the possible apoptosis induction in tumor cell cultures. The measurements were done by flow cytometry (Figure 2).

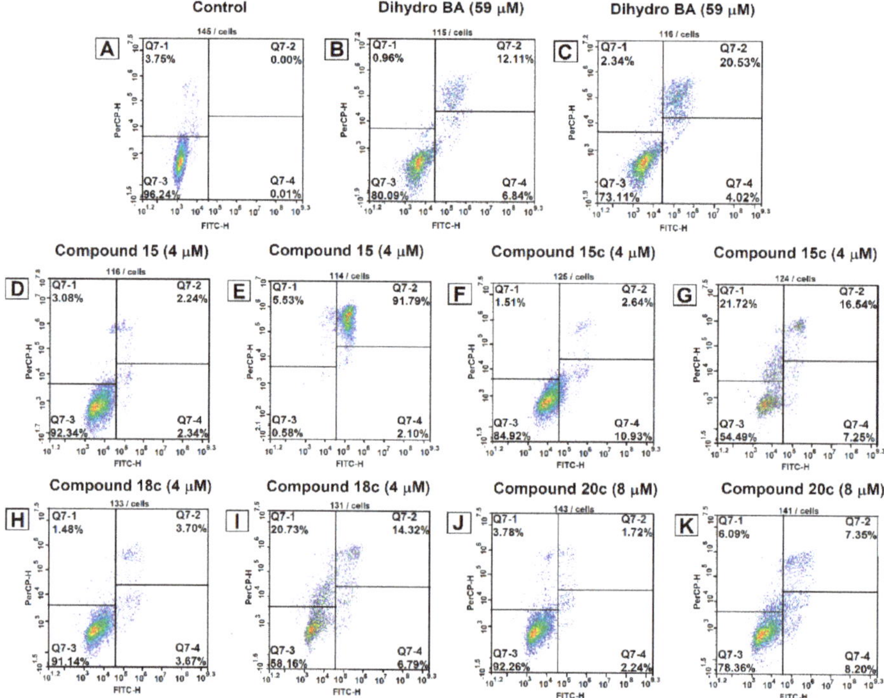

Figure 2. AnnexinV/7-AAD staining upon induction of apoptosis in Jurkat cells. Cells were treated with compounds at their IC$_{50}$ concentration for 24 and 48 h. Then, the cells were harvested, stained with Annexin V/7-AAD and analyzed by flow cytometry. The experiments were performed three times, and the results of the representative experiments are shown. The first cytogram represents an untreated cell sample (**A**); after incubation with dihydrobetulinic acid for 24 h (**B**) and for 48 h (**C**); after incubation with compound **15** for 24 h (**D**) and for 48 h (**E**); after incubation with compound **15c** for 24 h (**F**) and for 48 h (**G**); after incubation with compound **18c** for 24 h (**H**) and for 48 h (**I**); after incubation with compound **20c** for 24 h (**J**) and for 48 h (**K**). Q7-1, necrotic cells; Q7-2, late apoptotic cells; Q7-3, living cells; Q7-4, early apoptotic cells.

The highest percentage of late apoptosis (91.7%) was detected upon the treatment of Jurkat cells with the test compound **15** at IC$_{50}$ concentration exposure for 48 h as depicted in Figure 2E. Compounds **15c**, **18c** and **20c** also showed apoptotic mode of cell death on Jurkat cells line, but, in this case, the apoptotic effect of these guanidine derivatives was notably weaker. Thus, after treatment of Jurkat cells with compound **15c** at IC$_{50}$ concentration (4 µM) the number of vital cells is decreased from

96.2% (control) to 54.5%. Total number of apoptotic cells population constituted 23.7% (7.2% and 16.5% of early and late apoptotic cells, respectively) and number of necrotic cells was 21.7%. Comparable results were obtained with the guanidine derivative of ursolic acid **18c** (6.8, 14.3, and 20.7% of early, late apoptotic cells and necrotic cells, respectively). Treatment of the Jurkat cells with **20c** resulted in about 15.5% apoptotic cells and 6.1% necrotic cells, with 78.4% of the cells still being considered vital (Figure 2K). Next, we analyzed the ability of dihydrobetulinic acid to stimulate apoptosis. Our results showed that dihydrobetulinic acid triggers apoptosis in Jurkat cells at higher doses as compared to derivatives **15**, **15c**, and **18c**. The number of apoptotic cells on treatment with dihydrobetulinic acid (59 μM) for 48 h constituted around 24% (4.0% of early-stage and 20.5% of secondary necrotic/late-stage apoptotic), while the number of vital cells was 73.1% (Figure 2C).

In summary, our results indicate that the apoptosis is induced in Jurkat cells by all test compounds. However, we observed a higher rate of necrosis after **15c**, **18c** and **20c** incubation compared to derivative **15**.

DNA flow cytometry was also used to analyze the cell cycle kinetics in Jurkat cells pre-incubated with dihydrobetulinic acid and derivatives **15**, **15c**, **18c**, or **20c** at their IC$_{50}$ concentration for 24 and 48 h (Figure 3).

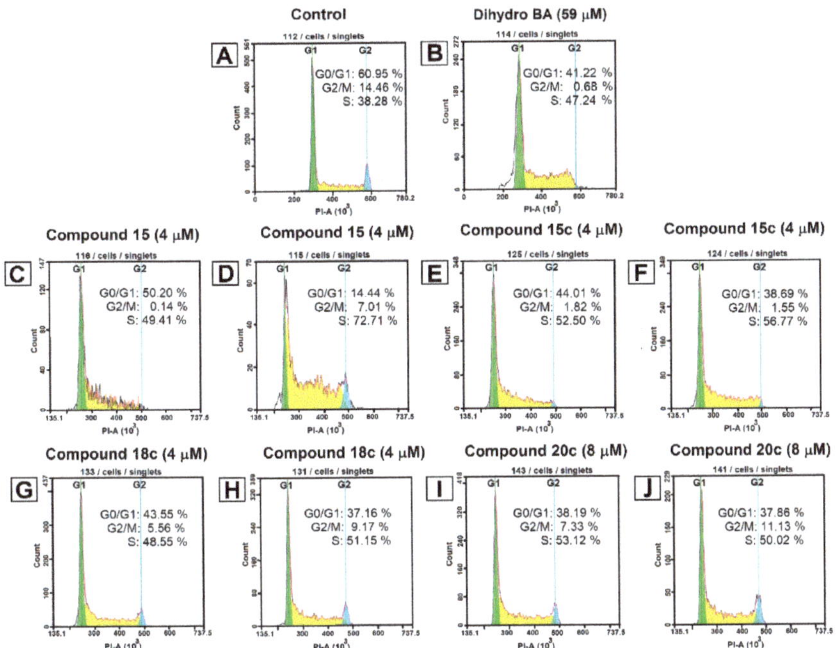

Figure 3. Cell cycle analysis in Jurkat cells. Jurkat cells treated with tested compounds at the IC$_{50}$ concentration for 24 and 48 h. The cells were trypsinized, harvested and washed three times with ice-PBS for PI-stained DNA content detected by flow cytometry. The experiments were performed three times, and the results of the representative experiments are shown. The first cytogram represents an untreated cell sample (**A**); after incubation with dihydrobetulinic acid for 48 h (**B**); after incubation with compound **15** for 24 h (**C**) and for 48 h (**D**); after incubation with compound **15c** for 24 h (**E**) and for 48 h (**F**); after incubation with compound **18c** for 24 h (**G**) and for 48 h (**H**); after incubation with compound **20c** for 24 h (**I**) and for 48 h (**J**).

The distribution of DNA content in Jurkat cells reveals whether cell proliferation is arrested at one checkpoint. The results of all experiments showed in significant S-phase arrest in cells after treating with these compounds. Thus, the ratio of cells in the S phase increased from 38.3% (control) to 72.7% in cells treated with **15** (4 μM for 48 h) and increased to 47.2–56.8% in cells treated with dihidrobetulinic acid, **15c**, **18c**, or **20c**. The appropriate decrease of the number of cells in the G2/M phase was observed. For example, the treatment of Jurkat cells with **15** and **15c** resulted in a decrease of cells in the G2/M phase from 14.5% (control) to 7.0 and 1.5% (Figure 3D,F). Considering these results, we assume that dihydrobetulinic acid and compounds **15**, **15c**, **18c**, or **20c** are able to trigger programmed cell death, including apoptotic mechanisms and cell cycle arrest in the S-phase.

3. Materials and Methods

3.1. Chemistry

IR spectra (thin films or solutions in $CHCl_3$) were obtained with use of a Vertex 70v spectrometer (Bruker, Karlsruhe, Germany). ^{1}H- and ^{13}C-NMR spectra were recorded in $CDCl_3$, in MeOD or in d_6-DMSO with Me_4Si as the internal standard on an AVANCE–500 instrument (500.13 (^{1}H), 125.78 MHz (^{13}C), 470.59 MHz (^{19}F)) or on an AVANCE-400 (400.13 (^{1}H), 100.62 MHz (^{13}C), 376.50 MHz (^{19}F)) (Bruker). Mass spectra of new compounds were recorded on a Bruker–Autoflex III spectrometer (MALDI TOF, positive ion mode, sinapic acid as the matrices). Optical rotation was determined on a 141 polarimeter (Perkin–Elmer, Beaconsfield, UK). Specific rotation $[\alpha]_D$ is expressed in $(deg \cdot mL)/(g \cdot dm)^{-1}$; the concentration of the solution c is expressed in g/100 mL. Elemental analysis was carried out on a 1106 analyzer (Carlo Erba, Milan, Italy). TLC was carried out on Sorbfil plates (Sorbpolimer, Krasnodar, Russia) in hexane–EtOAc and chloroform–methanol, spots were visualized with anisaldehyde. Silica gel L (KSKG grade, 50–160 μm) was employed for column chromatography. All reagents and solvents were of the purest grade available, and generally were used without further treatment. The starting compounds ursolic, oleanolic acids and reagents: sodium borohydride ($NaBH_4$), acetyl chloride, 10% Pd/C, oxalyl chloride, 1,2-diaminoethane, 1,4-diaminobutane, tris(2-aminoethyl)amine, 1,4-bis(3-aminopropyl)piperazine, tris(hydroxymethyl)aminomethane, 1,3-di-Boc-2-(trifluoromethylsulfonyl)guanidine, triethylamine (Et_3N), dimethylaminopyridine (DMAP), trifluoroacetic acid (TFA) were purchased from Acros Organics (Geel, Belgium). Dihydrobetulonic and dihydrobetulinic acids were obtained from betulin according to the known procedures [41]. Acetates of oleanolic, ursolic, dihydrobetulonic and dihydrobetulinic acids were synthesized according to the typical procedures. Mono-Boc-protected bis-aminopropylpiperazine and compounds **19–23** were prepared by a reported methods [43,45]. NMR ^{1}H and ^{13}C spectra of all new compounds are in Supplementary Materials.

3.1.1. General Procedure for the Synthesis of Amines **4–7**

Oxalyl chloride (0.13 mL, 1.5 mmol) was added with stirring to a solution of compounds **1** or **3** (0.5 mmol) in dry CH_2Cl_2 (5 mL) precooled to 0 °C, and stirring of the reaction mixture was continued at room temperature for 2 h. Then the solvent and excess oxalyl chloride were removed under vacuum. The amine (1.5 mmol) was dissolved in dry CH_2Cl_2 (2 mL) and under vigorous stirring were added Et_3N (0.2 mL, 1.5 mmol) and a solution of the acid chloride of **1** or **3** (0.5 mmol) in dry CH_2Cl_2 (4 mL). The mixture was stirred for 24 h (monitoring by TLC). The mixture was then poured into cold H_2O and extracted with CH_2Cl_2 (2 × 15 mL each). The organic phases washed brine and were dried over Na_2SO_4 and evaporated under reduced pressure. The residue was chromatographed on silica gel, using CH_2Cl_2/MeOH 30:1→1:1 (*v/v*), to obtain pure compounds **4–7**.

N-(4-Aminobuthyl)-3-oxolupane-28-amide (**4**), White powder, 78% yield; mp 170–172 °C (EtOH); $[\alpha]_D^{19}$: +4° (*c* 0.23, CH_2Cl_2); IR (CHCl$_3$) ν_{max} 1641, 1702 (C=O), 3357 (NH) cm^{-1}; ^{1}H-NMR (500 MHz, CDCl$_3$) δ: 5.86 (t, *J* = 5.5 Hz, 1H, CONH), 3.32–3.20 (m, 2H, H-1′), 2.76 (t, *J* = 6.5 Hz, 2H, H-4′), 2.55–1.14 (m, 26H, CH, CH$_2$ in pentacyclic skeleton, 4H, H-2′, H-3′), 1.08, 1.03, 0.98, 0.97, 0.94 (all s, 3H each, H-23–H-27), 0.87 (d, *J* = 6.5 Hz, 3H, H-29), 0.75 (d, *J* = 7.0 Hz, 3H, H-30); ^{13}C-NMR (125 MHz, CDCl$_3$) δ: 218.2 (C-3), 176.3 (C-28), 56.0 (C-17), 55.0 (C-5), 49.8 (C-9), 49.5 (C-19), 47.4 (C-4), 44.2 (C-18), 42.7 (C-14), 41.6 (C-4′), 40.7 (C-8), 39.6 (C-1), 39 (C-22), 38.8 (C-1′), 37.7 (C-13), 36.9 (C-10), 34.2 (C-2), 33.8 (C-16), 33.6 (C-7), 30.6 (C-3′), 29.9 (C-20), 29.4 (C-15), 27.2 (C-23), 27.0 (C-2′), 26.6 (C-12), 23.1 (C-29), 23.0 (C-21),

21.5 (C-24), 21.1 (C-11), 19.7 (C-6), 16.0 (C-25), 15.9 (C-26), 14.6 (C-30), 14.5 (C-27); Anal. Calcd. for $C_{34}H_{58}N_2O_2$: C, 77.51, H, 11.10. Found: C, 78.03, H, 11.02%. MS: *m/z* 527.45 [M + H]$^+$ (calcd. for $C_{34}H_{58}N_2O_2$, 526.45).

3β-N-(2-Aminoethyl)-3-O-acetyl-lupane-28-amide (**5**), White powder, 74% yield; mp 120–122 °C (EtOH); $[\alpha]_D^{19}$ −8° (*c* 0.24, CH$_2$Cl$_2$); IR (CHCl$_3$) ν_{max} 1647, 1735 (C=O), 3367 (NH) cm^{-1}; ^{1}H-NMR (500 MHz, CDCl$_3$) δ: 6.14 (t, *J* = 5.5 Hz, 1H, CONH), 4.48–4.44 (m, 1H, H-3), 3.35–3.21 (m, 2H, H-1′), 2.81 (t, *J* = 6.0 Hz, 2H, H-2′), 2.43–0.97 (m, 25H, CH, CH$_2$ in pentacyclic skeleton), 2.01 (s, 3H, CH$_3$CO−), 0.94, 0.92, 0.85, 0.84, 0.83, 0.82 (all s, 3H each, H-23–H-27 and H-29), 0.77 (d, *J* = 10.0 Hz, 1H, H-5), 0.73 (d, *J* = 7.0 Hz, 3H, H-30); ^{13}C-NMR (125 MHz, CDCl$_3$) δ: 176.7 (C-28), 171.0 (COCH$_3$), 80.9 (C-3), 56.2 (C-17), 55.4 (C-5), 50.3 (C-9), 49.5 (C-19), 44.3 (C-18), 42.6 (C-14), 41.8 (C-2′), 41.7 (C-1′), 40.8 (C-8), 38.7 (C-1), 38.4 (C-22), 37.8 (C-4), 37.6 (C-13), 37.1 (C-10), 34.4 (C-7), 33.6 (C-16), 29.9 (C-20), 29.5 (C-15), 27.9 (C-23), 27.0 (C-2), 23.7 (C-12), 23.1 (C-21, C-29), 21.3 (COCH$_3$), 21.0 (C-11), 18.2 (C-6), 16.5 (C-25), 16.2 (C-24, C-26), 14.6 (C-30), 14.5 (C-27); Anal. Calcd. for $C_{34}H_{58}N_2O_3$: C, 75.23, H, 8.84. Found: C, 75.74, H, 8.79%. MS: *m/z* 543.40 [M + H]$^+$ (calcd. for $C_{34}H_{58}N_2O_3$, 542.44).

3β-N-(4-Aminobuthyl)-3-O-acetyl-lupane-28-amide (**6**), White powder, 58% yield; mp 104–106 °C (EtOH); $[\alpha]_D^{19}$ −11° (*c* 0.69, CHCl$_3$); IR (CHCl$_3$) ν_{max} 1646, 1734 (C=O), 3367 (NH) cm^{-1}; ^{1}H-NMR (500 MHz, CDCl$_3$) δ: 5.86 (t, *J* = 5.5 Hz, 1H, CONH), 4.48–4.45 (m, 1H, H-3), 3.30–3.17 (m, 2H, H-1′), 2.73 (t, *J* = 6.5 Hz, 2H, H-4′), 2.47–0.98 (m, 25H, CH, CH$_2$ in pentacyclic skeleton, 4H, H-2′, H-3′), 2.03 (s, 3H, CH$_3$CO−), 0.95, 0.93, 0.92, 0.84, 0.83, 0.82, (all s, 3H each, H-23–H-27 and H-29), 0.78 (d, *J* = 10.0 Hz, 1H, H-5), 0.73 (d, *J* = 7.0 Hz, 3H, H-30); ^{13}C-NMR (125 MHz, CDCl$_3$) δ: 176.3 (C-28), 171.0 (COCH$_3$), 80.9 (C-3), 56.0 (C-17), 55.4 (C-5), 50.3 (C-9), 49.5 (C-19), 44.3 (C-18), 42.6 (C-14), 41.6 (C-4′), 40.8 (C-8), 39.0 (C-1′), 38.7 (C-22), 38.4 (C-1), 37.8 (C-4), 37.6 (C-13), 37.1 (C-10), 34.4 (C-7), 33.6 (C-16), 30.7 (C-3′), 29.9 (C-20), 29.4 (C-15), 27.9 (C-23), 27.2 (C-2′), 27.0 (C-2), 23.7 (C-12), 23.0 (C-21, C-29), 21.3 (COCH$_3$), 21.0 (C-11), 18.2 (C-6), 16.5 (C-25), 16.2 (C-26, C-24), 14.6 (C-30), 14.5 (C-27); Anal. Calcd. for $C_{36}H_{62}N_2O_3$: C, 75.74, H, 10.95. Found: C, 76.01, H, 10.89%. MS: *m/z* 609.43 [M + K]$^+$ (calcd. for $C_{36}H_{62}N_2O_3$, 570.48).

3β-N-[2-(N,N′-bis-Aminoethyl)-aminoethyl]-3-O-acetyl-lupane-28-amide (**7**), White powder, 69% yield; mp 108–110 °C (EtOH); $[\alpha]_D^{19}$ −5° (*c* 0.16, C$_2$H$_5$OH); IR (CHCl$_3$) ν_{max} 1648, 1735 (C=O), 3441 (NH) cm^{-1}; ^{1}H-NMR (400 MHz, MeOD) δ: 4.49–4.45 (m, 1H, H-3), 3.33–3.27 (m, 2H, H-1′), 2.74–2.72 (m, 4H, H-4′, H-42″), 2.60–2.58 (m, 6H, H-2′, H-3′, H-3″), 2.32–0.84 (m, 26H, CH, CH$_2$ in pentacyclic skeleton), 2.05 (s, 3H, CH$_3$CO−), 1.02, 1.00, 0.93, 0.90, 0.89, 0.88 (all s, 3H each, H-23–H-27 and H-29), 0.79 (d, *J* = 6.4 Hz, 3H, H-30); ^{13}C-NMR (100 MHz, MeOD) δ: 179.3 (C-28), 172.8 (COCH$_3$), 82.5 (C-3), 57.7 (C-17), 57.4 (C-5), 57.0 (C-3′, C-3″), 55.2 (C-2′), 51.9 (C-9), 51.1 (C-19), 45.5 (C-18), 43.8 (C-14), 42.2 (C-8), 40.2 (C-22), 39.9 (C-1), 39.8 (C-4′, C-4″), 39.0 (C-4), 38.8 (C-1′), 38.4 (C-13), 38.3 (C-10), 35.8 (C-7), 34.1 (C-16), 31.4 (C-20), 30.9 (C-15), 28.7 (C-23), 28.5 (C-2), 24.8 (C-12), 24.2 (C-29), 23.8 (C-21), 22.4 (COCH$_3$), 21.4 (C-11), 19.4 (C-6), 17.2 (C-25), 17.0 (C-24, C-26), 15.3 (C-30, C-27); Anal. Calcd. for $C_{38}H_{68}N_4O_3$: C, 72.56, H, 10.90. Found: C, 73.04, H, 10.87%. MS: *m/z* 629.56 [M + H]$^+$ (calcd. for $C_{38}H_{68}N_4O_3$, 628.53).

3β-N-{[3-(3-Aminopropyl)piperazinyl]propyl}-3-O-acetyl-lupane-28-amide (**8a**), The acid chloride of **3** was synthesized according to the general procedure for synthesis of amines **4**–**7**. Then the *N-tert*-butoxycarbonyl-1,4-bis(3-aminopropyl)piperazine (0.45 g, 1.5 mmol) and Et$_3$N (0.25 mL, 1.8 mmol) were added to acid chloride of **3** (0.50 g, 1 mmol) in dry CH$_2$Cl$_2$ (10 mL). The mixture was stirred at room temperature for 16 h (monitoring by TLC), then was poured into cold H$_2$O and extracted with CH$_2$Cl$_2$ (2 × 15 mL). The organic phases washed brine, dried over Na$_2$SO$_4$ and evaporated under reduced pressure to obtain compound 8. Then the compound 8 was dissolved in CH$_2$Cl$_2$ (7 mL), acidified with TFA 10% (*v/v*) in CH$_2$Cl$_2$ (17 mL) for deprotection reaction and stirred for around 5 h. The reaction was quenched using saturated aqueous potassium carbonate solution (20 mL). The aqueous phase was extracted with CH$_2$Cl$_2$ (2 × 15 mL). The organic phases washed brine, dried over Na$_2$SO$_4$ and evaporated under reduced pressure. The residue was chromatographed on silica gel, using CH_2Cl_2/MeOH 50:1→5:1, to obtain pure compound 8a. White powder, 78% yield; mp 119–121 °C (EtOH); $[\alpha]_D^{21}$ −7.4° (*c* 0.5, $CHCl_3$); IR ($CHCl_3$) ν_{max} 1735 (C=O), 3361 (NH) cm^{-1}; ^{1}H-NMR (500 MHz, $CDCl_3$) δ: 6.79 (t, *J* = 5.5 Hz, 1H, CONH), 4.47–4.44 (m, 1H, H-3), 3.31–3.25 (m, 2H, H-1′), 2.78–2.77 (m, 2H, H-1″), 2.50–2.42 (m, 12H, H-3′–H-5′, H-3″–H-5″), 2.41–0.98 (m, 25H, CH, CH_2 in pentacyclic skeleton, 4H, H-2′, H-2″), 2.02 (s, 3H, CH_3CO–), 0.93, 0.92, 0.91, 0.83, 0.82, 0.81 (3H each, all s, H-23–H-27 and H-29), 0.77 (d, *J* = 10.5 Hz, 1H, H-5), 0.72 (d, *J* = 6.5 Hz, 3H, H-30); ^{13}C-NMR (125 MHz, $CDCl_3$) δ: 176.3 (C-28), 171.0 (COCH$_3$), 80.9 (C-3), 57.8 (C-3′), 56.7 (C-3″), 55.9 (C-17), 55.4 (C-5), 53.5 (C-5′, C-5″), 53.3 (C-4′), 53.3 (C-4″), 50.3 (C-9), 49.6 (C-19), 44.1 (C-18), 42.6 (C-14), 40.9 (C-1″), 40.8 (C-8), 39.3 (C-1′), 38.8 (C-22), 38.4 (C-1), 37.8 (C-4), 37.4 (C-13), 37.1 (C-10), 34.4 (C-7), 33.7 (C-16), 29.9 (C-20), 29.8 (C-2′), 29.7 (C-15), 27.9 (C-23), 26.9 (C-2), 25.3 (C-2″), 23.7 (C-12), 23.0 (C-21, C-29), 21.3 (COCH$_3$), 21.0 (C-11), 18.2 (C-6), 16.5 (C-25), 16.3 (C-24), 16.2 (C-26), 14.6 (C-30), 14.5 (C-27); Anal. Calcd. for $C_{42}H_{74}N_4O_3$: C, 73.85, H, 10.92. Found: C, 74.13, H, 10.84%. MS: *m/z* 683.52 [M + H]$^+$ (calcd. for $C_{42}H_{74}N_4O_3$, 682.58).

3.1.2. General Procedure for the Synthesis of Compounds 15, 18–21

Acid chlorides of **3**, **16**, **17** was synthesized according to the general procedure for synthesis of amines **4–7**. Then the acid chlorides of **3**, **16**, **17** (1 mmol) was dissolved in a mixture of pyridine (4 mL), CH_2Cl_2 (1 mL) and DMAP (0.09 g, 0.7 mmol) was added. After complete dissolution of DMAP, a solution containing of TRIS (tris(hydroxymethyl)aminomethane) (0.24 g, 2 mmol) in pyridine (0,5 mL) was added. The mixture was stirred for 10 h at room temperature and the solvent was removed rapidly under vacuum. The residue was chromatographed on silica gel, using CH_2Cl_2/MeOH 30:1→1:1, to obtain pure compounds **15**, **18–21**.

3β-[2-Amino-3-hydroxy-2-(hydroxymethyl)propyl]-3-O-acetyl-lupane-28-oate (**15**), White powder, 16% yield; mp 134–136 °C (EtOH); $[\alpha]_D^{21}$ −13.1° (*c* 0.51, $CHCl_3$); IR ($CHCl_3$) ν_{max} 1731 (C=O), 3366 (OH), 3446 (NH) cm^{-1}; ^{1}H-NMR (400 MHz, $CDCl_3$) δ: 4.49–4.45 (m, 1H, H-3), 4.09, 4.03 (both d, *J* = 11.6 Hz, 1H each, H-1′), 3.52 (br s, 4H, H-3′, H-4′), 2.70 (br s, 4H, NH_2, OH), 2.05 (s, 3H, CH_3CO–), 2.24–0.81 (m, 26H, CH, CH_2 in pentacyclic skeleton), 0.95, 0.91, 0.85, 0.84, 0.83, 0.82 (all s, 3H each, H-23–H-27 and H-29), 0.76 (d, *J* = 6.8 Hz, 3H, H-30); ^{13}C-NMR (100 MHz, $CDCl_3$) δ: 177.0 (C-28), 171.1 (COCH$_3$), 80.9 (C-3), 64.9 (C-1′), 64.2 (C-3′, C-4′), 57.3 (C-17, C-2′), 55.4 (C-5), 50.2 (C-9), 48.8 (C-19), 44.2 (C-18), 42.6 (C-14), 40.7 (C-8), 38.4 (C-22), 38.1 (C-1), 37.8 (C-4), 37.5 (C-13), 37.1 (C-10), 34.3 (C-7), 32.0 (C-16), 29.7 (C-15, C-20), 27.9 (C-23), 26.9 (C-2), 23.7 (C-12), 23.0 (C-29), 22.8 (C-21), 21.3 (COCH$_3$), 20.9 (C-11), 18.2 (C-6), 16.5 (C-25), 16.2 (C-24), 16.1 (C-26), 14.7 (C-30), 14.6 (C-27); Anal. Calcd. for $C_{36}H_{61}NO_6$: C, 71.60, H, 10.18. Found: C, 72.09, H, 10.11%. MS: *m/z* 604.42 [M + H]$^+$ (calcd. for $C_{36}H_{61}NO_6$, 603.45).

3β-[2-Amino-3-hydroxy-2-(hydroxymethyl)propyl]-3-O-acetylurs-12-en-28-oate (**18**), White powder, 15% yield; mp 129–131 °C (EtOH); $[\alpha]_D^{18}$ +40° (*c* 0.74, $CHCl_3$); IR ($CHCl_3$) ν_{max} 1721 (C=O), 3444 (NH), 3468 (OH) cm^{-1}; ^{1}H-NMR (400 MHz, $CDCl_3$) δ: 5.25 (br s, 1H, H-12), 4.51–4.47 (m, 1H, H-3), 4.02, 3.93 (both d, *J* = 12 Hz, 1H each, H-1′), 3.48 (br s, 4H, H-3′, H-4′), 2.68 (br s, 4H, NH_2, OH), 2.19 (d, *J* = 11.2 Hz, 1H, H-18), 2.05 (s, 3H, CH_3CO–), 1.99–0.81 (m, 22H, CH, CH_2 in pentacyclic skeleton), 1.08, 0.96, 0.94, 0.88, 0.87, 0.86, 0.74 (all s, 3H each, H-23–H-27, H-29 and H-30); ^{13}C-NMR (100 MHz, $CDCl_3$) δ: 178.1 (C-28), 171.1 (COCH$_3$), 138.3 (C-13), 125.6 (C-12), 80.9 (C-3), 65.7 (C-1′), 64.4 (C-3′, C-4′), 56.6 (C-2′), 55.3 (C-5), 53.0 (C-18), 48.6 (C-17), 47.4 (C-9), 42.1 (C-14), 39.5 (C-4, C-8), 38.9 (C-19), 38.3 (C-20), 37.7 (C-1), 37.1 (C-22), 36.8 (C-10), 32.9 (C-7), 30.6 (C-21), 28.1 (C-23), 27.9 (C-15), 24.3 (C-16, C-27), 23.5 (C-2, C-11), 21.3 (COCH$_3$), 21.1 (C-30), 18.2 (C-6), 17.2 (C-29), 17.0 (C-26), 16.7 (C-24), 15.5 (C-25); Anal. Calcd. for $C_{36}H_{59}NO_6$: C, 71.84, H, 9.88. Found: C, 72.13, H, 9.83%. MS: *m/z* 602.40 [M + H]$^+$ (calcd. for $C_{36}H_{59}NO_6$, 601.43).

3β-N-[(1′,1′,1′-tris-Hidroxymethyl)methyl]-3-O-acetyl-ursolamide (**19**), White powder, 24% yield; mp 203–205 °C (EtOH); $[\alpha]_D^{18}$ +26.9° (*c* 0.74, CH_3OH); IR (CH_3OH) ν_{max} 1734 (C=O), 2923, 2871, 2852 (OH), 3352 (NH) cm^{-1}; ^{1}H-NMR (500 MHz, MeOD) δ: 5.38 (br s, 1H, H-12), 4.50–4.47 (m, 1H, H-3), 3.66, 3.58 (both d, *J* = 10 Hz, 3H each, H-2′–H-4′), 2.12–0.88 (m, 23H, CH, CH_2 in pentacyclic skeleton), 2.05 (s, 3H, CH_3CO–), 1.18, 1.04, 0.99, 0.95, 0.93, 0.92, 0.91 (all s, 3H each, H-23–H-27, H-29 and H-30);

^{13}C-NMR (125 MHz, CDCl$_3$) δ: 178.5 (C-28), 170.7 (COCH$_3$), 138.1 (C-13), 125.8 (C-12), 80.4 (C-3), 61.8 (C-1′), 60.8 (C-2′–C-4′), 55.1 (C-5), 53.6 (C-18), 48.3 (C-17), 47.4 (C-9), 42.3 (C-14), 39.5 (C-4, C-8), 38.9 (C-19), 38.4 (C-20), 37.7 (C-22, C-1), 36.8 (C-10), 33 (C-7), 30.9 (C-21), 28.3 (C-23), 27.8 (C-15), 24.5 (C-16), 23.7 (C-2), 23.4 (C-11, C-27), 21.5 (COCH$_3$), 21.4 (C-30), 18.2 (C-6), 17.5 (C-29, C-26), 17.1 (C-24), 15.7 (C-25); Anal. Calcd. for C$_{36}$H$_{59}$NO$_6$: C, 71.84, H, 9.88. Found: C, 72.09, H, 9.79%. MS: *m/z* 602.41 [M + H]$^+$ (calcd. for C$_{36}$H$_{59}$NO$_6$, 601.43).

3β-[2-Amino-3-hydroxy-2(hydroxymethyl)propyl]-3-O-acetylolean-12-en-28-oate (**20**), White powder, 23% yield; mp 141–143 °C (EtOH); [α]$_D^{18}$ +49° (*c* 0.75, CHCl$_3$); IR (CHCl$_3$) ν_{max} 1618, 1727 (C=O), 3447 (NH) cm^{-1}; ^{1}H-NMR (500 MHz, CDCl$_3$) δ: 5.30 (br s, 1H, H-12), 4.51–4.48 (m, 1H, H-3), 4.04, 3.99 (both d, *J* = 11.5 Hz, 1H each, H-1′), 3.50 (br s, 4H, H-3′, H-4′), 2.84 (d, *J* = 10.0 Hz, 1H, H-18), 2.68–2.60 (m, 4H, NH$_2$, OH), 2.05 (s, 3H, CH$_3$CO–), 2.01–0.83 (m, 22H, CH, CH$_2$ in pentacyclic skeleton), 1.14, 0.94, 0.93, 0.92, 0.87, 0.86, 0.73 (all s, 3H each, H-23–H-27, H-29 and H-30); ^{13}C-NMR (125 MHz, CDCl$_3$) δ: 178.2 (C-28), 171.1 (COCH$_3$), 143.7 (C-13), 122.6 (C-12), 80.9 (C-3), 65.7 (C-1′), 64.6 (C-3′, C-4′), 56.6 (C-2′), 55.3 (C-5), 47.5 (C-9), 47.2 (C-17), 45.7 (C-19), 41.8 (C-14), 41.5 (C-18), 39.3 (C-8), 38.1 (C-1), 37.7 (C-4), 36.9 (C-10), 33.8 (C-22), 33.0 (C-30), 32.8 (C-7), 32.7 (C-21), 30.7 (C-20), 28.0 (C-23), 27.6 (C-15), 25.8 (C-27), 23.6 (C-29), 23.5 (C-11), 23.4 (C-2), 23.1 (C-16), 21.3 (COCH$_3$), 18.2 (C-6), 17.1 (C-26), 16.7 (C-24), 15.4 (C-25); Anal. Calcd. for C$_{36}$H$_{59}$NO$_6$: C, 71.84, H, 9.88. Found: C, 72.04, H, 9.82%. MS: *m/z* 602.40 [M + H]$^+$ (calcd. for C$_{36}$H$_{59}$NO$_6$, 601.43).

3β-N-[(1′,1′,1′-tris-Hydroxymethyl)methyl]-3-O-acetyl-oleanolamide (**21**), White powder, 32% yield; mp 249–251 °C (EtOH); [α]$_D^{18}$ +33.9° (*c* 0.61, CH$_3$OH); IR (CH$_3$OH) ν_{max} 1732(C=O), 2945, 2927 (OH), 3353 (NH) cm^{-1}; ^{1}H-NMR (500 MHz, *d$_6$*-DMSO) δ: 6.62 (s, 1H, NH), 5.24 (br s, 1H, H-12), 4.97 (t, *J* = 5 Hz, 3H, –OH), 4.42–4.38 (m, 1H, H-3), 3.49–3.42 (m, 6H, H-2′–H-4′), 2.61 (d, *J* = 10.0 Hz, 1H, H-18), 2.00 (s, 3H, CH$_3$CO–), 1.97–0.84 (m, 22H, CH, CH$_2$ in pentacyclic skeleton), 1.91, 0.90, 0.88, 0.87, 0.82, 0.81, 0.78 (all s, 3H each, H-23–H-27, H-29 and H-30); ^{13}C-NMR (125 MHz, *d$_6$*-DMSO) δ: 178.3 (C-28), 170.6 (COCH$_3$), 143.7 (C-13), 122.6 (C-12), 80.4 (C-3), 61.8 (C-1′), 60.8 (C-2′–C-4′), 55 (C-5), 47.4 (C-9), 46.8 (C-17), 46.6 (C-19), 42.0 (C-14), 41.9 (C-18), 39.5 (C-8), 38.2 (C-1), 37.7 (C-4), 36.9 (C-10), 34.1 (C-22), 33.3 (C-7), 33.2 (C-30), 32.7 (C-21), 30.8 (C-20), 28.2 (C-23), 27.4 (C-15), 25.8 (C-27), 23.8 (C-29), 23.7 (C-2), 23.4 (C-11, C-16), 21.4 (COCH$_3$), 18.2 (C-6), 17.4 (C-26), 17.1 (C-24), 15.6 (C-25); Anal. Calcd. for C$_{36}$H$_{59}$NO$_6$: C, 71.84, H, 9.88. Found: C, 72.11, H, 9.85%. MS: *m/z* 602.39 [M + H]$^+$ (calcd. for C$_{36}$H$_{59}$NO$_6$, 601.43).

3.1.3. General Procedure for the Guanilation of Amines 4–7, 8a, 15, 18 and 20

The amine (0.5 mmol) is added neat to a solution of 1,3-di-Boc-2-(trifluoromethyl-sulfonyl)guanidine (0.18 g, 0.45 mmol) or (0.60 g, 0.9 mmol) for amine **7** and Et$_3$N (0.07 mL, 0.5 mmol) in CH$_2$Cl$_2$ (at r.t.) of compounds **4–7, 8a** or in CHCl$_3$ (at reflux) of compounds **15, 18, 20** (10 mL). The mixture was stirred for 2–12 h (TLC monitoring CH$_2$Cl$_2$/MeOH 20:1). After completion of the reaction, the mixture is diluted with CH$_2$Cl$_2$ and washed witH-NH$_4$Cl, NaHCO$_3$ and brine. After drying with sodium sulfate and filtering the solvent is removed under reduced pressure. The residue was chromatographed on silica gel, using hexane/EtOAc 10:1→1:1, to obtain pure compounds **9–13, 15a, 18a** and **20a**.

N-[4-tert-Butyloxycarbonyl buthylguanidine]-3-oxo-lupane-28-amide (**9**), White powder, 88% yield; mp 158–160 °C (EtOH); [α]$_D^{23}$ +3.4° (*c* 0.59, CH$_2$Cl$_2$); IR (CHCl$_3$) ν_{max} 1637, 1718 (C=O), 3337 (NH) cm^{-1}; ^{1}H-NMR (500 MHz, CDCl$_3$) δ: 11.52 (s, 1H, NH in Boc), 8.30 (br s, 1H, NH–C=N), 5.70 (br s, 1H, CONH), 3.48–3.41 (m, 2H, H-2′), 3.37–3.20 (m, 2H, H-1′), 1.51, 1.48 (both s, 9H each, CH$_3$ in Boc), 2.54–1.15 (m, 26H, CH, CH$_2$ in pentacyclic skeleton, 4H, H-2′, H-3′), 1.08, 1.03, 0.98, 0.97, 0.95 (all s, 3H each, H-23–H-27), 0.87 (d, *J* = 6.5 Hz, 3H, H-29), 0.75 (d, *J* = 6.5 Hz, 3H, H-30); ^{13}C-NMR (125 MHz, CDCl$_3$) δ: 218.1 (C-3), 176.3 (C-28), 163.6 (C=N), 156.1, 153.3 (CONH-Boc), 83.0, 79.2 (C in Boc), 56.0 (C-17), 55.0 (C-5), 49.8 (C-9), 49.5 (C-19), 47.3 (C-4), 44.1 (C-18), 42.6 (C-14), 40.7 (C-8), 40.5 (C-4′), 39.6 (C-1), 38.7 (C-22, C-1′), 37.6 (C-13), 36.9 (C-10), 34.1 (C-2), 33.8 (C-16), 33.6 (C-7), 29.9 (C-20), 29.4 (C-15), 28.4, 28.0 (CH$_3$ in Boc), 27.7 (C-23), 27.0 (C-2′, C-3′), 26.6 (C-12), 23.0 (C-21,

C-29), 21.5 (C-24), 21.0 (C-11), 19.6 (C-6), 16.0 (C-25), 15.9 (C-26), 14.6 (C-30), 14.5 (C-27); Anal. Calcd. for $C_{45}H_{76}N_4O_6$: C, 70.27, H, 9.96. Found: C, 70.62, H, 9.91%. MS: m/z 791.56 [M + Na]$^+$ (calcd. for $C_{45}H_{76}N_4O_6$, 768.58).

3β-N-(2-tert-Butyloxycarbonylethylguanidine)-3-O-acetyl-lupane-28-amide (**10**), White powder, 86% yield; mp 176–177 °C (EtOH); $[\alpha]_D^{23}$ +3.8° (*c* 0.56, CHCl$_3$); IR (CHCl$_3$) ν_{max} 1724 (C=O), 3329 (NH) cm^{-1}; ^{1}H-NMR (500 MHz, CDCl$_3$) δ: 11.53 (s, 1H, NH in Boc), 8.65 (br s, 1H, NH–C=N), 6.88 (br s, 1H, CONH), 4.50–4.47 (m, 1H, H-3), 3.69–3.54 (m, 2H, H-2′), 3.47–3.35 (m, 2H, H-1′), 2.05 (s, 3H, CH$_3$CO–), 1.51, 1.50 (both s, 9H each, CH$_3$ in Boc), 2.46–0.96 (m, 25H, CH, CH$_2$ in pentacyclic skeleton), 0.94, 0.90, 0.87, 0.86, 0.85, 0.84 (all s, 3H each, H-23–H-27 and H-29), 0.79 (d, *J* = 9.5 Hz, 1H, H-5), 0.75 (d, *J* = 7.0 Hz, 3H, H-30); ^{13}C-NMR (125 MHz, CDCl$_3$) δ: 176.9 (C-28), 171.0 (<u>C</u>OCH$_3$), 163.1 (C=N), 153.5, 153.0 (CONH-Boc), 83.5 (C in Boc), 80.9 (C-3), 77.3 (C in Boc), 56.0 (C-17), 55.4 (C-5), 50.3 (C-9), 49.7 (C-19), 44.2 (C-18), 42.5 (C-14), 41.2 (C-2′), 40.7 (C-8), 39.7 (C-1′), 38.6 (C-22), 38.4 (C-1), 37.8 (C-4), 37.5 (C-13), 37.1 (C-10), 34.3 (C-7), 33.4 (C-16), 29.8 (C-20), 29.4 (C-15), 28.3, 28.0 (CH$_3$ in Boc), 27.9 (C-23), 26.9 (C-2), 23.7 (C-12), 23.1 (C-29), 23.0 (C-21), 21.3 (CO<u>C</u>H$_3$), 21.0 (C-11), 18.3 (C-6), 16.5 (C-25), 16.2 (C-24), 16.1 (C-26), 14.7 (C-30), 14.6 (C-27); Anal. Calcd. for $C_{45}H_{76}N_4O_7$: C, 68.84, H, 9.76. Found: C, 69.23, H, 9.69%. MS: m/z 807.54 [M + Na]$^+$ (calcd. for $C_{45}H_{76}N_4O_7$, 784.57).

3β-N-(4-tert-Butyloxycarbonylbutylguanidine)-3-O-acetyl-lupane-28-amide (**11**), White powder, 82% yield; mp 148–150 °C (EtOH); $[\alpha]_D^{17}$ −4° (*c* 0.52, CHCl$_3$); IR (CHCl$_3$) ν_{max} 1640, 1720 (C=O), 3288, 3337, 3410 (NH) cm^{-1}; ^{1}H-NMR (500 MHz, CDCl$_3$) δ: 11.48 (s, 1H, NH in Boc), 8.30 (t, *J* = 5.0 Hz, 1H, NH–C=N), 5.73 (t, *J* = 5.5 Hz, 1H, CONH), 4.46–4.43 (m, 1H, H-3), 3.42–3.37 (m, 2H, H-2′), 3.31–3.16 (m, 2H, H-1′), 2.01 (s, 3H, CH$_3$CO–), 1.47, 1.46 (both s, 9H each, CH$_3$ in Boc), 2.45–0.94 (m, 25H, CH, CH$_2$ in pentacyclic skeleton, 4H, H-2′, H-3′), 0.91, 0.90, 0.83, 0.82, 0.81, 0.80 (all s, 3H each, H-23–H-27 and H-29), 0.75 (d, *J* = 9.5 Hz, 1H, H-5), 0.71 (d, *J* = 7.0 Hz, 3H, H-30); ^{13}C-NMR (125 MHz, CDCl$_3$) δ: 176.2 (C-28), 170.9 (<u>C</u>OCH$_3$), 163.6 (C=N), 156.1, 153.3 (CONH-Boc), 83.0 (C in Boc), 81.0 (C-3), 79.2 (C in Boc), 56.0 (C-17), 55.4 (C-5), 50.3 (C-9), 49.5 (C-19), 44.2 (C-18), 42.6 (C-14), 40.8 (C-8), 40.5 (C-4′), 38.7 (C-1′, C-22), 38.4 (C-1), 37.8 (C-4), 37.5 (C-13), 37.1 (C-10), 34.4 (C-7), 33.6 (C-16), 29.9 (C-3′), 29.4 (C-20), 28.3, 28.0 (CH$_3$ in Boc), 27.8 (C-15), 27.3 (C-23), 26.9 (C-2′), 26.5 (C-2), 23.7 (C-12), 23.0 (C-21, C-29), 21.3 (CO<u>C</u>H$_3$), 21.0 (C-11), 18.2 (C-6), 16.5 (C-25), 16.2 (C-24, C-26), 14.6 (C-30), 14.5 (C-27); Anal. Calcd. for $C_{47}H_{80}N_4O_7$: C, 69.42, H, 9.92. Found: C, 69.74, H, 9.85%. MS: m/z 835.51 [M + Na]$^+$ (calcd. for $C_{47}H_{80}N_4O_7$, 812.60).

3β-N-[2-(N,N′-bis-tert-Butyloxycarbonylethylgyanidine)-aminoethyl]-3-O-acetyl-lupane-28-amide (**12**), White powder, 87% yield; mp 188–190 °C (EtOH); $[\alpha]_D^{19}$ +1.02° (*c* 0.96, CHCl$_3$); IR (CHCl$_3$) ν_{max} 1641, 1722 (C=O), 3335, 3442 (NH) cm^{-1}; ^{1}H-NMR (500 MHz, CDCl$_3$) δ: 11.52 (s, 2H, NH in Boc), 8.52 (br s, 2H, NH–C=N), 6.35 (t, 1H, *J* = 5.5 Hz, CONH), 4.49–4.46 (m, 1H, H-3), 3.53–3.43 (m, 4H, H-4′, H-4″), 3.41–3.22 (m, 2H, H-1′), 2.65–2.54 (m, 6H, H-2′, H-3′, H-3″), 2.04 (s, 3H, CH$_3$CO–), 1.50, 1.48 (both br s, 18H each, CH$_3$ in Boc), 2.32–0.98 (m, 25H, CH, CH$_2$ in pentacyclic skeleton), 0.93, 0.92, 0.85, 0.84, 0.83, 0.82 (all s, 3H each, H-23–H-27 and H-29), 0.78 (d, *J* = 9.0 Hz, 1H, H-5), 0.73 (d, *J* = 6.5 Hz, 3H, H-30); ^{13}C-NMR (125 MHz, CDCl$_3$) δ: 176.7 (C-28), 171.0 (<u>C</u>OCH$_3$), 163.5 (C=N), 155.9, 153.2 (CONH-Boc), 82.9 (C in Boc), 81.0 (C-3), 79.2 (C in Boc), 55.9 (C-17), 55.5 (C-5), 54.8 (C-2′), 53.8 (C-3′, C-3″), 50.4 (C-9), 49.6 (C-19), 43.8 (C-18), 42.5 (C-14), 40.8 (C-8), 39.0 (C-4′, C-4″), 38.9 (C-22), 38.4 (C-1), 38.0 (C-13), 37.8 (C-1′), 37.2 (C-4), 37.1 (C-10), 34.4 (C-7), 33.4 (C-16), 29.8 (C-20), 29.5 (C-15), 28.3, 28.1 (CH$_3$ in Boc), 28.0 (C-23), 27.0 (C-2), 23.7 (C-12), 23.1 (C-29), 23.0 (C-21), 21.3 (CO<u>C</u>H$_3$), 21.0 (C-11), 18.3 (C-6), 16.5 (C-25), 16.4 (C-24), 16.2 (C-26), 14.7 (C-30), 14.4 (C-27); Anal. Calcd. for $C_{60}H_{104}N_8O_{11}$: C, 64.72; H, 9.41. Found: C, 65.02; H, 9.34%. MS: m/z 1135.71 [M + Na]$^+$ (calcd. for $C_{60}H_{104}N_8O_{11}$, 1112.78).

3β-N-{[3-(3-tert-Butyloxycarbonylpropylgyanidine)piperazinyl]propyl}-3-O-acetyl-lupane-28-amide (**13**), White powder, 60% yield; mp 131–134 °C (EtOH); $[\alpha]_D^{21}$ −7.9° (*c* 0.57, CHCl$_3$); IR (CHCl$_3$) ν_{max} 1640, 1722 (C=O), 3289, 3333 (NH) cm^{-1}; ^{1}H-NMR (500 MHz, CDCl$_3$) δ: 11.50 (br s, 1H, NH in Boc), 8.53 (br s, 1H, NH–C=N), 6.86 (br s, 1H, CONH), 4.47–4.45 (m, 1H, H-3), 3.51–3.42 (m, 2H, H-1″), 3.32–3.27 (m, 2H, H-1′), 2.48–2.41 (m,

14H, H-3′–H-5′, H-2″–H-5″), 2.06 (s, 3H, CH$_3$CO–), 1.49 (br s, 18H, CH$_3$ in Boc), 2.33–0.99 (m, 25H, CH, CH$_2$ in pentacyclic skeleton, 2H, H-2′), 0.93, 0.92, 0.85, 0.84, 0.83, 0.81 (all s, 3H each, H-23–H-27 and H-29), 0.78 (d, J = 10.5 Hz, 1H, H-5), 0.73 (d, J = 6.5 Hz, 3H, H-30); ^{13}C-NMR (125 MHz, CDCl$_3$) δ: 176.4 (C-28), 171.0 (<u>C</u>OCH$_3$), 163.7 (C=N), 156.1, 153.0 (CONH-Boc), 82.8 (C in Boc), 80.9 (C-3), 79.2 (C in Boc), 56.4 (C-3′), 55.9 (C-17), 55.9 (C-3″), 55.4 (C-5), 53.1 (C-4′, C-4″, C-5′, C-5″), 50.3 (C-9), 49.6 (C-19), 44.1 (C-18), 42.6 (C-14), 40.8 (C-8), 39.8 (C-22, C-1″), 39.2 (C-1′), 38.8 (C-1), 38.4 (C-4), 37.8 (C-10), 37.4 (C-13), 34.5 (C-7), 33.7 (C-16), 29.8 (C-20), 29.4 (C-15), 28.3, 28.1 (CH$_3$ in Boc), 27.9 (C-23), 26.9 (C-2), 26.9 (C-2′), 25.9 (C-2″), 23.7 (C-12), 23.1 (C-21), 23.0 (C-29), 21.3 (CO<u>C</u>H$_3$), 21.0 (C-11), 18.2 (C-6), 16.5 (C-25), 16.3 (C-24), 16.2 (C-26), 14.6 (C-30), 14.5 (C-27); Anal. Calcd. for C$_{53}$H$_{92}$N$_6$O$_7$: C, 68.79, H, 10.02. Found: C, 69.04, H, 9.96%. MS: m/z 947.58 [M + Na]$^+$ (calcd. for C$_{53}$H$_{92}$N$_6$O$_7$, 924.70).

3β-[2-tert-Butyloxycarbonylguanidine-3-hydroxy-2-(hydroxymethyl)propyl]-3-O-acetyl-lupane-28-oate (**15a**), White powder, 63% yield; mp 106–108 °C (EtOH); [α]$_D^{21}$ −5.6° (*c* 0.48, CHCl$_3$); IR (CHCl$_3$) ν_{max} 1655, 1714 (C=O), 3271, 3437 (NH) cm^{-1}; ^{1}H-NMR (400 MHz, CDCl$_3$) δ: 11.47 (s, 1H, NH in Boc), 9.05 (s, 1H, NH–C=N), 4.49–4.45 (m, 1H, H-3), 4.25 (br s, 2H, H-1′), 3.82-3.77, 3.57–3.53 (both m, 2H each, H-3′, H-4′), 2.05 (s, 3H, CH$_3$CO–), 1.49, 1.47 (both br s, 9H each, CH$_3$ in Boc), 2.30–0.98 (m, 25H, CH, CH$_2$ in pentacyclic skeleton), 0.94, 0.90, 0.88, 0.86, 0.85, 0.83 (all s, 3H each, H-23–H-27 and H-29), 0.79 (d, J = 9.6 Hz, 1H, H-5), 0.75 (d, J = 6.4 Hz, 3H, H-30); ^{13}C-NMR (100 MHz, CDCl$_3$) δ: 176.3 (C-28), 171.1 (<u>C</u>OCH$_3$), 161.8 (C=N), 155.7, 152.8 (CONH-Boc), 83.8 (C in Boc), 80.9 (C-3), 80.1 (C in Boc), 62.9 (C-1′), 62.8 (C-4′), 62.8 (C-3′), 61.9 (C-2′), 57.3 (C-17), 55.4 (C-5), 50.2 (C-9), 49.0 (C-19), 44.0 (C-18), 42.5 (C-14), 40.7 (C-8), 38.4 (C-22), 38.0 (C-1), 37.8 (C-4), 37.1 (C-13, C-10), 34.3 (C-7), 31.8 (C-16), 29.8 (C-20), 29.7 (C-15), 28.1 (CH$_3$ in Boc), 28.0 (CH$_3$ in Boc, C-23), 26.9 (C-2), 23.7 (C-12), 23.0 (C-29), 22.7 (C-21), 21.3 (CO<u>C</u>H$_3$), 20.9 (C-11), 18.2 (C-6), 16.5 (C-25), 16.2 (C-24), 15.9 (C-26), 14.7 (C-30), 14.6 (C-27); Anal. Calcd. for C$_{47}$H$_{79}$N$_3$O$_{10}$: C, 66.72, H, 9.41. Found: C, 67.13, H, 9.36%. MS: m/z 868.45 [M + Na]$^+$ (calcd. for C$_{47}$H$_{79}$N$_3$O$_{10}$, 845.58).

3β-[2-tert-Butyloxycarbonylguanidine-3-hydroxy-2-(hydroxymethyl)propyl]-3-O-acetyl-urs-12-en-28-oate (**18a**), White powder, 75% yield; mp 118–120 °C (EtOH); [α]$_D^{21}$ +31.3° (*c* 0.53, CHCl$_3$); IR (CHCl$_3$) ν_{max} 1653, 1729 (C=O), 3271, 3443 (NH) cm^{-1}; ^{1}H-NMR (500 MHz, CDCl$_3$) δ: 11.50 (s, 1H, NH in Boc), 9.06 (s, 1H, NH–C=N), 5.26 (br s, 1H, H-12), 4.51–4.48 (m, 1H, H-3), 4.13 (br s, 2H each, H-1′), 3.82–3.76 (m, 2H, H-3′), 3.54–3.52 (m, 2H, H-4′), 2.26 (d, J = 11.0 Hz, 1H, H-18), 2.04 (s, 3H, CH$_3$CO–), 1.99–0.81 (m, 22H, CH, CH$_2$ in pentacyclic skeleton), 1.50, 1.47 (both br s, 9H each, CH$_3$ in Boc), 1.07, 0.94, 0.93, 0.86, 0.85, 0.83, 0.75 (all s, 3H each, H-23–H-27, H-29 and H-30); ^{13}C-NMR (125 MHz, CDCl$_3$) δ: 177.4 (C-28), 171.0 (<u>C</u>OCH$_3$), 161.8 (C=N), 155.6, 152.8 (CONH-Boc), 138.1 (C-13), 125.8 (C-12), 83.7 (C in Boc), 80.9 (C-3), 79.9 (C in Boc), 63.4 (C-1′), 62.9 (C-3′), 62.6 (C-4′), 61.6 (C-2′), 55.3 (C-5), 52.8 (C-18), 48.5 (C-17), 47.5 (C-9), 42.0 (C-14), 39.5 (C-8), 39.0 (C-4), 38.7 (C-19), 38.3 (C-20), 37.7 (C-1), 36.8 (C-22), 36.5 (C-10), 32.9 (C-7), 30.6 (C-21), 28.1 (CH$_3$ in Boc), 28.0 (C-15), 28 (C-23), 24.1 (C-16), 23.6 (C-2), 23.5 (C-27), 23.3 (C-11), 21.3 (C-30), 21.0 (CO<u>C</u>H$_3$), 18.2 (C-6), 17.0 (C-29), 16.9 (C-26), 16.7 (C-24), 15.5 (C-25); Anal. Calcd. for C$_{47}$H$_{77}$N$_3$O$_{10}$: C, 66.87, H, 9.19. Found: C, 67.27, H, 9.14%. MS: m/z 866.43 [M + Na]$^+$ (calcd. for C$_{47}$H$_{77}$N$_3$O$_{10}$, 843.56).

3β-[2-tert-Butyloxycarbonylguanidine-3-hydroxy-2-(hydroxymethyl)propyl]-3-O-acetyl-olean-12-en-28-oate (**20a**), White powder, 79% yield; mp 140–142 °C (EtOH); [α]$_D^{19}$ +30.4° (*c* 0.56, CHCl$_3$); IR (CHCl$_3$) ν_{max} 1618, 1727 (C=O), 3434 (NH) cm^{-1}; ^{1}H-NMR (400 MHz, CDCl$_3$) δ: 11.49 (s, 1H, NH in Boc), 9.03 (s, 1H, NH–C=N), 5.30 (br s, 1H, H-12), 5.10 (br s, 1H, OH), 4.49–4.45 (m, 1H, H-3), 4.26, 4.15 (both d, J = 11.6 Hz, 1H each, H-1′), 3.77, 3.54 (both d, J = 12.0 Hz, J = 11.6 Hz, 2H each, H-3′, H-4′), 2.85 (d, J = 10.0 Hz, 1H, H-18), 2.03 (s, 3H, CH$_3$CO–), 1.99–0.80 (m, 22H, CH, CH$_2$ in pentacyclic skeleton), 1.51, 1.48 (both br s, 9H each, CH$_3$ in Boc), 1.12, 0.91, 0.90, 0.89, 0.85, 0.84, 0.71 (3H each, all s, H-23–H-27, H-29 and H-30); ^{13}C-NMR (100 MHz, CDCl$_3$) δ: 177.4 (C-28), 171.0 (<u>C</u>OCH$_3$), 161.8 (C=N), 155.6, 152.8 (CONH-Boc), 138.1 (C-13), 125.8 (C-12), 83.7 (C in Boc), 80.9 (C-3), 79.9 (C in Boc), 63.4 (C-1′), 62.9 (C-3′), 62.6 (C-4′), 61.6 (C-2′), 55.3 (C-5), 52.8 (C-18), 48.5 (C-17), 47.5 (C-9), 42.0 (C-14), 39.5 (C-8), 39.0 (C-4), 38.7 (C-19), 38.3 (C-20), 37.7 (C-1), 36.8 (C-22), 36.5 (C-10), 32.9 (C-7), 30.6 (C-21), 28.1 (CH$_3$

in Boc), 28.0 (C-15, C-23), 24.1 (C-16), 23.6 (C-2), 23.5 (C-27), 23.3 (C-11), 21.3 (C-30), 21.0 (COC$\underline{H_3}$), 18.2 (C-6), 17.0 (C-29), 16.9 (C-26), 16.7 (C-24), 15.5 (C-25); Anal. Calcd. for $C_{47}H_{77}N_3O_{10}$: C, 66.87, H, 9.19. Found: C, 67.24, H, 9.14%. MS: *m/z* 866.47 [M + Na]$^+$ (calcd. for $C_{47}H_{77}N_3O_{10}$, 843.56).

3.1.4. General Procedure for the Synthesis of Compounds **9a–13a**, **15b**, **18b** and **20b**

Compounds **9–13** and **15a**, **18a**, **20a** (0.2 mmol) in 1 mL of dry CH_2Cl_2 were treated with TFA (1 mL) and the mixture was stirred for 4–6 h at room temperature (TLC control, hexane:EtOAc, 1:1, *v/v*). The solution was evaporated to dryness to obtain pure compounds **9a–13a**, **15b**, **18b** and **20b**.

N-(4-Butylgyanidine)-3-oxolupane-28-amide trifluoroacetate (**9a**), White powder, 96% yield; mp 142–144 °C (EtOH); $[\alpha]_D^{19}$ −0.6° (*c* 0.34, CHCl$_3$); IR (CHCl$_3$) ν_{max} 1669 (C=O), 3206, 3437 (NH) cm^{-1}; ^{19}F-NMR (376.50 MHz, CDCl$_3$) δ: −75.91; ^{1}H-NMR (400 MHz, CDCl$_3$) δ: 11.13 (br s, 1H, NH in Boc), 7.50 (br s, 1H, NH–C=N), 6.82 (br s, 2H, NH$_2$), 6.39 (br s, 1H, CONH), 3.26 (m, 4H, H-1′, H-4′), 2.45–1.19 (m, 26H, CH, CH$_2$ in pentacyclic skeleton, 4H, H-2′, H-3′), 1.07, 1.01, 0.96, 0.92, 0.90 (all s, 3H each, H-23–H-27), 0.85 (d, *J* = 6.0 Hz, 3H, H-29), 0.75 (d, *J* = 6.0 Hz, 3H, H-30); ^{13}C-NMR (100 MHz, CDCl$_3$) δ: 220.4 (C-3), 178.8 (C-28), 161.4 (q, $J_{C,F}$ = 37 Hz), 157.2 (C=N), 117.5 (q, $J_{C,F}$ = 288 Hz), 56.4 (C-17), 54.8 (C-5), 49.6 (C-9), 49.2 (C-19), 47.4 (C-4), 44.4 (C-18), 42.7 (C-14), 40.9 (C-1, C-4′), 40.7 (C-8), 38.8 (C-22), 38.1 (C-1′), 38.0 (C-13), 36.8 (C-10), 34.2 (C-2), 33.6 (C-16), 33.3 (C-7), 29.9 (C-20), 29.7 (C-3′), 29.4 (C-15), 27.0 (C-2′), 26.7 (C-23), 25.5 (C-12), 23.0 (C-21), 22.9 (C-29), 20.9 (C-11, C-24), 19.6 (C-6), 15.8 (C-25), 15.6 (C-26), 14.5 (C-30), 14.4 (C-27); Anal. Calcd. for $C_{37}H_{61}F_3N_4O_4$: C, 65.08, H, 9.00. Found: C, 65.52, H, 8.94%. MS: *m/z* 569.43 [M + H]$^+$ (calcd. for $C_{35}H_{60}N_4O_2$, 568.47).

3β-N-(2-Ethylgyanidine)-3-O-acetyl-lupane-28-amide trifluoroacetate (**10a**), White powder, 93% yield; mp 132–134 °C (EtOH); $[\alpha]_D^{19}$ −5.6° (*c* 0.31, CHCl$_3$); IR (CHCl$_3$) ν_{max} 1671 (C=O), 3351 (NH) cm^{-1}; ^{19}F-NMR (470.59 MHz, CDCl$_3$) δ: −75.80; ^{1}H-NMR (500 MHz, CDCl$_3$) δ: 10.63 (br s, 1H, NH in Boc), 8.05 (br s, 1H, NH–C=N), 7.00 (br s, 2H, NH$_2$), 6.79 (br s, 1H, CONH), 4.48–4.45 (m, 1H, H-3), 3.39–3.28 (m, 4H, H-1′, H-2′), 2.09 (s, 3H, CH$_3$CO–), 2.31–1.01 (m, 25H, CH, CH$_2$ in pentacyclic skeleton), 0.95, 0.92, 0.89, 0.86, 0.85, 0.83 (all s, 3H each, H-23–H-27 and H-29), 0.79 (d, *J* = 11.0 Hz, 1H, H-5), 0.75 (d, *J* = 6.0 Hz, 3H, H-30); ^{13}C-NMR (125 MHz, CDCl$_3$) δ: 179.6 (C-28), 171.6 ($\underline{C}$OCH$_3$), 161.2 (q, $J_{C,F}$ = 37 Hz), 157.9 (C=N), 117.2 (q, $J_{C,F}$ = 287 Hz), 81.2 (C-3), 56.4 (C-17), 55.4 (C-5), 50.2 (C-9), 49.3 (C-19), 44.3 (C-18), 42.6 (C-14), 40.7 (C-1, C-2′), 40.6 (C-8), 38.6 (C-1′), 38.4 (C-22), 37.8 (C-4, C-13), 37.0 (C-10), 34.2 (C-7), 33.1 (C-16), 29.9 (C-20), 29.3 (C-15), 27.9 (C-23), 26.9 (C-2), 23.6 (C-12), 23.0 (C-21, C-29), 21.3 (COC$\underline{H_3}$), 21.0 (C-11), 18.1 (C-6), 16.4 (C-25), 16.0 (C-24), 15.9 (C-26), 14.5 (C-27, C-30); Anal. Calcd. for $C_{37}H_{61}F_3N_4O_5$: C, 63.59, H, 8.80. Found: C, 63.88, H, 8.73%. MS: *m/z* 585.62 [M + H]$^+$ (calcd. for $C_{35}H_{60}N_4O_3$, 584.47).

3β-N-(4-Butylgyanidine)-3-O-acetyl-lupane-28-amide trifluoroacetate (**11a**), White powder, 95% yield; mp 148–150 °C (EtOH); $[\alpha]_D^{17}$ −12° (*c* 0.49, CHCl$_3$); IR (CHCl$_3$) ν_{max} 1672 (C=O), 3207, 3354 (NH) cm^{-1}; ^{19}F-NMR (470.59 MHz, CDCl$_3$) δ: −77.19; ^{1}H-NMR (500 MHz, MeOD) δ: 4.48–4.45 (m, 1H, H-3), 3.28–3.15 (m, 4H, H-1′, H-4′), 2.04 (s, 3H, CH$_3$CO–), 2.61–1.04 (m, 25H, CH, CH$_2$ in pentacyclic skeleton, 4H, H-2′, H-3′), 1.02, 0.98, 0.92, 0.90, 0.89, 0.88 (3H each, all s, H-23–H-27 and H-29), 0.84 (d, *J* = 11.0 Hz, 1H, H-5), 0.79 (d, *J* = 7.0 Hz, 3H, H-30); ^{13}C-NMR (125 MHz, MeOD) δ: 179.6 (C-28), 172.9 ($\underline{C}$OCH$_3$), 161.2 (q, $J_{C,F}$ = 37 Hz), 158.8 (C=N), 117.2 (q, $J_{C,F}$ = 287 Hz), 82.6 (C-3), 57.5 (C-17), 57.0 (C-5), 51.9 (C-9), 51.1 (C-19), 45.5 (C-18), 43.8 (C-14), 42.2 (C-8, C-4′), 40.0 (C-22), 39.8 (C-1), 39.4 (C-1′), 39.0 (C-4), 38.9 (C-13), 38.4 (C-10), 35.8 (C-7), 34.1 (C-16), 31.3 (C-20), 30.7 (C-3′), 28.7 (C-15), 28.5 (C-23), 28.1 (C-2′), 27.4 (C-2), 24.8 (C-12), 24.2 (C-29), 23.7 (C-21), 22.5 (C-11), 21.3 (COC$\underline{H_3}$), 19.4 (C-6), 17.2 (C-25), 17.0 (C-24, C-26), 15.3 (C-30), 15.2 (C-27); Anal. Calcd. for $C_{39}H_{65}F_3N_4O_5$: C, 64.44, H, 9.01. Found: C, 64.87, H, 8.94%. MS: *m/z* 613.48 [M + H]$^+$ (calcd. for $C_{37}H_{64}N_4O_3$, 612.50).

3β-N-[2-(N,N′-bis-Ethylgyanidine)-aminoethyl]-3-O-acetyl-lupane-28-amide trifluoroacetate (**12a**), White powder, 96% yield; mp 116–118 °C (EtOH); $[\alpha]_D^{17}$ −10.5° (*c* 0.2, C$_2$H$_5$OH); IR (CHCl$_3$) ν_{max} 1681 (C=O), 3199, 3362 (NH) cm^{-1}; ^{19}F-NMR (470.59 MHz, MeOD) δ: −76.98; ^{1}H-NMR (500 MHz, MeOD) δ:

4.48–4.44 (m, 1H, H-3), 3.73–3.70 (m, 4H, H-4′, H-4″), 3.56–3.53 (m, 2H, H-1′), 3.46 (t, *J* = 6.5 Hz, 4H, H-3′, H-3″), 3.29–3.26 (m, 2H, H-2′), 2.04 (s, 3H, CH_3CO–), 2.52–1.06 (m, 25H, CH, CH_2 in pentacyclic skeleton), 1.02, 0.97, 0.92, 0.90, 0.88, 0.87 (all s, 3H each, H-23–H-27 and H-29), 0.85 (d, *J* = 11.5 Hz, 1H, H-5), 0.79 (d, *J* = 6.5 Hz, 3H, H-30); ^{13}C-NMR (125 MHz, MeOD) δ: 181.2 (C-28), 173.0 (COCH$_3$), 161.7 (q, $J_{C,F}$ = 37 Hz), 159.1 (C=N), 117.5 (q, $J_{C,F}$ = 288 Hz), 35.6 (C-4″), 53.4 (C-3″), 82.6 (C-3), 57.6 (C-17), 56.9 (C-5), 53.9 (C-2′), 53.4 (C-3′), 51.8 (C-9), 51.0 (C-19), 45.5 (C-18), 43.8 (C-14), 42.2 (C-8), 39.7 (C-1), 39.7 (C-22), 39.0 (C-13, C-4), 38.4 (C-1′), 37.8 (C-10), 35.6 (C-4′, C-7), 33.8 (C-16), 31.3 (C-20), 30.7 (C-15), 28.6 (C-23), 28.4 (C-2), 24.8 (C-12), 24.1 (C-21), 23.5 (C-29), 22.3 (C-11), 21.3 (COCH$_3$), 19.4 (C-6), 17.1 (C-25), 17.0 (C-24), 16.9 (C-26), 15.2 (C-30), 15.1 (C-27); Anal. Calcd. for $C_{44}H_{74}F_6N_8O_7$: C, 56.16, H, 7.93. Found: C, 56.63, H, 7.86%. MS: *m*/*z* 713.59 [M + H]$^+$ (calcd. for $C_{40}H_{72}N_8O_3$, 712.57).

3β-N-{[3-(3-Propylguanidine)piperazinyl]propyl}-3-O-acetyl-lupane-28-amide trifluoroacetate (**13a**), White powder, 96% yield; mp 92–94 °C (EtOH); [α]$_D^{19}$ +8° (*c* 0.54, CHCl$_3$); IR (CHCl$_3$) ν_{max} 1673, 1773 (C=O), 3190, 3367 (NH) cm^{-1}; ^{19}F-NMR (470.59 MHz, MeOD) δ: −77.25; ^{1}H-NMR (500 MHz, MeOD) δ: 4.48–4.45 (m, 1H, H-3), 3.49–3.40 (m, 4H, H-2″, H-3″), 3.31–3.22 (m, 8H, H-4′, H-4″, H-5′, H-5″), 3.13–3.10 (m, 2H, H-1″), 3.00–2.98 (m, 2H, H-1′), 2.58–0.84 (m, 26H, CH, CH_2 in pentacyclic skeleton, 4H, H-2′, H-3′), 2.08 (s, 3H, CH_3CO–), 1.02, 0.99, 0.92, 0.89, 0.86, 0.84 (all s, 3H each, H-23–H-27 and H-29), 0.80 (d, *J* = 7.0 Hz, 3H, H-30); ^{13}C-NMR (125 MHz, MeOD) δ: 180.2 (C-28), 173.0 (COCH$_3$), 161.7 (q, $J_{C,F}$ = 37 Hz), 158.9 (C=N), 117.5 (q, $J_{C,F}$ = 288 Hz), 82.6 (C-3), 57.6 (C-17), 57.0 (C-5), 56.0 (C-3″), 55.2 (C-3′), 51.9 (C-9), 51.0 (C-19), 50.1 (C-4′, C-4″, C-5′, C-5″), 45.5 (C-18), 43.8 (C-14), 42.2 (C-8), 39.9 (C-1′), 39.8 (C-1″), 39.6 (C-1, C-22), 39.0 (C-4), 38.9 (C-13), 38.4 (C-10), 35.7 (C-7), 34.0 (C-16), 31.3 (C-20), 30.7 (C-15), 28.6 (C-23), 28.5 (C-2″), 25.6 (C-2′), 24.9 (C-2), 24.2 (C-12), 23.6 (C-21, C-29), 22.4 (C-11), 21.3 (COCH$_3$), 19.4 (C-6), 17.1 (C-25), 17.0 (C-24), 16.9 (C-26), 15.2 (C-30), 15.1 (C-27); Anal. Calcd. for $C_{45}H_{77}F_3N_6O_5$: C, 64.41, H, 9.25. Found: C, 64.88, H, 9.19. MS: *m*/*z* 747.58 [M + Na]$^+$ (calcd. for $C_{43}H_{76}N_6O_3$, 724.60).

3β-[2-Guanidine-3-hydroxy-2-(hydroxymethyl)propyl]-3-O-acetyl-lupane-28-oate trifluoroacetate (**15b**), White powder, 95% yield; mp 126–129 °C (EtOH); [α]$_D^{19}$ −9.5° (*c* 0.34, CHCl$_3$); IR (CHCl$_3$) ν_{max} 1620, 1698 (C=O), 3437 (NH) cm^{-1}; ^{19}F-NMR (470.59 MHz, CDCl$_3$) δ: −75.97; ^{1}H-NMR (500 MHz, MeOD) δ: 4.50–4.47 (m, 1H, H-3), 4.31 (2H, m, H-1′), 3.75 (m, 4H, H-3′, H-4′), 2.18–0.99 (m, 26H, CH, CH_2 in pentacyclic skeleton), 2.08 (s, 3H, CH_3CO–), 0.98, 0.97, 0.96, 0.87, 0.86 (all s, 3H each, H-23–H-27), 0.89 (d, *J* = 6.0 Hz, 3H, H-29), 0.76 (d, *J* = 6.0 Hz, 3H, H-30); ^{13}C-NMR (125 MHz, MeOD) δ: 177.4 (C-28), 172.3 (COCH$_3$), 161.4 (q, $J_{C,F}$ = 37 Hz), 157.5 (C=N), 117.5 (q, $J_{C,F}$ = 288 Hz), 81.7 (C-3), 63.1 (C-3′, C-4′), 62.0 (C-2′), 60.2 (C-1′), 57.5 (C-17), 55.4 (C-5), 50.2 (C-9), 48.8 (C-19), 44.1 (C-18), 42.6 (C-14), 40.7 (C-8), 38.4 (C-22), 38.2 (C-1), 37.8 (C-4, C-13), 37.0 (C-10), 34.2 (C-7), 31.9 (C-16), 29.7 (C-20), 29.0 (C-15), 27.9 (C-23), 26.8 (C-2), 23.6 (C-12), 22.9 (C-29, C-21), 21.3 (COCH$_3$), 20.9 (C-11), 18.1 (C-6), 16.4 (C-25), 16.1 (C-24), 15.8 (C-26), 14.6 (C-27, C-30); Anal. Calcd. for $C_{39}H_{64}F_3N_3O_8$: C, 61.64; H, 8.49. Found: C, 62.34; H, 8.43%. MS: *m*/*z* 646.39 [M + H]$^+$ (calcd. for $C_{37}H_{63}N_3O_6$, 645.47).

3β-[2-Guanidine-3-hydroxy-2-(hydroxymethyl)propyl]-3-O-acetyl-urs-12-en-28-oate trifluoroacetate (**18b**), White powder, 98% yield; mp 124–126 °C (EtOH); [α]$_D^{19}$ +30.0° (*c* 0.72, CH$_2$Cl$_2$); IR (CHCl$_3$) ν_{max} 1619, 1682 (C=O), 3438 (NH) cm^{-1}; ^{19}F-NMR (470.59 MHz, MeOD) δ: −77.17; ^{1}H-NMR (500 MHz, MeOD) δ: 5.29 (br s, 1H, H-12), 4.49–4.46 (m, 1H, H-3), 4.31, 4.17 (both d, *J* = 11.5 Hz, 1H each, H-1′), 3.75 (m, 4H, H-3′, H-4′), 2.24 (d, *J* = 11.0 Hz, 1H, H-18), 2.04 (s, 3H, CH_3CO–), 2.13–0.87 (m, 22H, CH, CH_2 in pentacyclic skeleton), 1.15, 1.01, 0.97, 0.98, 0.92, 0.90, 0.81 (3H each, all s, H-23–H-27, H-29 and H-30); ^{13}C-NMR (125 MHz, MeOD) δ: 179.2 (C-28), 173.0 (COCH$_3$), 161.4 (q, $J_{C,F}$ = 37 Hz), 159.2 (C=N), 140.1 (C-13), 127.2 (C-12), 117.5 (q, $J_{C,F}$ = 288 Hz), 82.5 (C-3), 64.3 (C-3′), 64.2 (C-4′), 62.7 (C-2′), 62.2 (C-1′), 56.8 (C-5), 54.4 (C-18), 49.5 (C-17), 48.6 (C-9), 43.3 (C-14), 40.9 (C-8), 40.4 (C-4, C-19), 39.6 (C-20), 38.8 (C-1), 38.1 (C-22), 38.0 (C-10), 34.2 (C-7), 31.7 (C-21), 29.2 (C-15), 28.7 (C-23), 25.5 (C-16), 24.6 (C-2), 24.5 (C-27), 24.4 (C-11), 21.6 (C-30), 21.3 (COCH$_3$), 19.4 (C-6), 17.9 (C-29), 17.8 (C-26), 17.3 (C-24), 16.2 (C-25); Anal. Calcd. for $C_{39}H_{62}F_3N_3O_8$: C, 61.80, H, 8.25. Found: C, 62.33, H, 8.19%. MS: *m*/*z* 644.47 [M + H]$^+$ (calcd. for $C_{37}H_{61}N_3O_6$, 643.46).

3β-[2-Guanidine-3-hydroxy-2-(hydroxymethyl)propyl]-3-O-acetyl-olean-12-en-28-oate trifluoroacetate (**20b**), White powder, 97% yield; mp 130–132 °C (EtOH); $[\alpha]_D^{17}$ +26.6° (*c* 0.53, C_2H_5OH); IR ($CHCl_3$) ν_{max} 1619, 1682 (C=O), 3438 (NH) cm^{-1}; ^{19}F-NMR (470.59 MHz, $CDCl_3$) δ: −77.24; ^{1}H-NMR (500 MHz, MeOD) δ: 5.31 (br s, 1H, H-12), 4.49–4.46 (m, 1H, H-3), 4.33, 4.20 (d, *J* = 11.5 Hz, 1H each, H-1′), 3.75 (m, 4H, H-3′, H-4′), 2.89 (d, 1H, *J* = 9.5 Hz, H-18), 2.09–0.88 (m, 22H, CH, CH_2 in pentacyclic skeleton), 2.04 (s, 3H, CH_3CO–), 1.19, 1.00, 0.97, 0.95, 0.90, 0.89, 0.79 (all s, 3H each, H-23–H-27, H-29 and H-30); ^{13}C-NMR (125 MHz, MeOD) δ: 178.9 (C-28), 173.0 ($\underline{C}OCH_3$), 161.4 (q, $J_{C,F}$ = 37 Hz), 159.2 (C=N), 145.0 (C-13), 124.1 (C-12), 117.5 (q, $J_{C,F}$ = 288 Hz), 82.6 (C-3), 64.3 (C-4′), 64.2 (C-3′), 62.7 (C-2′), 62.3 (C-1′), 56.9 (C-5), 48.5 (C-9, C-17), 47.1 (C-19), 43.0 (C-14), 42.0 (C-18), 40.8 (C-8), 39.5 (C-1), 38.9 (C-4), 38.3 (C-10), 34.9 (C-22), 33.9 (C-30), 33.7 (C-7), 33.6 (C-21), 31.7 (C-20), 28.9 (C-15), 28.7 (C-23), 26.6 (C-27), 24.7 (C-11, C-29), 24.3 (C-2), 24.2 (C-16), 21.3 ($CO\underline{C}H_3$), 19.5 (C-6), 17.9 (C-26), 17.3 (C-24), 16.1 (C-25); Anal. Calcd. for $C_{39}H_{62}F_3N_3O_8$: C, 61.80, F, 7.52, H, 8.25. Found: C, 62.12, H, 8.21%. MS: *m/z* 644.57 [M + H]$^+$ (calcd. for $C_{37}H_{61}N_3O_6$, 643.46).

3.1.5. General Procedure for the Synthesis of Compounds **9b–12b**, **15c**, **18c** and **20c**

The compounds **9a–12a**, **15b**, **18b**, **20b** (0.2 g) were dissolved in 2 mL MeOH and 5M HCl was added dropwise until the precipiate formed. The solution was evaporated to dryness and this procedure was repeated three times. The precipitate which formed was filtered off and washed with water to pH = 7. The salts **9b–12b** and **15c**, **18c**, **20c** were obtained as white solids with a quantitative yield.

N-(4-Butylgyanidine)-3-oxolupane-28-amide dihydrochloride (**9b**), White powder, 87% yield; mp 176–178 °C (EtOH); $[\alpha]_D^{21}$ −11° (*c* 0.29, C_2H_5OH); IR ($CHCl_3$) ν_{max} 1721 (C=O), 3342 (NH) cm^{-1}; ^{1}H-NMR (500 MHz, d_6-DMSO) δ: 7.71 (br s, 1H, NH), 7.59 (br s, 1H, CONH), 3.10–3.01 (m, 4H, H-1′, H-4′), 2.61–1.04 (m, 26H, CH, CH_2 in pentacyclic skeleton, 4H, H-2′, H-3′), 0.99, 0.93, 0.91, 0.87, 0.86 (3H each, all s, H-23–H-27), 0.81 (d, *J* = 6.5 Hz, 3H, H-29), 0.71 (d, *J* = 7.0 Hz, 3H, H-30); ^{13}C-NMR (125 MHz, d_6-DMSO) δ: 217.1 (C-3), 176.1 (C-28), 157.4 (C=N), 55.7 (C-17), 54.3 (C-5), 49.6 (C-9), 49.5 (C-19), 47.0 (C-4), 43.8 (C-18), 42.6 (C-14), 40.9 (C-4′), 40.3 (C-8), 39.7 (C-1), 38.1 (C-22, C-1′), 37.0 (C-13), 36.8 (C-10), 34.1 (C-2), 33.8 (C-16), 32.7 (C-7), 30.0 (C-20), 29.4 (C-15), 27.2 (C-3′), 27.0 (C-2′), 26.9 (C-23), 26.4 (C-12), 23.6 (C-29), 23.1 (C-21), 21.7 (C-11), 21.2 (C-24), 19.7 (C-6), 16.2 (C-25), 16.1 (C-26), 15.0 (C-30), 14.6 (C-27); Anal. Calcd. for $C_{35}H_{62}Cl_2N_4O_2$: C, 65.50, Cl, 11.05, H, 9.74. Found: C, 65.97, Cl, 11.78, H, 9.68%. MS: *m/z* 569.49 [M + H]$^+$ (calcd. for $C_{35}H_{60}N_4O_2$, 568.47).

3β-N-(2-Ethylgyanidine)-3-O-acetyl-lupane-28-amide hydrochloride (**10b**), White powder, 82% yield; mp 192–194 °C (EtOH); $[\alpha]_D^{17}$ −16° (*c* 0.23, C_2H_5OH); IR ($CHCl_3$) ν_{max} 1652, 1716 (C=O), 3155, 3327 (NH) cm^{-1}; ^{1}H-NMR (500 MHz, MeOD) δ: 4.48–4.45 (m, 1H, H-3), 3.43–3.16 (m, 4H, H-1′, H-2′), 2.04 (s, 3H, CH_3CO–), 2.60–0.81 (m, 25H, CH, CH_2 in pentacyclic skeleton), 1.02, 0.98, 0.92, 0.90, 0.88, 0.87 (all s, 3H each, H-23–H-27 and H-29), 0.80 (d, *J* = 6.5 Hz, 3H, H-30); ^{13}C-NMR (125 MHz, MeOD) δ: 180.8 (C-28), 173.0 ($\underline{C}OCH_3$), 159.0 (C=N), 82.6 (C-3), 57.6 (C-17), 57.0 (C-5), 51.1 (C-9, C-19), 45.5 (C-18), 42.7 (C-14, C-1′), 42.2 (C-8), 39.8 (C-1), 39.7 (C-2′), 39.4 (C-22), 39.0 (C-4), 38.9 (C-13), 38.4 (C-10), 35.7 (C-7), 34.0 (C-16), 31.4 (C-20), 30.7 (C-15), 28.6 (C-23), 28.5 (C-2), 24.8 (C-12), 24.1 (C-21), 23.6 (C-29), 22.4 (C-11), 21.3 ($CO\underline{C}H_3$), 19.4 (C-6), 17.1 (C-25), 16.9 (C-24, C-26), 15.2 (C-30), 15.1 (C-27); Anal. Calcd. for $C_{35}H_{61}ClN_4O_3$: C, 67.66, Cl, 5.71, H, 9.90. Found: C, 67.99, Cl, 5.40, H, 9.84%. MS: *m/z* 585.54 [M + H]$^+$ (calcd. for $C_{35}H_{60}N_4O_3$, 584.47).

3β-N-(4-Butylgyanidine)-3-O-acetyl-lupane-28-amide hydrochloride (**11b**), White powder, 79% yield; mp 156–158 °C (EtOH); $[\alpha]_D^{22}$ −14.5° (*c* 0.53, C_2H_5OH); IR ($CHCl_3$) ν_{max} 1645, 1716 (C=O), 3168, 3338 (NH) cm^{-1}; ^{1}H-NMR (500 MHz, MeOD) δ: 7.45 (br s, 1H, NH), 4.48–4.45 (m, 1H, H-3), 3.23–3.18 (m, 4H, H-1′, H-4′), 2.04 (s, 3H, CH_3CO–), 2.61–0.84 (m, 26H, CH, CH_2 in pentacyclic skeleton, 4H, H-2′, H-3′), 1.01, 0.98, 0.92, 0.89, 0.88, 0.86 (all s, 3H each, H-23–H-27 and H-29), 0.79 (d, *J* = 7.0 Hz, 3H, H-30); ^{13}C-NMR (125 MHz, MeOD) δ: 179.6 (C-28), 172.9 ($\underline{C}OCH_3$), 158.8 (C=N), 82.6 (C-3), 57.5 (C-17), 57.0 (C-5), 51.9 (C-9), 51.1 (C-19), 45.5 (C-18), 43.8 (C-14), 42.4 (C-8), 42.2 (C-4′), 40.0 (C-22, C-1′), 39.8 (C-1), 39.0 (C-4), 38.9 (C-13), 38.4 (C-10), 35.8 (C-7), 34.1 (C-16), 31.4 (C-20), 30.8 (C-3′), 28.6 (C-15), 28.2 (C-23,

C-2'), 27.4 (C-2), 24.8 (C-12), 24.4 (C-21), 24.2 (C-29), 22.5 (C-11), 21.3 (CO$\underline{C}$H$_3$), 19.4 (C-6), 17.1 (C-25), 17 (C-24), 16.9 (C-26), 15.2 (C-30), 15.1 (C-27); Anal. Calcd. for C$_{37}$H$_{65}$ClN$_4$O$_3$: C, 68.43, Cl, 5.46, H, 10.09. Found: C, 68.88; Cl, 5.12; H, 10.02%. MS: *m/z* 614.51 [M + 2H]$^+$ (calcd. for C$_{37}$H$_{64}$N$_4$O$_3$, 612.50).

3β-N-[2-(N,N'-bis-Ethylgyanidine)-aminoethyl]-3-O-acetyl-lupane-28-amide dihydrochloride (**12b**), White powder, 88% yield; mp 198–200 °C (EtOH); [α]$_D^{17}$ −16° (*c* 0.29, C$_2$H$_5$OH); IR (CHCl$_3$) ν_{max} 1637, 1672, 1734 (C=O), 3162, 3271, 3324 (NH) cm^{-1}; ^{1}H-NMR (500 MHz, MeOD) δ: 4.48–4.45 (m, 1H, H-3), 3.90–3.83 (m, 4H, H-4', H-4''), 3.66–3.59 (m, 6H, H-2', H-3', H-3''), 3.38–3.33 (m, 2H, H-1'), 2.04 (s, 3H, CH$_3$CO–), 2.51–0.85 (m, 26H, CH, CH$_2$ in pentacyclic skeleton), 1.03, 0.99, 0.93, 0.90, 0.88, 0.87 (all s, 3H each, H-23–H-27 and H-29), 0.80 (d, *J* = 6.5 Hz, 3H, H-30); ^{13}C-NMR (125 MHz, MeOD) δ: 181.0 (C-28), 172.9 (CO$\underline{C}$H$_3$), 158.9 (C=N), 82.5 (C-3), 57.7 (C-17), 56.9 (C-5), 53.7 (C-3', C-3''), 53.6 (C-2'), 51.8 (C-9), 51.0 (C-19), 45.5 (C-18), 43.8 (C-14), 42.2 (C-8), 39.7 (C-22), 39.0 (C-1, C-13), 38.4 (C-4), 37.7 (C-10, C-1'), 35.7 (C-7), 35.0 (C-4', C-4''), 33.9 (C-16), 31.3 (C-20), 30.8 (C-15), 28.6 (C-23), 28.4 (C-2), 24.8 (C-12), 24.2 (C-21), 23.6 (C-29), 22.4 (C-11), 21.3 (CO$\underline{C}$H$_3$), 19.4 (C-6), 17.1 (C-25, C-24), 16.9 (C-26), 15.2 (C-30), 15.1 (C-27); Anal. Calcd. for C$_{40}$H$_{74}$Cl$_2$N$_8$O$_3$: C, 61.13, Cl, 9.02, H, 9.49. Found: C, 61.55, Cl, 11.40, H, 9.42%. MS: *m/z* 713.51 [M + H]$^+$ (calcd. for C$_{40}$H$_{72}$N$_8$O$_3$, 712.57).

3β-[2-Guanidine-3-hydroxy-2-(hydroxymethyl)propyl]-3-O-acetyl-lupane-28-oate hydrochloride (**15c**), White powder, 68% yield; mp 136–138 °C (EtOH); [α]$_D^{22}$ −14.5° (*c* 0.53, C$_2$H$_5$OH); IR (CHCl$_3$) ν_{max} 1673, 1733 (C=O), 3325 (NH) cm^{-1}; ^{1}H-NMR (500 MHz, MeOD) δ: 4.48–4.45 (m, 1H, H-3), 4.31 (2H, m, H-1'), 3.75 (m, 4H, H-3', H-4'), 2.04 (s, 3H, CH$_3$CO–), 2.29–0.85 (m, 26H, CH, CH$_2$ in pentacyclic skeleton), 1.03, 1.00, 0.98, 0.91, 0.88, 0.87 (all s, 3H each, H-23–H-27 and H-29), 0.81 (d, *J* = 6.5 Hz, 3H, H-30); ^{13}C-NMR (125 MHz, MeOD) δ: 177.1 (C-28, $\underline{C}$OCH$_3$), 159.1 (C=N), 82.6 (C-3), 64.1 (C-3'), 64.1 (C-4'), 62.8 (C-2'), 61.8 (C-1'), 58.7 (C-17), 56.9 (C-5), 51.8 (C-9), 50.4 (C-19), 45.8 (C-18), 43.9 (C-14), 42.1 (C-8), 39.7 (C-22), 39.6 (C-1), 39.0 (C-4), 38.4 (C-10, C-13), 35.6 (C-7), 33.2 (C-16), 31.2 (C-20), 30.9 (C-15), 28.6 (C-23), 28.4 (C-2), 24.8 (C-12), 23.9 (C-21), 23.5 (C-29), 22.3 (C-11), 21.3 (CO$\underline{C}$H$_3$), 19.4 (C-6), 17.1 (C-25), 16.9 (C-24), 16.8 (C-26), 15.2 (C-30), 15.1 (C-27); Anal. Calcd. for C$_{37}$H$_{64}$ClN$_3$O$_6$: C, 65.13, Cl, 5.20, H, 9.45. Found: C, 65.6 4, Cl, 4.50; H, 9.39%. MS: *m/z* 647.47 [M + 2H]$^+$ (calcd. for C$_{37}$H$_{63}$N$_3$O$_6$, 645.47).

3β-[2-Guanidine-3-hydroxy-2-(hydroxymethyl)propyl]-3-O-acetyl-urs-12-en-28-oate hydrochloride (**18c**), White powder, 62% yield; mp 142–144 °C (EtOH); [α]$_D^{19}$ +37° (*c* 0.51, C$_2$H$_5$OH); IR (CHCl$_3$) ν_{max} 1675, 1733 (C=O), 3186, 3335 (NH) cm^{-1}; ^{1}H-NMR (500 MHz, MeOD) δ: 7.07 (s, 1H, NH), 5.30 (br s, 1H, H-12), 4.50–4.47 (m, 1H, H-3), 4.29, 4.23 (both d, *J* = 11.5 Hz, 1H each, H-1'), 3.75 (m, 4H, H-3', H-4'), 2.25 (d, *J* = 11.5 Hz, 1H, H-18), 2.05 (s, 3H, CH$_3$CO–), 2.14–0.88 (m, 22H, CH, CH$_2$ in pentacyclic skeleton), 1.16, 1.02, 1.00, 0.92, 0.91, 0.90, 0.81 (all s, 3H each, H-23–H-27, H-29 and H-30); ^{13}C-NMR (125 MHz, MeOD) δ: 178.6 (C-28), 173.0 ($\underline{C}$OCH$_3$), 159.2 (C=N), 139.6 (C-13), 127.3 (C-12), 82.6 (C-3), 64.2 (C-3', C-4'), 62.9 (C-2'), 62.3 (C-1'), 56.8 (C-5), 54.5 (C-18), 49.3 (C-17), 49.0 (C-9), 43.4 (C-14), 41.0 (C-8), 40.5 (C-4, C-19), 39.6 (C-20), 38.9 (C-1), 38.2 (C-22), 38.0 (C-10), 34.2 (C-7), 31.8 (C-21), 29.3 (C-15), 28.8 (C-23), 25.5 (C-16), 24.7 (C-2), 24.5 (C-27), 24.4 (C-11), 21.7 (C-30), 21.3 (CO$\underline{C}$H$_3$), 19.5 (C-6), 17.9 (C-29), 17.8 (C-26), 17.4 (C-24), 16.2 (C-25); Anal. Calcd. for C$_{37}$H$_{62}$ClN$_3$O$_6$: C, 65.32, Cl, 5.21, H, 9.19. Found: C, 65.74, Cl, 4.51, H, 9.14%. MS: *m/z* 644.34 [M + H]$^+$ (calcd. for C$_{37}$H$_{61}$N$_3$O$_6$, 643.46).

3β-[2-Guanidine-3-hydroxy-2-(hydroxymethyl)propyl]-3-O-acetyl-olean-12-en-28-oate hydrochloride (**20c**), White powder, 53% yield; mp 124–128 °C (EtOH); [α]$_D^{19}$ +34° (*c* 0.34, C$_2$H$_5$OH); IR (CHCl$_3$) ν_{max} 1662, 1733 (C=O), 3344 (NH) cm^{-1}; ^{1}H-NMR (500 MHz, MeOD) δ: 5.32 (br s, 1H, H-12), 4.49–4.46 (m, 1H, H-3), 4.33, 4.20 (both d, *J* = 11.5 Hz, 1H each, H-1'), 3.76 (m, 4H, H-3', H-4'), 2.89 (d, *J* = 9.5 Hz, 1H, H-18), 2.13–0.90 (m, 22H, CH, CH$_2$ in pentacyclic skeleton), 2.05 (s, 3H, CH$_3$CO–), 1.20, 1.00, 0.98, 0.96, 0.94, 0.91, 0.88 (all s, 3H each, H-23–H-27, H-29 and H-30); ^{13}C-NMR (125 MHz, MeOD) δ: 178.8 (C-28), 172.9 ($\underline{C}$OCH$_3$), 159.1 (C=N), 145.0 (C-13), 124.1 (C-12), 82.6 (C-3), 64.2 (C-4'), 64.1 (C-3'), 62.8 (C-2'), 62.3 (C-1'), 56.8 (C-5), 49.0 (C-9), 48.5 (C-17), 47.1 (C-19), 43.0 (C-14), 42.9 (C-18), 40.7 (C-8), 39.5 (C-1),

38.9 (C-4), 38.2 (C-10), 34.9 (C-22), 33.9 (C-30), 33.7 (C-7, C-21), 31.7 (C-20), 28.9 (C-15), 28.7 (C-23), 26.6 (C-27), 24.7 (C-11, C-29), 24.3 (C-2), 24.2 (C-16), 21.3 (COC$\underline{H}_3$), 19.5 (C-6), 17.9 (C-26), 17.3 (C-24), 15.6 (C-25); Anal. Calcd. for $C_{37}H_{62}ClN_3O_6$: C, 65.32, Cl, 5.21, H, 9.19. Found: C, 65.74, Cl, 4.80, H, 9.12%. MS: m/z 644.44 $[M + H]^+$ (calcd. for $C_{37}H_{61}N_3O_6$, 643.46).

3β-N-(4-Butylgyanidine)-3-hydroxy-lupane-28-amide (**14**), To a solution of the compound **11b** (0.33 g, 0.5 mmol) in MeOH (4 mL) and THF (4 mL) was added 4 N NaOH (4 mL). The reaction mixture was stirred at room temperature for 24 h (monitoring by TLC) and then neutralized with 20% HCl. The solution was dried under vacuum and reconstituted with CH_2Cl_2. The organic layer was washed with brine and dried over anhydrous $MgSO_4$ and concentrated under reduced pressure to obtain pure compound 14. White powder, 79% yield; mp 260–262 °C (EtOH); $[\alpha]_D^{19}$ −10° (*c* 0.24, DMSO); IR (CHCl$_3$) ν_{max} 1731 (C=O), 3366 (NH) cm^{-1}; ^{1}H-NMR (500 MHz, DMSO-d$_5$) δ: 7.73 (br s, 1H, NH), 7.58 (br s, 1H, CONH), 4.29 (br s, 1H, OH), 3.09–2.98 (m, 5H, H-3, H-1', H-4'), 2.19–1.02 (m, 26H, CH, CH$_2$ in pentacyclic skeleton, 4H, H-2', H-3'), 0.89, 0.87, 0.84, 0.78, 0.66 (all s, 3H each, H-23–H-27), 0.81 (d, J = 6.5 Hz, 3H, H-29), 0.72 (d, J = 6.5 Hz, 3H, H-30); ^{13}C-NMR (125 MHz, DMSO-d$_5$) δ: 176.1 (C-28), 157.4 (C=N), 77.2 (C-3), 55.7 (C-17), 55.4 (C-5), 50.3 (C-9), 49.6 (C-19), 43.8 (C-18), 42.5 (C-14), 40.9 (C-8), 40.8 (C-4'), 39.0 (C-22), 38.8 (C-1), 38.6 (C-4), 38.1 (C-1'), 37.2 (C-13), 36.9 (C-10), 34.6 (C-7), 32.8 (C-16), 30.0 (C-20), 29.4 (C-15), 28.6 (C-23), 27.4 (C-2), 27.2 (C-3'), 27.0 (C-2'), 26.4 (C-12), 23.6 (C-29), 23.2 (C-21), 21.1 (C-11), 18.5 (C-6), 16.4 (C-25), 16.3 (C-24, C-26), 15.0 (C-30), 14.7 (C-27); Anal. Calcd. for $C_{35}H_{62}N_4O_2$: C, 73.63, H, 10.95. Found: C, 73.74, H, 10.88%. MS: m/z 593.31 $[M + Na]^+$ (calcd. for $C_{35}H_{62}N_4O_2$, 570.49).

3.2. Biology

3.2.1. Cell Culturing

Cells (Jurkat, K562, U937, HeLa, HEK293 and normal Fibroblasts) were purchased from Russian Cell Culture Collection (Institute of Cytology of the Russian Academy of Sciences, Saint Petersburg, Russia) and cultured according to standard mammalian tissue culture protocols and sterile technique. Human cell lines HEK293 and HeLa were obtained from the HPA Culture Collections (Salisbury, UK). All cell lines used in the study were tested and shown to be free of mycoplasma and viral contamination.

HEK293, HeLa cell lines and fibroblasts were cultured as monolayers and maintained in Dulbecco's modified Eagle's medium (DMEM, Gibco BRL, Waltham, MA, USA) supplemented with 10% foetal bovine serum and 1% penicillin-streptomycin solution at 37 °C in a humidified incubator under a 5% CO_2 atmosphere.

Cells were maintained in RPMI 1640 (Jurkat, K562, U937) (Gibco, Thermo Fisher Scientific, Waltham, MA, USA) supplemented with 4 mM glutamine, 10% FBS (Sigma, Burlington, MA, USA) and 100 units/mL penicillin-streptomycin (Sigma). All types of cells were grown in an atmosphere of 5% CO_2 at 37 °C. The cells were subcultures at 2–3 days intervals. Adherent cells (HEK293, HeLa, fibroblasts) were suspended using trypsin/EDTA and counted after they have reached 80% confluency. Cells were then seeded in 24 well plates at 5×104 cells per well and incubated overnight. Jurkat, K562, U937 cells were subcultured at 2 day intervals with a seeding density of 1×105 cells per 24 well plates in RPMI with 10% FBS.

3.2.2. Cytotoxicity Assay

Viability (Live/dead) assessment was performed by staining cells with 7-AAD (7-Aminoactinomycin D) (Biolegend, San Diego, CA, USA). Cells were treated of test compounds with six different concentrations (1, 5, 10, 15, 30 and 60 μM). After treatment, cells were harvested, washed 1–2 times with phosphate-buffered saline (PBS) and centrifuged at $400\times g$ for 5 min. Cell pellets were resuspended in 200 μL of flow cytometry staining buffer (PBS without Ca^{2+} and Mg^{2+}, 2.5% FBS) and stained with 5 μL of 7-AAD staining solutionfor 15 min at room temperature in the dark. Samples were acquired on NovoCyteTM 2000 FlowCytometry

System (ACEA, San Diego, CA, USA) equipped with 488 nm argon laser. Detection of 7-AAD emission was collected through a 675/30 nm filter in FL4 channel.

3.2.3. Viability and Apoptosis

Apoptosis was determined by flow cytometric analysis of Annexin V and 7-aminoactinomycin D staining. Briefly, 200 μL of Guava Nexin reagent (Millipore, Bedford, MA, USA) was added to 5×105 cells in 200 μL, and the cells were incubated with the reagent for 20 min at room temperature in the dark. The plates were treated with compounds **15**, **15c**, **18c**, **20c** and dihydrobetulinic acid at IC_{50} concentration (4, 8 and 59 μM) for 24 h and 48 h. At the end of incubation, the cells were analyzed on NovoCyteTM 2000 FlowCytometry System (ACEA). Different states of cell death were defined as follows: normal cells are localized in the lower-left quadrant (Annexin V^-/PI^-); early apoptotic cells are in the lower-right quadrant (Annexin V^+/PI^-); late apoptotic cells and necrotic cells are in the upper-right quadrant (Annexin V^+/PI^+); and necrotic cells are in the upper-left quadrant (Annexin V^-/PI^+).

3.2.4. Cell Cycle Analysis

Cell cycle was analyzed using the method of propidium iodide staining. Briefly, cells were plated in 24-well round bottom plates at density 10×105 cells per well, centrifuged at $450 \times g$ for 5 min, and fixed with ice-cold 70% ethanol for 24 h at 0 °C. Cells were then washed with PBS and incubated with 250 μL of Guava Cell Cycle Reagent (Millipore, Burlington, MA, USA) for 30 min at room temperature in the dark. Samples were analyzed on NovoCyte^{TM} 2000 FlowCytometry System (ACEA, San Diego, CA, USA).

4. Conclusions

Novel betulinic, ursolic, and oleanolic acid derivatives, containing a guanidine moiety have been designed and synthesized in an attempt to develop potent antitumor agents. These compounds and their precursors, monoamine, diamine and triamine derivatives, were tested for cytotoxic activity on various human tumor cell lines. Guanidine-functionalized triterpenoids demonstrated higher cytotoxicity in Jurkat cells, compared with original triterpenoic acids. Most of the tested guanidine derivatives showed higher IC_{50} values than amines, but were less toxic to human fibroblasts. The lead molecules—dihydrobetulinic acid amine **15**, its guanidine derivative **15c**, and guanidinium salts of ursolic and oleanolic acids **18c** and **20c** were selected for extended biological testing by using flow cytometry analysis. Our results showed that the antitumor activity of compounds **15**, **15c**, and **18c** is caused by apoptotic processes and induction of cell cycle arrest in the S-phase. Nevertheless, addition information concerning the molecular mechanisms and targets of these triterpene acid derivatives is needed.

Supplementary Materials: The following are available online: ^{1}H-NMR and ^{13}C-NMR spectra of all new compounds.

Author Contributions: Supervision, U.D. and V.O.; validation and writing-review & editing, A.S. and L.D.; performing the chemistry experiments R.K. and D.N.; performing the biology experiments M.Y., L.D. and V.D.; The manuscript was prepared through the contributions L.D., V.D. and D.N.

Funding: This work was performed under financial support from the Russian Science Foundation (Grant 16-13-10051).

Conflicts of Interest: The authors declare no conflict of interest.

References

1. Hill, R.A.; Connolly, J.D. Triterpenoids. *Nat. Prod. Rep.* **2017**, *34*, 90–122. [CrossRef] [PubMed]
2. Dzubak, P.; Hajduch, M.; Vydra, D.; Hustova, A.; Kvasnica, M.; Biedermann, D.; Markova, L.; Urban, M.; Sarek, J. Pharmacological activities of natural triterpenoids and their therapeutic implications. *Nat. Prod. Rep.* **2006**, *23*, 394–411. [CrossRef] [PubMed]

3. Sarek, J.; Kvasnica, M.; Vlk, M.; Urban, M.; Dzubak, P.; Hajduch, M. The potential of triterpenoids in the treatment of melanoma. In *Research on Melanoma: A Glimpse into Current Directions and Future Trends*; Murph, M., Ed.; IntechOpen: Rijeka, Croatia, 2011; pp. 125–158.

4. Sheng, H.; Sun, H. Synthesis, biology and clinical significance of pentacyclic triterpenes: A multi-target approach to prevention and treatment of metabolic and vascular diseases. *Nat. Prod. Rep.* **2011**, *28*, 543–593. [CrossRef] [PubMed]

5. Salvador, J.A.R.; Moreira, V.M.; Goncalves, B.M.F.; Lealab, A.S.; Jing, Y. Ursane-type pentacyclic triterpenoids as useful platforms to discover anticancer drugs. *Nat. Prod. Rep.* **2012**, *29*, 1463–1479. [CrossRef] [PubMed]

6. Chen, H.; Gao, Y.; Wang, A.; Zhou, X.; Zheng, Y.; Zhou, J. Evolution in medicinal chemistry of ursolic acid derivatives as anticancer agents. *Eur. J. Med. Chem.* **2015**, *92*, 648–655. [CrossRef] [PubMed]

7. Cichewicz, R.H.; Kouzi, S.A. Chemistry, biological activity, and chemotherapeutic potential of betulinic acid for the prevention and treatment of cancer and HIV infection. *Med. Res. Rev.* **2004**, *24*, 90–114. [CrossRef] [PubMed]

8. Csuk, R. Betulinic acid and its derivatives: A patent review (2008–2013). *Expert Opin. Ther. Pat.* **2014**, *24*, 913–923. [CrossRef] [PubMed]

9. Fulda, S.; Galluzzi, L.; Kroemer, G. Targeting mitochondria for cancer therapy. *Nat. Rev. Drug Discov.* **2010**, *9*, 447–464. [CrossRef] [PubMed]

10. Damle, A.A.; Pawar, Y.P.; Narkar, A.A. Anticancer activity of betulinic acid on MCF-7 tumors in nude mice. *Indian J. Exp. Biol.* **2013**, *51*, 485–491. [PubMed]

11. Mullauer, F.B.; Bloois, L.; Daalhuisen, J.B.; Brink, M.S.T.; Storm, G.; Medema, J.P.; Schiffelers, R.M.; Kessler, J.H. Betulinic acid delivered in liposomes reduces growth of human lung and colon cancers in mice without causing systemic toxicity. *Anticancer Drugs.* **2011**, *22*, 223–233. [CrossRef] [PubMed]

12. Pathak, A.K.; Bhutani, M.; Nair, A.S.; Ahn, K.S.; Chakraborty, A.; Kadara, H.; Guha, S.; Sethi, G.; Aggarwal, B.B. Ursolic acid inhibits STAT3 activation pathway leading to suppression of proliferation and chemosensitization of human multiple myeloma cells. *Clin. Cancer Res.* **2007**, *5*, 943–955. [CrossRef] [PubMed]

13. Villar, V.H.; Vögler, O.; Barceló, F.; Broto, J.M.; Serra, J.M.; Gutiérrez, V.R.; Alemany, R. Down-regulation of AKT signalling by ursolic acid induces intrinsic apoptosis and sensitization to doxorubicin in soft tissue sarcoma. *PLoS ONE* **2016**, *11*, e0155946. [CrossRef] [PubMed]

14. Lin, C.; Wen, X.; Sun, H. Oleanolic acid derivatives for pharmaceutical use: A patent review. *Expert Opin. Ther. Pat.* **2016**, *26*, 643–655. [CrossRef] [PubMed]

15. Hussain, H.; Green, I.R.; Ali, I.; Khan, I.A.; Ali, Z.; Al-Sadi, A.M.; Ahmed, I. Ursolic acid derivatives for pharmaceutical use: A patent review (2012–2016). *Expert Opin. Ther. Pat.* **2017**, *27*, 1061–1072. [CrossRef] [PubMed]

16. Biedermann, D.; Eignerova, B.; Hajduch, M.; Sarek, J. Synthesis and evaluation of biological activity of the quaternary ammonium salts of lupane-, oleanane-, and ursane-type acids. *Synthesis* **2010**, *22*, 2839–3848. [CrossRef]

17. Kataev, V.E.; Strobykina, Y.; Zakharova, L.Y. Quaternary ammonium derivatives of natural terpenoids. Synthesis and properties. *Russ. Chem. Bull.* **2014**, *63*, 1884–1900. [CrossRef]

18. Serafim, T.L.; Carvalho, F.S.; Bernardo, T.C.; Pereira, G.C.; Perkins, E.; Holy, J.; Krasutsky, D.A.; Kolomitsyna, O.N.; Krasutsky, P.A.; Oliveira, P.J. New derivatives of lupine triterpenoids disturb breast cancer mitochondria and induce cell death. *Bioorg. Med. Chem.* **2014**, *22*, 6270–6287. [CrossRef] [PubMed]

19. Bernardo, T.C.; Cunha-Oliveira, T.; Serafim, T.L.; Holy, J.; Krasutsky, D.; Kolomitsyna, O.; Krasutsky, P.; Moreno, A.M.; Oliveira, P.J. Dimethylaminopyridine derivatives of lupine triterpenoids cause mitochondrial disruption and induce the permeability transition. *Bioorg. Med. Chem.* **2013**, *21*, 7239–7249. [CrossRef] [PubMed]

20. Tsepaeva, O.V.; Nemtarev, A.V.; Abdullin, T.I.; Grigor'eva, L.R.; Kuznetsova, E.V.; Akhmadishina, R.A.; Ziganshina, L.E.; Cong, H.H.; Mironov, V.F. Design, synthesis, and cancer cell growth inhibitory activity of triphenylphosphonium derivatives of the triterpenoid betulin. *J. Nat. Prod.* **2017**, *80*, 2232–2239. [CrossRef] [PubMed]

21. Ye, Y.; Zhang, T.; Yuan, H.; Li, D.; Lou, H.; Fan, P. Mitochondria-targeted lupanetriterpenoid derivatives and their selective apoptosis-Inducing anticancer mechanisms. *J. Med. Chem.* **2017**, *60*, 6353–6363. [CrossRef] [PubMed]

22. Strobykina, I.Y.; Belenok, M.G.; Semenova, M.N.; Semenov, V.V.; Babaev, V.M.; Rizvanov, I.K.; Mironov, V.F.; Kataev, V.E. Triphenylphosphonium cations of the diterpenoid isosteviol: Synthesis and antimitotic activity in a sea urchin embryo model. *J. Nat. Prod.* **2015**, *78*, 1300–1308. [CrossRef] [PubMed]
23. Spivak, A.Y.; Nedopekina, D.A.; Shakurova, E.R.; Khalitova, R.R.; Gubaidullin, R.R.; Odinokov, V.N.; Dzhemilev, U.M.; Bel'skii, Y.P.; Bel'skaya, N.V.; Stankevich, S.A.; et al. Synthesis of lupine triterpenoids with triphenylphosphonium substituents and studies of their antitumor activity. *Russ. Chem. Bull.* **2013**, *62*, 188–198. [CrossRef]
24. Spivak, A.Y.; Nedopekina, D.A.; Khalitova, R.R.; Gubaidullin, R.R.; Odinokov, V.N.; Bel'skii, Y.P.; Bel'skaya, N.V.; Khazanov, V.A. Triphenylphosphonium cations of betulinic acid derivatives: Synthesis and antitumor activity. *Med. Chem. Res.* **2017**, *26*, 518–531. [CrossRef]
25. Nedopekina, D.A.; Gubaidullin, R.R.; Odinokov, V.N.; Maximchik, P.V.; Zhivotovsky, B.; Bel'skii, Y.P.; Khazanov, V.A.; Manuylova, A.V.; Gogvadze, V.; Spivak, A.Y. Mitohondria-targeted betulinic and ursolic acid derivatives: Synthesis and anticancer activity. *Med. Chem. Commun.* **2017**, *8*, 1934–1945. [CrossRef] [PubMed]
26. Saczewski, F.; Balewski, L. Biological activities of guanidine compounds. *Expert Opin. Ther. Pat.* **2009**, *19*, 1417–1448. [CrossRef] [PubMed]
27. Wexselblatt, E.; Esko, J.D.; Tor, Y. On guanidinium and cellular uptake. *J. Org. Chem.* **2014**, *79*, 6766–6774. [CrossRef] [PubMed]
28. Pantos, A.; Tsogas, I.; Paleos, C.M. Guanidinium group: A versatile moiety inducing transport and multicompartmentalization in complementary membranes. *Biochim. Biophys. Acta* **2008**, *1778*, 811–823. [CrossRef] [PubMed]
29. Castagnolo, D.; Schenone, S.; Botta, M. Guanylated diamines, triamines, and polyamines: Chemistry and biological properties. *Chem. Rev.* **2011**, *111*, 5247–5300. [CrossRef] [PubMed]
30. Sibrian-Vazquez, M.; Nesterova, I.V.; Jensen, T.J.; Vicente, M.G. Mitochondria targeting by guanidine- and biguanidine-porphyrin photosensitizers. *Bioconjug. Chem.* **2008**, *19*, 705–713. [CrossRef] [PubMed]
31. Blanchet, M.; Borselli, D.; Brunel, J.M. Polyamine derivatives: A revival of an old neglected scaffold to fight resistant Gram-negative bacteria? *Future Med. Chem.* **2016**, *8*, 963–973. [CrossRef] [PubMed]
32. Nowotarski, S.L.; Woster, P.M.; Casero, R.A. Polyamines and cancer: Implications for chemotherapy and chemoprevention. *Expert Rev. Mol. Med.* **2013**, *15*, e3. [CrossRef] [PubMed]
33. Fujiwara, T.; Hasegawa, S.; Hirashima, N.; Nakanishi, M.; Ohwada, T. Gene transfection activities of amphiphilic steroid-polyamine conjugates. *Biochim. Biophys. Acta* **2000**, *1468*, 396–402. [CrossRef]
34. Brycki, B.; Koenig, H.; Kowalczyk, I.; Pospieszny, T. Synthesis, spectroscopic and theoretical studies of new dimeric quaternary alkylammonium conjugates of sterols. *Molecules* **2014**, *19*, 9419–9434. [CrossRef] [PubMed]
35. Vida, N.; Svobodova, H.; Rarova, L.; Drasar, P.; Saman, D.; Cvacka, J.; Wimmer, Z. Polyamine conjugates of stigmasterol. *Steroids* **2012**, *77*, 1212–1218. [CrossRef] [PubMed]
36. Brycki, B.; Koenig, H.; Pospieszny, T. Quaternary alkylammonium conjugates of steroids: Synthesis, molecular structure, and biological studies. *Molecules* **2015**, *20*, 20887–20900. [CrossRef] [PubMed]
37. Giniyatullina, G.V.; Flekhter, O.B.; Tolstikov, G.A. Synthesis of squalamine analogues on the basis of lupine triterpenoids. *Mendeleev Commun.* **2009**, *19*, 32–33. [CrossRef]
38. Kazakova, O.B.; Giniyatullina, G.V.; Medvedeva, N.I.; Tolstikov, G.A. Synthesis of a triterpene–spermidine conjugate. *Russ. J. Org. Chem.* **2012**, *48*, 1370–1373. [CrossRef]
39. Bildziukevich, U.; Vida, N.; Rárová, L.; Kolář, M.; Šaman, D.; Havlíček, L.; Drašar, P.; Wimmer, Z. Polyamine derivatives of betulinic acid and β-sitosterol: A comparative investigation. *Steroids* **2015**, *100*, 27–35. [CrossRef] [PubMed]
40. Wang, J.; Jiang, Z.; Xiang, L.; Li, Y.; Ou, M.; Yang, X.; Shao, J.; Lu, Y.; Lin, L.; Chen, J.; et al. Synergism of ursolic acid derivative US597 with 2-deoxy-D-glucose to preferentially induce tumor cell death by dual-targeting of apoptosis and glycolysis. *Sci. Rep.* **2014**, *4*, 5006. [CrossRef] [PubMed]
41. Kim, D.S.; Pezzuto, J.M.; Pisha, E. Synthesis of betulinic acid derivatives with activity against human melanoma. *Bioorg. Med. Chem. Lett.* **1998**, *8*, 1707–1712. [CrossRef]
42. You, Y.J.; Kim, Y.; Nam, N.H.; Ahn, B.Z. Synthesis and cytotoxic activity of A-ring modified betulinic acid derivatives. *Bioorg. Med. Chem. Lett.* **2003**, *13*, 3137–3140. [CrossRef]

43. Coulibaly, W.K.; Paquin, L.; Benie, A.; Bekro, Y.A.; Guevel, R.L.; Ravache, M.; Corlu, A.; Bazureau, J.P. Prospective study directed to the synthesis of unsymmetrical linked bis-5-arylidene rhodanine derivatives via "one-pot two steps" reactions under microwave irradiation with their antitumor activity. *Med. Chem. Res.* **2015**, *24*, 1653–1661. [CrossRef]
44. Feichtinger, K.; Zapf, C.; Sings, H.L.; Goodman, M. Diprotected triflyl-guanidines: A new class of guanidinylation reagents. *J. Org. Chem.* **1998**, *63*, 3804–3805. [CrossRef]
45. Kommera, H.; Kaluderovic, G.N.; Kalbitz, J.; Dräger, B.; Paschke, R. Small structural changes of pentacycliclupane type triterpenoid derivatives lead to significant differences in their anticancer properties. *Eur. J. Med. Chem.* **2010**, *45*, 3346–3353. [CrossRef] [PubMed]
46. Willmann, M.; Wacheck, V.; Buckley, J.; Nagy, K.; Thalhammer, J.; Paschke, R.; Triche, T.; Jansen, B.; Selzer, E. Characterization of NVX-207, a novel betulinic acid-derived anti-cancer compound. *Eur. J. Clin. Investig.* **2009**, *39*, 384–394. [CrossRef] [PubMed]
47. Bache, M.; Bernhardt, S.; Passin, S.; Wichmann, H.; Hein, A.; Zschornak, M.; Kappler, M.; Taubert, H.; Paschke, R.; Vordermark, D. Betulinic acid derivatives NVX-207 and B10 for treatment of glioblastoma—An in vitro study of cytotoxicity and radiosensitization. *Int. J. Mol. Sci.* **2014**, *15*, 19777–19790. [CrossRef] [PubMed]
48. Liebscher, G.; Vanchangiri, K.; Mueller, T.; Feige, K.; Cavalleri, J.M.; Paschke, R. In vitro anticancer activity of betulinic acid and derivatives thereof on equine melanoma cell lines from grey horses and *in vivo* safety assessment of the compound NVX-207 in two horses. *Chem. Biol. Interact.* **2016**, *246*, 20–29. [CrossRef] [PubMed]

Sample Availability: Samples of all compounds are available from the authors.

 *molecules*

Article

Synthesis of 9-Hydroxystearic Acid Derivatives and Their Antiproliferative Activity on HT 29 Cancer Cells

Natalia Calonghi [1,*], Carla Boga [2,*], Dario Telese [2], Silvia Bordoni [2], Giorgio Sartor [1], Chiara Torsello [1] and Gabriele Micheletti [2]

1   Department of Pharmacy and Biotechnology, University of Bologna, 40127 Bologna, Italy; giorgio.sartor@unibo.it (G.S.); chiara.torsello@virgilio.it (C.T.)
2   Department of Industrial Chemistry 'Toso Montanari', Alma Mater Studiorum Università di Bologna Viale Del Risorgimento 4, 402136 Bologna, Italy; dario.telese2@unibo.it (D.T.); silvia.bordoni@unibo.it (S.B.); gabriele.micheletti3@unibo.it (G.M.)
*   Correspondence: natalia.calonghi@unibo.it (N.C.); carla.boga@unibo.it (C.B.); Tel.: +39-051-2091231 (N.C.); +39-051-2093616 (C.B.)

Academic Editor: Luigi A. Agrofoglio
Received: 27 September 2019; Accepted: 12 October 2019; Published: 15 October 2019

Abstract: 9-Hydroxystearic acid (9-HSA) is an endogenous cellular lipid that possesses antiproliferative and selective effects against cancer cells. A series of derivatives were synthesized in order to investigate the effect of the substituent in position 9 and on the methyl ester functionality on the biological activity. The two separate enantiomers of methyl 9-hydroxystearate and of methyl 9-aminostearate showed antiproliferative activity against the HT29 cell line. This indicates the importance of position 9 groups being able to make hydrogen bonding with the molecular target. Further, this effect must be preserved when the carboxy group of 9-HSA is esterified. The biological tests showed that the amines, contrarily to methyl esters, resulted in cytotoxicity. A deep investigation on the effect of methyl (R)-9-hydroxystearate on HT29 cells showed an antiproliferative effect acting through the CDKN1A and MYCBP gene expression.

Keywords: 9-hydroxystearic acid; methyl 9-hydroxystearate; methyl 9-aminostearate; cancer

1. Introduction

9-Hydroxystearic acid (9-HSA, Figure 1) is an endogenous cellular lipid that possesses a natural negative regulatory activity of tumor cell proliferation. [1–3] Its content has been found greatly diminished in cancer cells with respect to the corresponding normal line [1] and its exogenous administration to adenocarcinoma (HT29), [4] osteosarcoma (U2OS, [5] or SaOS [6]) cancer cell lines results in a significant inhibition of the proliferation rate, as well as a significant increase of p53-independent p21 expression. [4] The expression of the cell cycle kinase inhibitor CDKN1A (P21) is induced in neoplastic cells by inhibitors of histone deacetylase 1 (HDAC1). Actually, the authors found that 9-HSA acts as a competitive inhibitor of the HDAC1 (histone deacetylase 1) isoform [7,8]. The studies of molecular docking [7] showed an interaction between the 9-HSA carboxy group and the zinc ion in the active site of HDAC1, by a mechanism similar to that already known for HDAC inhibitors such as valproate, butyrate, trichostatin A (TSA), and suberoylanilidehydroxamic acid (SAHA, Vorinostat) [9]. The molecular docking also predicted a more favorable binding energy for the HDAC1-(R)-9-HSA complex, with respect to the complex formed with the opposite enantiomer (S)-9-HSA. Enantiopure (R)-9-HSA is more easily accessible within the natural chiral pool since its precursor, (S)-dimorphecholic acid, is one of the main components of seeds of the genus Dimorphotheca plants [10]. Thus, both the (R)-9-HSA and its enantiomer were synthesized, and the docking prediction experimentally confirmed [11]. Based on the above, the importance of both the carboxy- and the hydroxyl- group in inducing the antiproliferative activity of 9-HSA is apparent.

Figure 1. The two enantiomeric forms of 9-hydroxystearic acid, an HDAC inhibitor.

Hence, this study made some modifications on these key positions to better enlighten the role played by these functionalities and to design, if possible, some structure-activity relationships. Herein, the results obtained are reported.

2. Results and Discussion

2.1. Synthesis of 9-HSA Derivatives

The 9-HSA derivatives synthesized and investigated are in Figure 2.

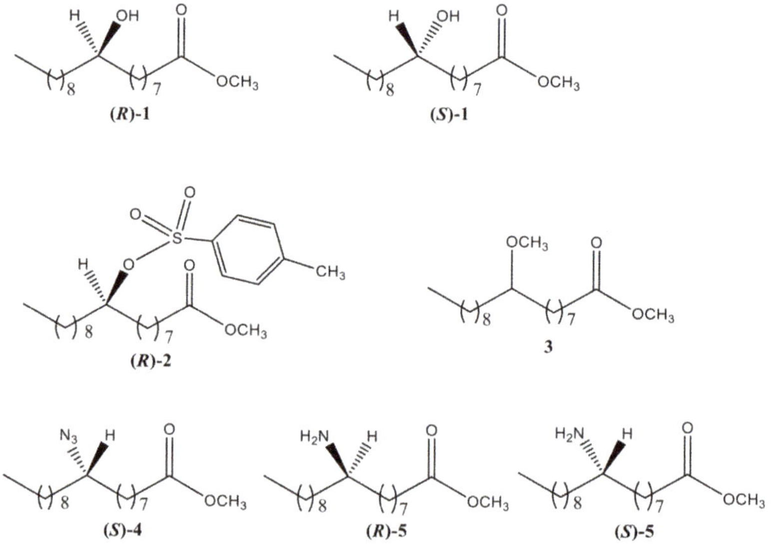

Figure 2. 9-HSA derivatives synthesized and biologically tested.

Compounds **(R)-2** and **3** bear an ester and an ether group bound to the C-9, respectively, and this difference might give information on the free hydroxyl functionality about the biological action. Further, this study planned to prepare the relative amine **(S)-5** isosteric structure, to gain information on the biological behavior of the methyl 9-hydroxystearate and methyl 9-aminostearate. This study decided to test also the biological activity of the azide **(S)-4** species, as precursor of the mentioned amine species.

All compounds shown in Figure 2 were obtained starting from methyl (9R)-9-hydroxystearate [**(R)-1**], which in turn was obtained from *Dimorphotheca sinuata* seeds, which is oil, as well as the seeds of other plants of the genus Dimorphotheca. These contain high amounts of (9S,10E,12E)-9-hydroxyoctadeca-10,12-dienoic acid [(S)-dimorphecholic acid] [11], a precursor of **(R)-1** and of **(R)-9-HSA**. The natural availability of (S)-dimorphecholic acid is a green and efficient way to obtain 9-HSA derivatives with the C-9 chiral carbon atom in enantiopure form, permitting all the difficulties exhibited by alternative synthetic methods [12] to be overcome. Scheme 1 shows the synthetic strategy adopted to prepare compounds **(R)-1, (R)-2, 3, (S)-4** and **(S)-5**.

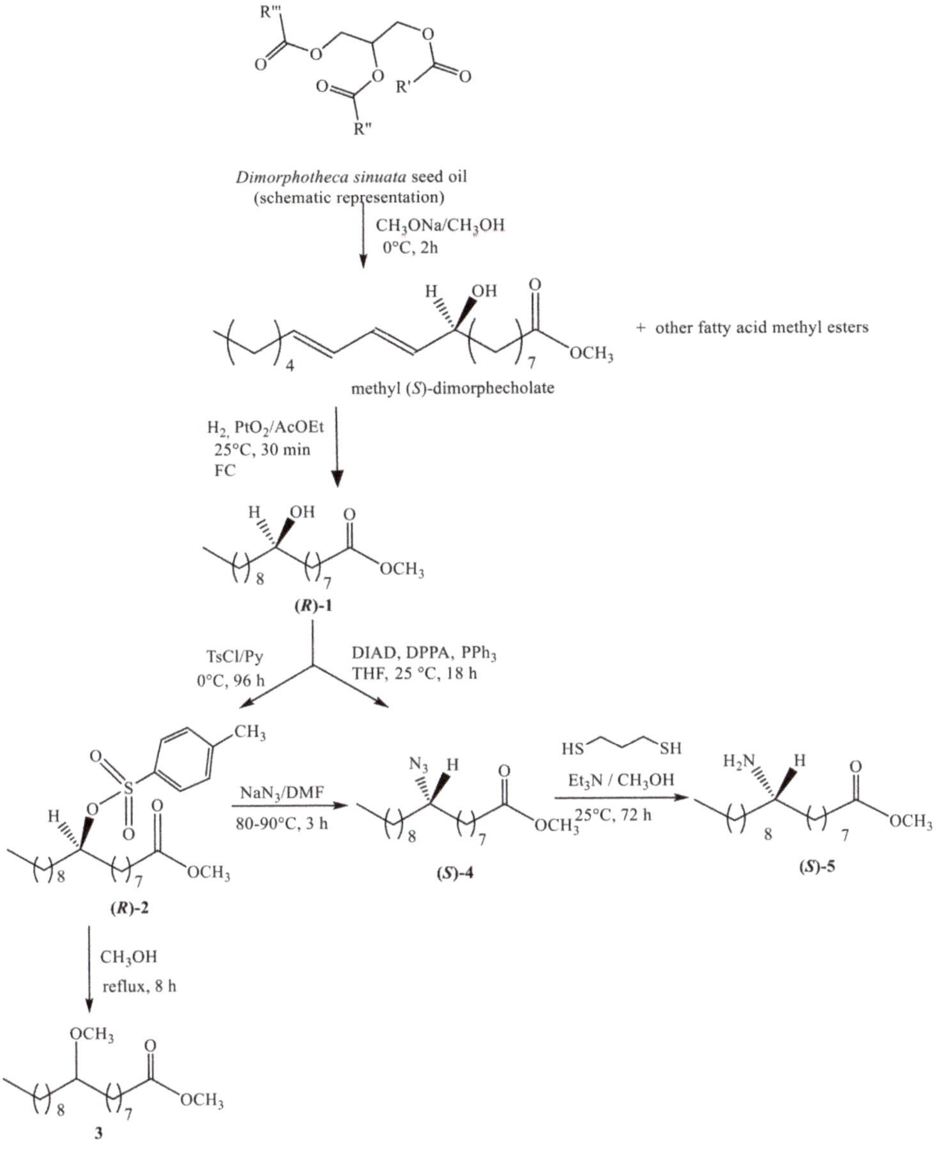

Scheme 1. Synthetic routes to 9-HSA derivatives.

The procedure appears to be the first reaction of transmethylation of the triglycerides, obtained from the extraction of the seeds and subsequent hydrogenation over the Adam's catalyst. After purification of the crude product by silica gel chromatography, the (9R)-methyl-9-hydroxystearate [**(R)-1**] was isolated and its optical purity was ascertained by ^{1}H NMR upon derivatization with (R)-O-acetylmandelic acid [10].

Compound **(R)-1** was then transformed by treatment with tosyl chloride into **(R)-2**. The reaction of **(R)-2** with refluxing methanol produced the methyl ether **3**. For the latter, the absolute configuration of the C-9 chiral center is not defined due to the possibility of occurrence, together with an inversion of the configuration, and also of a S_N1 mechanism producing racemization. The addition to **(R)-2** of sodium azide in *N,N*-dimethylformamide (DMF) at 80–90 °C produced the azide **(S)-4**. However, due to difficulties of the work-up and purification steps, an alternative method was adopted by reacting (**R**)-**2** with diphenylphosphoryl azide (DPPA) under Mitsunobu conditions yielding (**S**)-**4** in satisfactory yield (77%).

The last step was devoted to the reduction of the azide. The adoption of Bayley's method (1,3-propandithiol and Et_3N) [13] suffered a particularly hard work-up: After the separation of the amine from by-products through immobilization on the Dowex resin and subsequent release with triethylamine, a further treatment by liquid-liquid extraction was required and the desired amine was recovered in 13% yield. This encouraged the study to try the alternative synthetic route shown in Scheme 2 that gave, after purification, the amine (**S**)-**5** in 36% overall yield from (**R**) **1**.

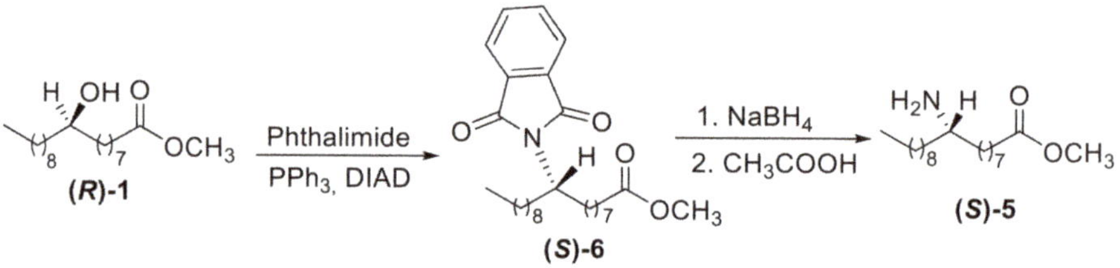

Scheme 2. Synthetic pathway to obtain aminoderivative (**S**)-**5** from (**R**)-**1**.

The enantiopurity of the new amine (**S**)-**5** was ascertained by reacting it with the acyl chloride of the (*S*)-*O*-acetylmandelic acid (**7**) (Scheme 3) and by calculating, from the ^{1}H-NMR spectrum of the crude, the ratio between the two diastereomeric amides (**S,S**)-**8** and (**R,S**)-**8**. The signals were falling at 3.669 and 3.665 ppm, belonging to the (**S,S**) and (**R,S**) diastereomer, respectively, from which a 9/1 *dr* was calculated, which was an indication of an almost complete inversion of the configuration during synthesis (see Figures S1–S17). This was confirmed by adding a little amount of the (**R,S**)-**8** diastereomer (obtained as described below) to the above mixture: A sensitive increase of the area of the peak at 3.665 ppm was observed.

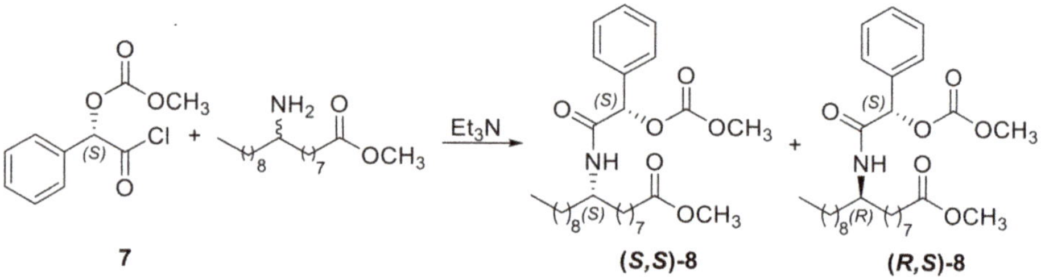

Scheme 3. Derivatization of the amine **5** with (*S*)-2-chloro-2-oxo-1-phenylethyl methyl carbonate (**7**) to determine the (**S,S**)-**8**/(**R,S**)-**8** diastereomeric ratio.

As reported in Scheme 1, several steps of the synthetic pathway imply the inversion of configuration on C-9 producing the compounds **4** and **5** with the (*S*) configuration. Since the two enantiomers showed distinct antiproliferative activity by the preliminary tests in the case of 9-HSA, this study planned to prepare the opposite enantiomers of those compounds that, in preliminary tests, were biologically active. Thus, (**S**)-**1** and (**R**)-**5** were prepared in order to compare the biological activity of the two enantiomers (Figure 2), and to evaluate the trend in biological activity ascribed to both the isosters with same configuration.

The procedure utilized is shown in Scheme 4: (**S**)-**1** was obtained by the inversion of the chiral center of (**R**)-**1** through the Mitsunobu reaction followed by the treatment of the intermediate **9** with methanolic KOH and esterification with BF_3/CH_3OH. Then, (**R**)-**5** was obtained from (**S**)-**1** by the same procedure depicted in Scheme 2 to obtain the opposite enantiomer.

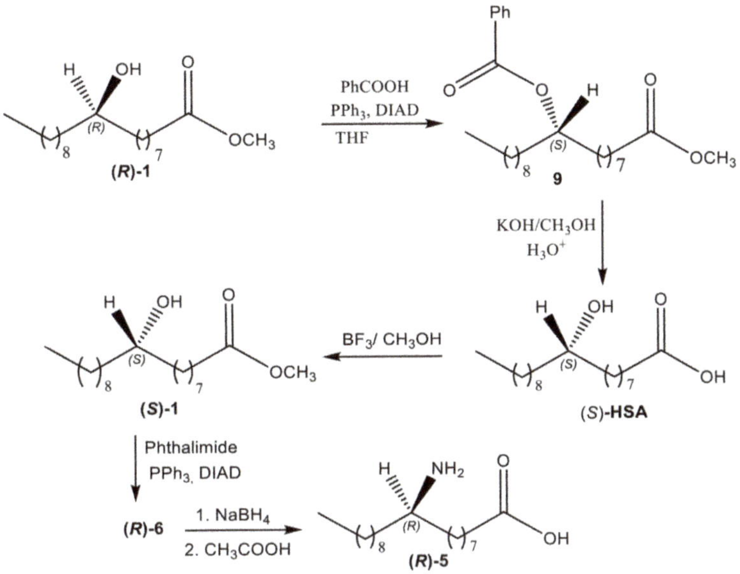

Scheme 4. Synthesis of (*S*)-1 and (*R*)-5.

The enantiopurity degree of **(S)-1** was measured [11,14] by reacting it with (*R*)-(−)-*O*-acetylmandelic acid: A diastereomeric ratio > 5/95 between the (*R*,*R*) and (*S*,*R*) diastereomers was calculated from the ^{1}H-NMR of the crude reaction mixture. Analogously, the **(R)-5** enantiopurity degree was evaluated as a 10/90 diastereomeric ratio, by analyzing ^{1}H-NMR spectrum of the reaction with the acyl chloride of (*S*)-*O*-acetylmandelic acid.

2.2. Biological Activity

2.2.1. Preliminary Biological Activity of (*R*)-**2**, **3**, (*S*)-**4**, and (*S*)-**5**

In order to gain information on the effect of the change of the group bound to C-9, preliminary biological tests on IC_{50} value for compounds **(R)-1**, **(R)-2**, **3**, **(S)-4**, and **(S)-5** were carried out. From the MTT tests, the only active compounds were the methyl ester **(R)-1** and the amino derivative **(S)-5** (see below). The absence of significant activity in the case of compounds functionalized on C-9 with groups not able to make hydrogen bonds might be considered as a strong indication of the importance of the presence of a group bearing hydrogen atoms bound to a heteroatom on this position. The results obtained with **(R)-1** and **(S)-5** encouraged this study to also synthesize the corresponding enantiomer in order to compare their biological effects.

2.2.2. Effect of Methyl (9*R*)-9-hydroxystearate [(*R*)-1], Methyl (9*S*)-9-hydroxystearate [(*S*)-1], Methyl (9*R*)-9-aminostearate [(*R*)-5] and Methyl (9*S*)-9-aminostearate [(*S*)-5] on Proliferation of HT29

The in vitro IC_{50} growth inhibitory concentration was determined for **(R)-1**, **(S)-1**, **(R)-5** and **(S)-5**, incubating the HT29 with increasing concentrations of the compounds for 24 h. The data obtained from the MTT analyses were examined to assess the concentration of compounds required for 50% inhibition of cell viability (IC_{50}): The values corresponded to 49 ± 1.3 μM for **(R)-1**, 51 ± 1.1 μM for **(S)-1**, 57 ± 2.1 μM for **(R)-5** and 43 ± 4.5 μM for **(S)-5** (Figure 3).

The effects on HT29 cell proliferation after treatment with **(R)-1** and **(S)-1** and with **(R)-5** and **(S)-5** are reported in Figure 4. In panel A, the effects of 49 μM **(R)-1** and 51 μM **(S)-1** are reported. The results for the **(R)-1** and **(S)-1** treatment show a significantly greater effect for the (*R*)-enantiomer, recalling the already reported behavior of the two enantiomers of 9-hydroxystearic acid. In panel B, the effects for (*R*)-5 and (*S*)-5 at the respective concentrations of 57 µM and 43 µM are reported: In this case, after 24 h of treatment, the cell proliferation strongly reduced, and after 48 h and 72 h of treatment, the cell numbers still diminished, indicating a continuous cytotoxic effect.

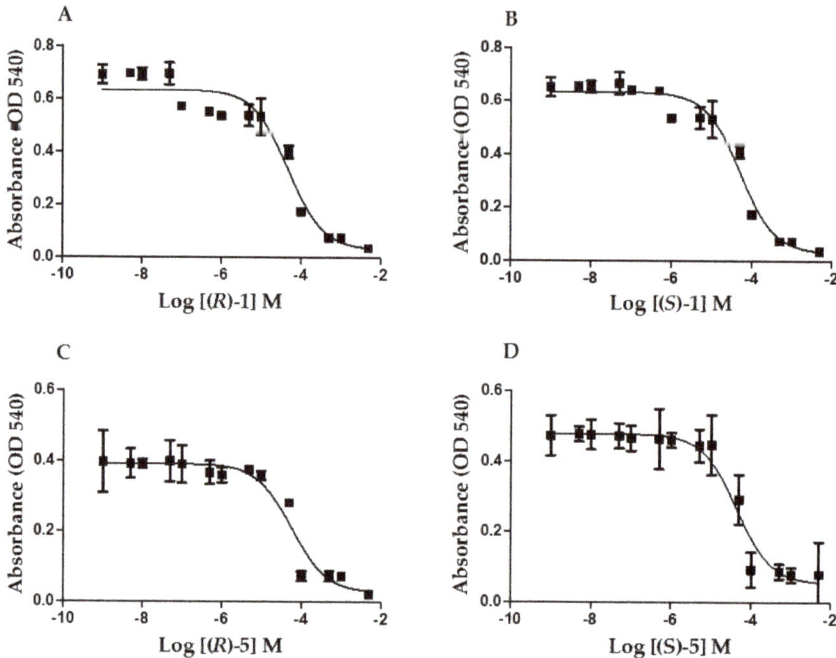

Figure 3. Dose response curves of HT29 cell viability upon treatment with different concentrations of (**A**) (*R*)-**1**, of (**B**) (*S*)-**1**, of (**C**) (*R*)-**5** and (**D**) (*S*)-**5**.

These findings prompted this study to gain more information on the biological effect of the methyl (9*R*)-9-hydroxystearate (*R*)-**1**.

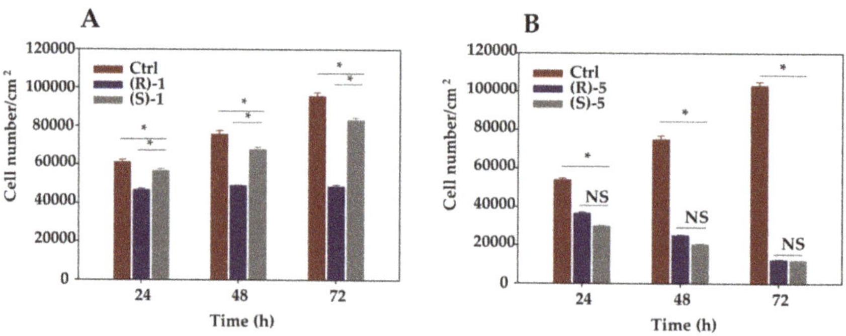

Figure 4. Effects on cell proliferation. (**A**) Antiproliferative effect of (*R*)-**1** and (*S*)-**1** on HT29 at 24, 48, and 72 h of treatment as compared with the control. (**B**) Antiproliferative effect of (*R*)-**5** or (*S*)-**5** in the same experimental conditions. The analysis was carried out by two-way analysis of variance (ANOVA) followed by the Bonferroni multiple comparisons. The differences of at least $p < 0.05$ were considered significant. Statistical analysis was carried out using Prism GraphPad software.

2.2.3. Cell Cycle Analysis

In order to assess whether the antiproliferative effect of (*R*)-1 was associated with the interference of the cell cycle progression, DNA profiles of cultured cells were examined by flow cytometry. The cells were exposed to the compound for 48 h prior to processing and analysis. As shown in Figure 5, the exposure to (***R***)-1 resulted in an increase in the number of the G0/G1 (80.18 ± 0.7% versus 73.91 ± 0.9%) phase, while reducing their number in the S (13.18 ± 0.5% versus 17.97 ± 0.2%) and G2/M (6.64 ± 0.2%versus 8.12 ± 0.1%) phases, thus indicating that the treatment leads to a cell cycle arrest. The flow cytometric data correlates with cell counting over time. Indeed, the administration of (***R***)-1 reduced the number of cells by 23% at 24 h, by 35% at 48 h and by 49% at 72 from the treatment, indicating that the cells did not duplicate anymore.

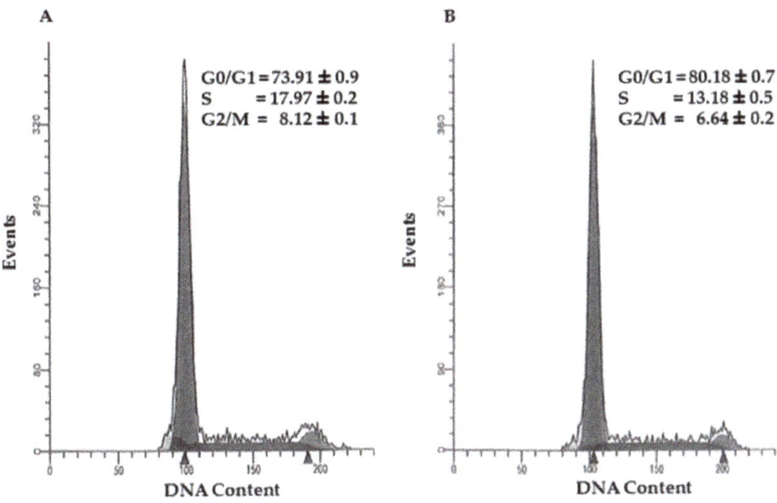

Figure 5. Effect of (***R***)-1 on cell cycle. The cells were incubated for 48 h with the vehicle (**A**) or with 49 μM of (***R***)-1 (**B**), afterward the cell cycle distribution was determined by flow cytometry. The two panels report the cytofluorimeter outputs obtained in one typical experiment repeated twice with similar results.

2.2.4. Effect of (*R*)-1 on Histone Acetylation

Since the above findings parallel those found for the parent (*R*)-9-HSA [11], an investigation of whether the molecular target might be histone deacetylase, as in the case of 9-HSA, was undertaken.

To identify the state of histones acetylation, these proteins were extracted from the control HT29 and treated for 6 h with 49 μM of (***R***)-1. Acetylation was detected by a western blot using an anti-acetyl lysine monoclonal antibody. The histone acetylation signals were quantified by densitometry and normalized on histone H3. As shown in Figure 6, the treatment with (***R***)-1 for 6 h increased histone H4 acetylation by 50%, while the acetylation status of histones H2/H3 did not change.

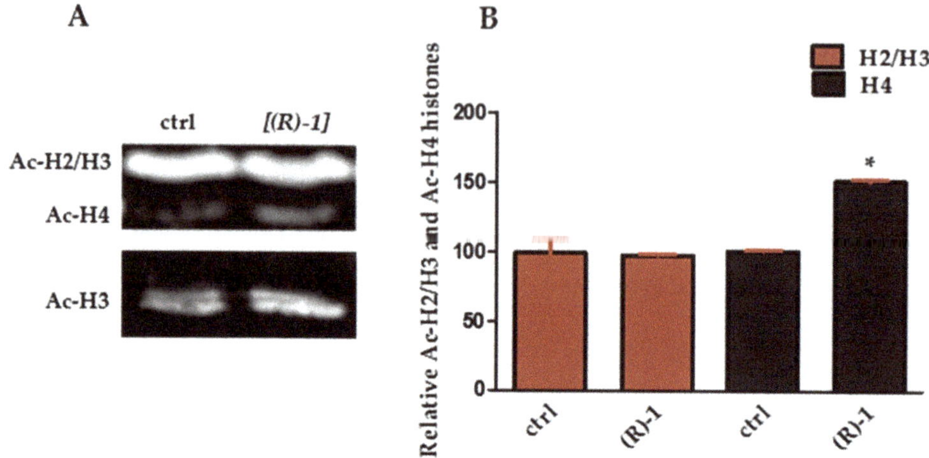

Figure 6. Effect of (*R*)-1 on histone H4 acetylation. (**A**) The cells were cultured for 6 h in the absence (control) or presence of 49 µM (*R*)-1. The cell nuclear extracts were prepared and subjected to a western blot analysis for acetylated (ac) histone H2/H3, histone H4 and histone H3 as loading controls. (**B**) A representative experiment, repeated three times with similar results and densitometric analysis of the bands (mean ± SEM; $n = 3$), are shown. * $p < 0.05$.

2.2.5. (*R*)-1 Induces Gene and Relative Protein Expression Modulations

The induction of p21 is one of the common phenomena observed after treatment with HDAC inhibitors such as TSA, NaBu, SAHA or 9-HSA [7,15–17]. Myc is a transcription factor whose activity is causally involved in cancers, principally through its capacity to drive tumor cell proliferation and promote angiogenesis, invasion, and metastasis [18–21]. The downregulation of Myc is an essential part of the anti-proliferative response to differentiation signals. Thus, a western blot analysis was performed to assess the expression of p21 and Myc in HT29 cells after (*R*)-1 treatment. As expected, p21 was induced by the treatment with (*R*)-1 compared to its expression level in the untreated control cells, while the Myc expression decreased significantly by (*R*)-1 treatment (Figure 7A,B).

The observed cytostatic effects of (*R*)-1 transcription levels of P21 and MYCBP (MYC) were analysed by quantitative Real Time-Polymerase Chain Reaction (RT-PCR. The genes were analysed after 6 h of treatment. The qRT-PCR was performed on cDNA of the control and (*R*)-1-treated cells, and the $\Delta\Delta C_T$ method was used with the GAPDH gene as the housekeeping gene. The relative transcription levels, expressed as a means of fold changes, are reported in Figure 7C. (*R*)-1 administration transcription significantly increases P21 gene at 6 h of treatment, while MYC is significantly reduced.

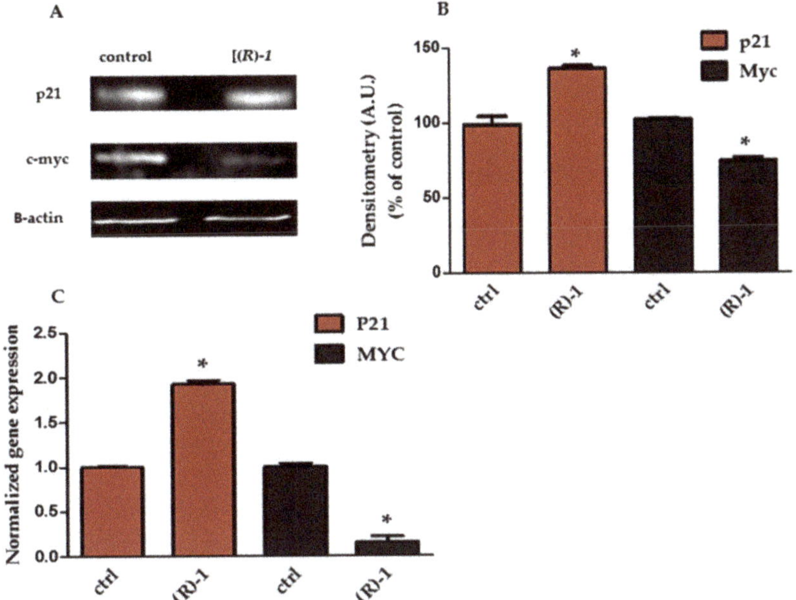

Figure 7. The effects of **(R)-1** on the expression of cell growth regulatory genes and proteins in HT29 colon cancer cell line. **A)** Representative western blots for p21 and MYC in HT29 cells following 6 h treatment with **(R)-1**. β-actin was used as the control. **B)** Densitometry from western blots for HT29 following 6 h treatment with **(R)-1**. Protein quantification was normalized to the β-actin band. The mean ± SEM, $n = 3$, and * $p < 0.05$. **C)** Changes in mRNA expression of P21 and MYC. GAPDH were used as a control gene. CT values were obtained, the data was normalized against GAPDH and the fold change was calculated by ΔCT method. The data are the mean ± SEM of three independent experiments. * $p < 0.05$.

3. Materials and Methods

3.1. Chemical Syntheses

Dimorphotheca sinuata L. seeds were bought from Galassi Sementi Srl (Gambettola, FC, Italy). The reagents used, unless stated otherwise, were purchased from Sigma-Aldrich (Milan, Italy). For flash chromatography (FC), silica gel 0.037–0.063 mm (Merck KGaA, Darmstadt, Germany) was used as the stationary phase. Thin layer chromatography (TLC) was carried out on silica gel 60 (Fluka Analytical, Buchs, Switzerland) and the spots were revealed, depending on the compound to be analyzed, using UV light, and an aqueous solution of $(NH_4)_6MoO_{24}$ (2.5%) and $(NH_4)_4Ce(SO_4)_4$ (4%) in 10% H_2SO_4, a solution of ninhydrin 0.013 M in *n*-butanol/acetic acid, or a basic solution of potassium permanganate. Anhydrous THF was freshy distilled over sodium benzophenone ketyl. Pyridine, trimethylamine and methanol were dried by standard methods. The hydrogenation was carried out on a Parr hydrogenator model 3911EKX at a hydrogen gas pressure of approximately 40 psi. The nuclear magnetic resonance spectra (^{1}H-NMR ^{13}C-NMR,) were recorded at 25 °C on Varian spectrometers Gemini 300, Mercury 400, or Inova 600 (Varian, Palo Alto, CA, USA). The signal multiplicities were established by DEPT-135 experiments. The chemical shifts were referenced to the solvent (CDCl$_3$, δ = 7.27 and 77.0 ppm for ^{1}H and ^{13}C-NMR, respectively). The GC-MS analyses were carried out by gas chromatograph directly interfaced with a mass selective detector (injection temperature: 250 °C; oven temperature was programmed as follows: 60 °C for 2 min, increased up to 260 °C at the rate of 20°/min, followed by 260 °C for 20 min; the carrier gas was helium, used at a flow rate of 1 mL; the transfer line temperature was 280 °C; the ionization was obtained by electron impact (EI); the acquisition range was 50–500 *m/z*). The ESI-MS spectra were recorded using a Waters 2Q 4000 instrument (Waters Corporation, Milford, MA, USA). The melting points were measured on a Büchi apparatus (Stone, Staffs, UK) and were not corrected. The Fourier transform infrared (FT-IR) spectra of the organic compounds were recorded using a Perkin-Elmer FT-IR MOD 1600 spectrophotometer (Norwalk, CT, USA).

3.1.1. Synthesis of Methyl (9*R*)-9-hydroxyoctadecanoate [(*R*)-1]

Dimorphotheca sinuata L. seeds (12.0 g) were ground and suspended in a $CHCl_3$/MeOH mixture (2:1 *v/v*, 200 mL) and the mixture was kept under nitrogen atmosphere, in the dark, and stirred for 24 h. After filtration and washing of the solid with $CHCl_3$ (3 × 20 mL), 20 mL of 0.1 M HCl and 0.1 M NaCl aqueous solution was added to the green solution. After extraction with $CHCl_3$ (3 × 15 mL), the combined organic layers were dried over anhydrous $MgSO_4$. After filtration and solvent removal in vacuo at 30 °C, 4.0 g of residue was recovered. The crude yellow oil was dissolved in CH_3OH (40 mL) under nitrogen atmosphere and the solution was kept at 0 °C by immersion in an external water/ice bath. Then, CH_3ONa (2.16 g) was added and the mixture was magnetically stirred at 0 °C for 2 h. The mixture was acidified to pH 4.0 and extracted with *n*-hexane (3 × 20 mL). The combined organic layers were dried over anhydrous $MgSO_4$. After filtration, the solvent was removed under reduced pressure and the residue (2.2 g) dissolved in 10 mL of ethyl acetate and transferred into the hydrogenation vessel. A catalytic amount of PtO_2 (Adam's catalyst) was added and the mixture was subjected to hydrogenation (40 psi H_2 pressure). The reaction course was monitored by ^{1}H-NMR spectroscopy. Usually, the hydrogenation is complete after approximately 40 min. The mixture was then filtered over celite and the solution concentrated under vacuum. Flash chromatography (eluent: petroleum ether/diethyl ether 7/3) of the residue gave **methyl (9*R*)-9-hydroxyoctadecanoate [(*R*)-1]** ($C_{19}H_{38}O_3$) as a white solid (0.74 g), m.p.: 50–51 °C (Lit. [22]: 50–51.5 °C). ^{1}H-NMR (600 MHz, $CDCl_3$): δ (ppm) = 3.67 (s, 3 H, OCH_3), 3.61–3.55 (m, 1 H, CHOH), 2.30 (t, 2 H, *J* = 7.6 Hz, CH_2COO), 1.62 (quint, 2 H, *J* = 7.4 Hz, CH_2CH_2COO), 1.48–1.21 (m, 27 H, incl. OH), 0.88 (t, 3 H, *J* = 7.0 Hz, CH_3). ^{13}C-NMR (100.6 MHz, $CDCl_3$) δ (ppm) = 174.1, 71.7, 51.3, 37.4, 37.3, 33.9, 31.8, 29.6, 29.56, 29.5, 29.4, 29.2, 29.1, 28.9, 25.5, 25.4, 24.8, 22.6, 14.05. IR ($CHCl_3$): 3427, 1728 cm^{-1}. MS (EI) *m/z* (%): 283 (M$^+$ − OCH_3, 2), 264 (4), 187(45), 159 (11), 158 (53), 155 (100), 129 (6), 115 (18), 109 (14), 87 (50), 74 (33), 69 (18), 55 (31). The enantiomeric excess was determined by derivatization with (*R*)-(−)-*O*-acetyl mandelic acid as previously reported [12,14] and was found to be 80%.

3.1.2. Synthesis of Methyl (*R*)-9-(tosyloxy)octadecanoate [(*R*)-2]

Methyl (9*R*)-9-Hydroxyoctadecanoate [(*R*)-1, 0.190 g (0.60 mmol], 0.607 g (3.18 mmol) of TsCl and 10 mL of anhydrous pyridine were kept at 0 °C for 1 h then at 4 °C. The reaction was monitored by TLC (*n*-hexane/AcOEt 1:1, R_F = 0.54) and, when it was complete (after approximately 5 days), it was quenched with water (40 mL). The mixture was extracted with Et_2O and the combined organic layers were acidified with HCl 1 M. The organic layer was dried over anhydrous $MgSO_4$, filtered and concentrated under reduced pressure. Compound (*R*)-2 ($C_{26}H_{44}O_5S$) was obtained as colorless oil in 70% yield. ^{1}H-NMR (300 MHz, $CDCl_3$): δ (ppm) = 0.87 (t, *J* = 6.9 Hz, 3 H, CH_3), 1.13–1.30 (m, 27 H, $(CH_2)_{13}$), 1.55 (m, 2 H, $CH_2CH_2COOCH_3$), 2.27 (t, *J* = 7.56 Hz, 2 H, CH_2COOCH_3), 2.42 (s, 3 H, $PhCH_3$), 3.65 (s, 3 H, $COOCH_3$), 4.52 (q, *J* = 6.1 Hz, 1 H $CHOSO_2$), 7.31(d, 2 H, *J* = 4 Hz, phenyl), 7.77 (d, *J* = 4.4 Hz, 2 H, phenyl). ^{13}C-NMR (75.44 MHz, $CDCl_3$): δ (ppm) = 174.3, 144.3, 134.8, 129.6, 127.7, 84.5, 51.4, 34.1 (two signals overlapped), 34.0, 31.9, 29.43, 29.38, 29.3(two signals overlapped), 29.1, 29.0, 28.9, 24.8, 24.7, 24.6, 22.6, 21.6, 14.1. ESI-MS (*m/z*): 481 [M + Na]$^+$, 469 [M + H]$^+$

3.1.3. Synthesis of Methyl 9-MethoxyOctadecanoate (3)

Compound (*R*)-2 (0.167 g, 0.357 mmol) was dissolved in CH_3OH (5 mL) and the solution was refluxed for 7 h. After removal of the solvent under reduced pressure, brine (20 mL) was added and the mixture was extracted with Et_2O (4 × 20 mL). The combined organic layers were dried over anhydrous

MgSO$_4$. After filtration and concentration, the residue was purified by FC (eluent: *n*-hexane/CH$_2$Cl$_2$ from 6:4 to 1:1 to 3:7 to 2:8 until CH$_2$Cl$_2$). Pure methyl 9-methoxyoctadecanoate (**3**, (C$_{20}$H$_{40}$O$_3$) was obtained in 55% (0.064 g, 0.195 mmol). ^{1}H-NMR (300 MHz, CDCl$_3$): δ (ppm) = 0.86 (t, *J* = 6.9 Hz, 3 H, CH$_3$), 1.24–1.32 (m, 27 H, (CH$_2$)$_{13}$), 1.60 (m, 2 H, CH$_2$CH$_2$COOCH$_3$), 2.29 (t, *J* = 7.49 Hz, 2 H, CH$_2$COOCH3), 3.10 (m, 1 H, CHOCH$_3$), 3.30 (s, 3 H, CHOCH$_3$), 3.65 (s, 3 H, COOCH$_3$). ^{13}C-NMR (150.80 MHz, CDCl$_3$): δ (ppm) = 174.3, 80.9, 56.3, 51.4, 34.1, 33.42, 33.41, 31.9, 29.9, 29.66, 29.65, 29.58, 29.3, 29.2, 29.1, 25.3, 25.26, 24.9, 22.7, 14.1. MS (EI) *m/z* (%): 327 (M$^+$ − 1, 0.5) 297 (3), 264 (14), 222 (6), 201 (100), 171 (50), 137 (29), 123 (8), 109 (19), 97 (43), 83 (72), 69 (57), 55 (83).

3.1.4. Synthesis of Methyl (9*S*)-9-azidooctadecanoate [(*S*)-**4**]

Method 1:

Compounds (*R*)-**2** (0.147 g, 0.314 mmol), NaN$_3$ (0.102 g, 1.57 mmol), and DMF (10 mL) were heated at reflux for 3 h. The mixture was treated once with brine (20 mL) then extracted with Et$_2$O (3 × 15 mL), then two folds with water (30 mL and 50 mL) and each time extracted with Et$_2$O (3 × 15 mL), finally with brine (20 mL) then extracted with Et$_2$O (3 × 15 mL). Each time the organic layer was dried over anhydrous MgSO$_4$ and filtered: The absence of DMF was checked. The combined organic layers were concentrated and the residue chromatographed on silica gel (eluent: 7:3 *n*-hexane/AcOEt). Compound (*S*)-**4** (C$_{19}$H$_{37}$N$_3$O$_2$, 0.064 g, 58%) was recovered as colorless oil. ^{1}H-NMR (300 MHz, CDCl$_3$): δ (ppm) = 0.88 (t, *J* = 6.6 Hz, 3 H, CH$_3$), 1.24–1.33 (m, 26 H, (CH$_2$)$_{13}$), 1.61 (m, 2 H, CH$_2$CH$_2$COOCH$_3$), 2.30 (t, *J* = 7.9 Hz, 2 H, CH$_2$COOCH$_3$), 3.21(q, *J* = 6.1 Hz, 1 H, CHN$_3$), 3.65 (s, 3 H, COOCH$_3$). ^{13}C-NMR (150.80 MHz, CDCl$_3$): δ (ppm) =174.3, 63.1, 51.4, 34.4, 34.3, 34.0, 31.9, 30.3, 29.5, 29.4, 29.3, 29.2, 29.1, 29.0, 26.1, 26.0, 24.9, 22.6, 14.1; IR-Neat (cm^{-1}): 2928.0, 2856.0, 2361.2, 2098.7, 1742.6, 1465.4, 1257.5, 1171.1; ESI-MS (*m/z*): 352 [M + Na]$^+$, 340 [M + H]$^+$

Method 2:

Compound (*R*)-**1** (0.230 g, 0.732 mmol) was dissolved in anhydrous THF (7.5 mL) and under nitrogen atmosphere. PPh$_3$ (0.398 g, 1.52 mmol) and DIAD 94% (300 μL, 1.52 mmol) were added and the yellow mixture was stirred. After a few minutes, DPPA 97% (330 μL, 1.52 mmol) was added. The solution became cloudy and after 18 h, had come clear. The crude was subjected to FC (eluent: *n*-hexane/AcOEt 9:1) and pure (*S*)-**4** (0.190 g, 77%) was recovered.

3.1.5. Synthesis of Methyl (9*S*)-9-aminooctadecanoate [(*S*)-**5**]

Compound (*R*)-**1** (0.222 g, 0.70 mmol) was dissolved in anhydrous THF (8.0 mL) in a flame dried apparatus immersed in an ice-bath and kept under nitrogen atmosphere. PPh$_3$ (0.103 g (0,70 mmol) and phthalimide (0.103 g (0,70 mmol) were added. Through a funnel, DIAD (0.2 mL, 0.70 mmol) in 2.0 mL of THF was added dropwise. After 12 h, the reaction appeared complete (TLC: petroleum light/diethyl ether 6:4). The reaction mixture was concentrated. FC on silica gel (light petroleum/diethyl ether: 20:1) of the residue gave 0.280 g (0.63 mmol, 90% yield) of the intermediate (9*S*)-9-(1,3-dioxo-1,3-dihydro-2*H*-isoindol-2-yl) octadecanoate [(*S*)-**6**] (C$_{27}$H$_{41}$NO$_4$) as a colorless oil:

^{1}H-NMR (400 MHz, CDCl$_3$), δ (ppm) = 7.83–7.79 (m, 2 H), 7.71–7.68 (m, 2 H), 4.26–4.06 (m, 1 H), 3.63 (s, 3 H, OCH$_3$), 2.25 (t, *J* = 7.43 Hz, 2 H), 2.11–1.97 (m, 2 H), 1.74–1.61 (m, 2 H), 1.61–1.49 (m, 2 H), 1.32–1.14 (m, 22 H), 0.84 (t, *J* = 6.97 Hz, 3 H, CH$_3$). ^{13}C-NMR (100.56 MHz, CDCl$_3$), δ (ppm) = 174.2, 168.8, 133.8, 131.8, 123.0, 52.2, 51.4, 34.0, 32.5, 32.4, 31.8, 29.45, 29.43, 29.23, 29.21, 29.04 (2 signals overlapped), 29.0, 26.6, 26.5, 24.8, 22.6, 14.1. ESI-MS (*m/z*): 444 [M + H]$^+$.

Sodium borohydride (0.06 g, 1.6 mmol) in 2-propanol (3.0 mL) was added to a solution of the compound (*S*)-**6** (0.131 g, 0.29 mmol) and the mixture was magnetically stirred at room temperature for 12 h. Glacial acetic acid (0.3 mL) was added and the mixture was heated at 80 °C for 2 h. After concentration, water (5 mL), and then saturated aqueous solution of NaHCO$_3$ (5 mL) were added. After extraction with ethyl acetate (3 × 10 mL), the combined organic layer was dried over anhydrous magnesium sulfate. After filtration and removal of the solvent under reduced pressure, the crude was subjected to FC (eluent: CH$_2$Cl$_2$/CH$_3$OH 10/1), the spots being evidenced with ninhydrin stain.

Compound **(S)-5** (0.027 g, 0.09 mmol) was obtained in a 30% yield. An alternative method was as follows: Hydrazine hydrate (0.16 mL, 3.2 mmol) was added to a solution of compound **(S)-6** (0.280 g, 0.63 mmol) in ethanol (5 mL) and the mixture was heated and refluxed for 3 h. The formation of a white solid was observed. After cooling, 1 M HCl was added until pH = 6. Then, the solution was filtered on a Buchner funnel. The mixture was treated with saturated solution of $NaHCO_3$ until reaching pH ~ 9 and then it was extracted with CH_2Cl_2 (4 × 10 mL) and dried over anhydrous $MgSO_4$. After filtration and removal of the solvent under reduced pressure, the light yellow oil was subjected to FC (eluent: CH_2Cl_2/CH_3OH 10/1), obtaining the product in 28% yield. Compound **(S)-5** ($C_{19}H_{39}NO_2$): ^{1}H-NMR (400 MHz, $CDCl_3$): δ (ppm) = 0.88 (t, J = 6.9 Hz, 3 H, CH_3), 1.24–1.32 (m, 26 H, $(CH_2)_{13}$), 1.58–1.65 (m, 2 H, $CH_2CH_2COOCH_3$), 2.29 (t, J = 7.7 Hz, 2 H, CH_2COOCH_3), 3.13 (m, 1 H, $CHNH_2$), 3.66 (s, 3 H, $COOCH_3$), 8.30 (br.s, 2 H, NH_2). ^{13}C-NMR (100.56 MHz $CDCl_3$): δ (ppm) = 174.2, 51.6, 51.4, 35.92, 35.86, 34.0, 31.9, 29.63, 29.55 (two signals overlapped), 29.4, 29.3, 29.12, 29.06, 25.8, 25.7, 24.9, 22.6, 14.1. MS (m/z): 282 (5), 267(1), 250(1), 207(3), 186(100), 156(99), 109(4), 97(2), 83(4), 70(5), 56(11). ESI-MS (m/z): 336 [M + Na]$^+$, 314 [M + H]$^+$; IR (cm^{-1}): 3423.5, 2950.0, 2870.0, 1710, 1635.8.

The synthesis of **(R)-5** was carried out with the same procedure starting from **(S)-1**.

3.1.6. Synthesis of (S)-2-Chloro-2-oxo-1-phenylethyl Methyl Carbonate (7)

In a flame-dried apparatus and under nitrogen atmosphere, (S)-O-acetylmandelic acid (1.046 g, 5,4 mmol) and anhydrous CH_2Cl_2 (10 mL) were added. To the solution, oxalyl chloride (0.47 mL, 5.4 mmol) dissolved in anhydrous CH_2Cl_2 (2 mL) were added dropwise. After 2 h at room temperature and under magnetic stirring, the solvent was distilled at room pressure. The residue was distilled at 0.01 mmHg and the colorless oil with b.p.= 102–103 °C found to be compound **7** ($C_{10}H_9ClO_4$).

^{1} H-NMR (300 MHz, $CDCl_3$), δ (ppm): 7.59–7.37 (m, 5 H), 6.09 (s, 1 H), 2.22 (s, 3 H). ^{13} C-NMR (75.44 MHz, $CDCl_3$), δ (ppm): 170.7, 169.8, 130.8, 130.2, 129.2, 128.4, 80.8, 20.4.

3.1.7. Methyl (9S)-9-{[(2S)-2-(Acetyloxy)-2-Phenylacetyl]amino}octadecanoate [(S,S)-8] and Methyl (9R)-9-{[(2S)-2-(Acetyloxy)-2-Phenylacetyl]amino}octadecanoate [(R,S)-8]

The optical purity of **(S)-5** was determined by ^{1}H-NMR analysis after derivatization with **7**, the acyl chloride of the (S)-(–)-O-acetylmandelic acid. Compound **7** (2.8 mg) in 0.5 mL of CD_2Cl_2 was put directly into the NMR spectroscopy tube and methyl 9(S)-9-aminostearate [**(S)-5**] (4 mg, 0.013 mmol) dissolved in CD_2Cl_2 (0.5 mL) was added. Then, triethylamine (2.7 μL) was added and the ^{1}H-NMR spectrum of the crude was recorded. A diastereomeric ratio of ~9:1 between methyl (9S)-9-{[(2S)-2-(acetyloxy)-2-phenylacetyl]amino}octadecanoate and methyl (9R)-9-{[(2S)-2-(acetyloxy)-2-phenylacetyl]amino}octadecanoate was calculated from the integration of the signals belonging to the methoxy group of the two diastereomers (see spectrum SI-17). This relative ratio was calculated from the spectrum of the crude reaction mixture. The product **(S,S)-8** was then purified by preparative TLC on silica gel eluting with CH_2Cl_2 and scraping the spot with RF ~ 0.21. The pure compound **(S,S)-8** was extracted with $CHCl_3$ and analyzed by NMR.

Methyl (9S)-9-{[(2S)-2-(acetyloxy)-2-phenylacetyl]amino}octadecanoate [**(S,S)-8**] ($C_{29}H_{47}NO_6$): ^{1}H-NMR (600 MHz, $CDCl_3$), δ (ppm): 7.45–7.41 (m, 2 H), 7.39–7.32 (m. 3H), 6.03 (s, 1 H), 5.66 (br. d., J = 9.44 Hz, 1 H, NH), 3.94–3.85 (m, 1 H. H-9), 3.669 (s, 3 H, OCH_3), 2.29 (t, J = 7.61 Hz, 2 H), 2.19 (s, 3 H, CH_3CO), 1.64–1.14 (m, 28 H), 0.88 (t, J = 7.04 Hz, 3 H, CH_3).

The same procedure was used to determine the optical purity of methyl (9R)-9-{[(2S)-2-(acetyloxy)-2-phenylacetyl]amino}octadecanoate [**(R,S)-8**] ($C_{29}H_{47}NO_6$): ^{1}H-NMR (600 MHz, $CDCl_3$), δ (ppm): 7.45–7.42 (m, 2 H), 7.39–7.33 (m. 3H), 6.03 (s, 1 H), 5.6 (br. d., J = 9.04 Hz, 1 H, NH), 3.95–3.85 (m, 1 H. H-9), 3.665 (s, 3 H, OCH_3), 2.29 (t, J = 7.61 Hz, 2 H), 2.17 (s, 3 H, CH_3CO), 1.64–1.14 (m, 28 H), 0.88 (t, J = 7.04 Hz, 3 H, CH_3).

In order to verify if some signals of the two diastereisomers could be separated in the ^{1}H-NMR spectrum, a mixture of **(S,S)-8** and **(R,S)-8** was prepared by mixing different amounts of the two species, previously purified by preparative TLC on silica gel (20×20 cm glass plates). The spectrum (600 MHz) showed separate methoxyl hydrogen signals (see Figures S1–S18).

Finally, to confirm the assignment of the signals at 3.669 and 3.665 ppm to the relative diastereomers, a small portion of purified **(R,S)-8** was added to a solution of a 9:1 diastereomeric mixture of **8**, and the peak at 3.665 ppm increased, as expected.

3.2. Cell Culture and Treatments

The human colon cancer cell line HT29 was obtained from American Type Culture Collection (Manassas, VA, USA). The cells were cultured in RPMI 1640 medium (Labtek Eurobio, Milan, Italy), supplemented with 10% FCS (Euroclone, Milan, Italy) and 2 mM L-glutamine (Sigma-Aldrich, Milan, Italy), at 37 °C and 5% CO_2 atmosphere. These conditions were used in all cell incubation steps for the experiments described below. **(R)-1**, **(S)-1**, **(R)-5** and **(S)-5** were dissolved in ethanol to obtain a 33 mM stock solution then diluted in culture medium to obtain the required concentrations. The cells were plated at a density of 1.0×10^4 cells/well in 6-well culture plates (Orange Scientific, Braine-l'Alleud, Belgium). Twenty-four hours after plating, **(R)-1** or **(S)-1** were added to a final concentration of 49 µM or 51 µM for each drug, and incubated for 24, 48, and 72 h. Moreover, HT29 cells were seeded onto 6-well plates, and after 24 h, they were treated with **(R)-5** at the concentration of 57 µM or with **(S)-5** at the concentration of 43 µM for 24, 48 and 72 h. The Trypan Blue exclusion dye method was used to determine the number of viable cells.

3.2.1. MTT Assay

To evaluate **(R)-1**, **(S)-1**, **(R)-5** and **(S)-5** activity, the cells were treated for 24 h without (control) or with concentrations between 1 nM–5 mM (1 nM, 5 nM, 10 nM, 50 nM, 100 nM, 500 nM, 1 µM, 5 µM, 10 µM, 50 µM, 100 µM, 500 µM, 1 mM and 5mM) of the test samples. The culture medium was removed and the cells were further incubated for 2 h with 0.2 mg/mL MTT in PBS. After removal of the medium, the cells were lysed with 0.1 mL of iso-propanol. The absorbance of the solubilized formazan pellet at 540 nm was determined by Victor2, Multilabel plate reader (Perkin Elmer, MI, Italy). The IC_{50} was determined from three different experiments of the dose-response curve by using GraphPadPrism 5 (GraphPad Software, Inc., La Jolla, CA, USA) fitting a symmetrical sigmoidal shaped curve.

3.2.2. Cell Cycle Analysis by Flow Cytometry

The HT29 cells were seeded in 25 cm^2 flasks at a density of 2×10^4 cells/cm^2. The effects on the cell cycle were studied 48 h after treatment with 49 µM **(R)-1**. To determine the cell cycle distribution at the end of incubation, HT29 cells were stained according to Busi et al. [23]. Briefly, 1×10^6 cells were pelleted and resuspended in trisodium citrate 0.1%, RNAse 0.1 mg/L, Igepal 0.01%, Propidium Iodide (PI) 50 µg/L. After 30 min at 37 °C in the dark, the cells were analyzed on a Coulter Epics Elite flow cytometer (Beckman Coulter) equipped with an argon ion laser tuned at 488 nm. PI fluorescence was collected on a linear scale at 600 nm and the DNA distribution was analyzed by the Modfit 5.0 software (ModFit Verity Software House http://www.vsh.com/).

3.2.3. Histone Extraction and Western Blot

HT29 cells were cultured with compounds **(R)-1** for 6 h, and the histone fraction was extracted. The cells were harvested using 0.11% trypsin and 0.02% EDTA, washed twice with PBS, and nuclei were isolated according to Amellem et al. [24]. The nuclear histones were extracted, and proteins were quantified using a protein assay kit (Bio-Rad, Hercules, CA, USA). The histones were examined by 15% SDS-PAGE and a western blot analysis against acetylated lysines using anti-acetylated lysine (Cell Signaling Technology, Danvers, MA #9441, USA). The detection of the immunoreactive bands was performed with a secondary antibody conjugated with horseradish peroxidase (Amersham, Uppsala, SE NA931V, Sweden) and developed with the enhanced chemiluminescence system Clarity Western

(Bio-Rad, Hercules, CA, USA), and the quantification was done by Fluor-S Max MultiImager (Bio-Rad) using the histone H3 signal as the control (Cell Signaling Technology, Beverly, MA #3638, USA).

3.2.4. Total Protein Extraction and Western Blot

HT29 cells were treated with 49 μM of (**R**)-**1** for 24 h. Cell were lysed in according to ref. 23. In brief, the cells were lysed for 20 min in an HEPES buffer and the protein concentration was determined by using the Bio-Rad protein assay method (Bio-Rad, Hercules, CA, USA). The proteins were resolved on a 10% density gel and immunoblotted with p21 (Cell Signaling Technology, Danvers, MA#2946, USA) or Myc (Cell Signaling Technology, Danvers, MA #5605, USA) or β-actin (Cell Signaling Technology, Danvers, MA #4967, USA) antibodies. The nitrocellulose membrane was incubated with secondary horseradish peroxidase-conjugated antibodies (GE Healthcare, Chicago, IL #NA931V or #NA9340, USA), developed as described in the previous section, and the quantification was done by Fluor-S Max MultiImager (Bio-Rad) using the β-actin signal as the control.

3.2.5. Quantitative Real Time-PCR Analysis

HT29 cells were seeded in 25 cm^2 flasks at a density of 2×10^4 cells/cm^2 and treated with (**R**)-**1** for 6 h. RT-PCR was assessed as previously described [23]. The gene expression was quantified by $\Delta\Delta C_T$ method, by using GAPDH as the housekeeping gene. The following primers were used: 5′-ATTTGGTCGTATTGGGCGCC-3′ (forward) and 5′-ACGGTGCCATGGAATTTGCC-3′ (reverse) for GAPDH detection, 5′-CCTAAGAGTGCTGGGCATTTT-3′ (forward) and 5′-TGAATTTCATAACCGCCTGTG-3′ (reverse) for P21 detection, 5′-TAGCTTCACCAACAGGAACT-3′ (forward) and 5′-AGCTCGAATTTCTTCCAGAT-3′ (reverse) for MYC detection.

3.3. Statistical Analysis

The data for the MTT, western blot, cell cycle and RT-PCR were analysed with a one-way analysis of variance (ANOVA) followed by Dunnett's multiple comparison test. The data of the cell growth were analysed with a two-way analysis of variance (ANOVA) followed by the Bonferroni multiple comparisons. The differences of at least $p < 0.05$ were considered significant. A statistical analysis was carried out using Prism GraphPad software.

4. Conclusions

The novel derivatives of (9R)-9-hydroxystearic acid (9-HSA) have been synthesized and the influence of the modification at the level of the substituents on C-9 and of the carboxy group have been studied on HT 29 colon cancer cell line.

The modification on the C-9 by ester (namely methyl (9R)-9-{[(4-methylphenyl)sulphonyl]oxy}octadecanoate, (**R**)-**2**), by ether (namely methyl 9-methoxyoctadecanoate, **3**), or azide group did not produce relevant antiproliferative activity on HT 29 cells. On the contrary, methyl (9S)-9-aminooctadecanoate [(**S**)-**5**] and its enantiomeric form (**R**)-**5** was active against the above colon tumor cells, but very toxic, as evidenced by the high cytotoxicity after 48 and 72 h of the treatment. This is an indication of the importance, for a biological effect, to have a group bound to the C-9 with hydrogen atoms able to interact with the active site of the biological target. When the modification was at the carboxy group of 9-HSA, as in case of the formation of methyl esters (**R**)-**1** and (**S**)-**1**, an antiproliferative effect was observed. In particular, (**R**)-**1** showed a more marked cytostatic effect compared to the (**S**)-**1** enantiomer. In fact, the number of cells dropped at 24 h and remained unchanged over time, suggesting a halt in cell growth. The anti-proliferative effect induced by (**R**)-**1** is characterized by an accumulation of cells in the G0 / G1 phase. The antiproliferative effect of the (**R**)-**1** is due to an increase of the p21WAF/CYP1 and a decrease of c-myc expression that is reflected on these two protein levels.

Supplementary Materials: The following are available online: Figures S1–S18: ^{1}H and ^{13}C-NMR spectra of compounds **1–8**.

Author Contributions: Conceptualization, C.B. and N.C.; investigation, C.B., N.C., D.T., C.T., G.M.; resources, C.B., N.C., G.S., S.B.; writing—original draft preparation, C.B., N.C.; writing—review and editing, S.B., G.S. All authors have read the final version of the manuscript.

Funding: This research received no external funding.

Acknowledgments: Work supported by Alma Mater Studiorum–Università di Bologna (RFO funds). Authors thank Luca Prati, Isabel Orlando, Arige Othman, and Luca Zuppiroli for running the mass spectra.

Conflicts of Interest: The authors declare no conflicts of interest.

References

1. Masotti, L.; Casali, E.; Gesmundo, N.; Sartor, G.; Galeotti, T.; Borrello, S.; Piretti, M.V.; Pagliuca, G. Lipid peroxidation in cancer cells: Chemical and physical studies. *Ann. N. Y. Acad. Sci.* **1989**, *551*, 47–58. [CrossRef] [PubMed]

2. Masotti, L.; Casali, E.; Gesmundo, N. Influence of hydroxystearic acid on in vitro cell proliferation. *Mol. Aspects Med.* **1993**, *14*, 209–215. [CrossRef]

3. Bertucci, C.; Hudaib, M.; Boga, C.; Calonghi, N.; Cappadone, C.; Masotti, L. Gas chromatography/mass apectrometry assay of endogenous cellular lipid peroxidation products: Quantitative analysis of 9- and 10-hydroxystearic acids. *Rapid Commun. Mass Spectrom.* **2002**, *16*, 859–864. [CrossRef] [PubMed]

4. Calonghi, N.; Cappadone, C.; Pagnotta, E.; Farruggia, G.; Buontempo, F.; Boga, C.; Brusa, G.L.; Santucci, M.A.; Masotti, L. 9-Hydroxystearic acid upregulates p21WAF1 in HT29 cancer cells. *Biochem. Biophys. Res. Commun.* **2004**, *314*, 138–142. [CrossRef]

5. Calonghi, N.; Pagnotta, E.; Parolin, C.; Molinari, C.; Boga, C.; Dal Piaz, F.; Brusa, G.L.; Santucci, M.A.; Masotti, L. Modulation of apoptotic signalling by 9-hydroxystearic acid in osteosarcoma cells. *Biochim. Biophys. Acta, Mol. Cell. Biol. Lipids* **2007**, *1771*, 139–146. [CrossRef]

6. Boanini, E.; Torricelli, P.; Boga, C.; Micheletti, G.; Cassani, M.C.; Fini, M.; Bigi, A. (9R)-9-Hydroxystearate-Functionalized Hydroxyapatite as Anti-Proliferative and Cytotoxic Agent towards Osteosarcoma Cells. *Langmuir* **2016**, *32*, 188–194. [CrossRef]

7. Calonghi, N.; Cappadone, C.; Pagnotta, E.; Boga, C.; Bertucci, C.; Fiori, J.; Tasco, G.; Casadio, R.; Masotti, L. Histone deacetylase 1: A target of 9-hydroxystearic acid in the inhibition of cell growth in human colon cancer. *J. Lipid Res.* **2005**, *46*, 1596–1603. [CrossRef]

8. Calonghi, N.; Pagnotta, E.; Parolin, C.; Tognoli, C.; Boga, C.; Masotti, L. 9-Hydroxystearic acid interferes with EGF signalling in a human colon adenocarcinoma. *Biochem. Biophys. Res. Commun.* **2006**, *342*, 585–588. [CrossRef]

9. West, A.C.; Johnstone, R.W. New and emerging HDAC inhibitors for cancer treatment. *J. Clin. Invest.* **2014**, *124*, 30–39. [CrossRef]

10. Earle, F.R.; Mikolajczak, K.L.; Wolff, I.A.; Barclay, A.S. Search for new industrial oils. X. Seed oils of the calenduleae. *J. Am. Oil Chem. Soc.* **1964**, *41*, 345–347. [CrossRef]

11. Parolin, C.; Calonghi, N.; Presta, E.; Boga, C.; Caruana, P.; Naldi, M.; Andrisano, V.; Masotti, L.; Sartor, G. Mechanism and stereoselectivity of HDAC I inhibition by (R)-9-hydroxystearic acid in colon cancer. *Biochim. Biophys. Acta* **2012**, *1821*, 1334–1340.

12. Ebert, C.; Felluga, F.; Forzato, C.; Foscato, M.; Gardossi, L.; Nitti, P.; Pitacco, G.; Boga, C.; Caruana, P.; Micheletti, G. Enzymatic kinetic resolution of hydroxystearic acids: A combined experimental and molecular modelling investigation. *J. Molec. Catal. B: Enzymatic* **2012**, *83*, 38–45. [CrossRef]

13. Bayley, H.; Standring, D.N.; Knowles, J.R. Propane-1,3-di-thiol: A selective reagent for the efficient reduction of alkyl and aryl azides to amines. *Tetrahedron Lett.* **1978**, *39*, 3633–3634. [CrossRef]

14. Boga, C.; Drioli, S.; Forzato, C.; Micheletti, G.; Nitti, P.; Prati, F. Easy route to enantiomerically enriched 7- and 8-hydroxy-stearic acids by olefin-metathesis-based approach. *Synlett* **2016**, *27*, 1354–1358.

15. Richon, V.M.; Sandhoff, T.W.; Rifkind, R.A.; Marks, P.A. Histone deacetylase inhibitor selectively induces p21WAF1 expression and gene-associated histone acetylation. *Proc. Natl. Acad. Sci. USA* **2000**, *97*, 10014–10019. [CrossRef] [PubMed]

16. Nakano, K.; Mizuno, T.; Sowa, Y.; Orita, T.; Yoshino, T.; Okuyama, Y.; Fujita, T.; Ohtani-Fujita, N.; Matsukawa, Y.; Tokino, T.; et al. Butyrate activates the WAF1/Cip1 gene promoter through Sp1 sites in a p53-negative human colon cancer cell line. *J. Biol. Chem.* **1997**, *272*, 22199–22206. [CrossRef] [PubMed]

17. Huang, L.; Sowa, Y.; Sakai, T.; Pardee, A.B. Activation of the p21WAF1/CIP1 promoter independent of p53 by the histone deacetylase inhibitor suberoylanilide hydroxamic acid (SAHA) through the Sp1 sites. *Oncogene* **2000**, *19*, 5712–5719. [CrossRef]

18. Dang, C.V. MYC, metabolism, cell growth, and tumorigenesis. *Cold Spring Harb. Perspect. Med.* **2013**, *3*, a014217. [CrossRef]

19. Lawlor, E.R.; Soucek, L.; Brown-Swigart, L.; Shchors, K.; Bialucha, C.U.; Evan, G.I. Reversible kinetic analysis of Myc targets in vivo provides novel insights into Myc-mediated tumorigenesis. *Cancer Res.* **2006**, *66*, 4591–4601. [CrossRef]

20. Sodir, N.M.; Swigart, L.B.; Karnezis, A.N.; Hanahan, D., Evan, G.I., Soucek, L. Endogenous Myc maintains the tumor microenvironment. *Genes Dev.* **2011**, *25*, 907–916. [CrossRef]

21. Wolfer, A.; Wittner, B.S.; Irimia, D.; Flavin, R.J.; Lupien, M.; Gunawardane, R.N.; Meyer, C.A.; Lightcap, E.S.; Tamayo, P.; Mesirov, J.P.; et al. MYC regulation of a "poor-prognosis" metastatic cancer cell state. *Proc. Natl. Acad. Sci. USA* **2010**, *107*, 3698–3703. [CrossRef] [PubMed]

22. Cochrane, C.; Harwood, H.J. Phase properties of mixtures of 9- and 10-oxo-octadecanoic acids and of 9- and 10-hydroxyoctadecanoic acids. *J. Org. Chem.* **1961**, *26*, 1278–1282. [CrossRef]

23. Busi, A.; Aluigi, A.; Guerrini, A.; Boga, C.; Sartor, G.; Calonghi, N.; Sotgiu, G.; Posati, T.; Corticelli, F.; Fiori, J.; et al. Unprecedented behavior of (9R)-9-hydroxystearic acid loaded keratin nanoparticles on cancer cell cycle. *Mol. Pharmaceutics* **2019**, *16*, 931–942. [CrossRef] [PubMed]

24. Amellem, O.; Stokke, T.; Sandvik, J.A.; Pettersen, E.O. The retinoblastoma gene product is reversibly dephosphorylated and bound in the nucleus in S and G2 phases during hypoxic stress. *Exp. Cell Res.* **1996**, *227*, 106–115. [CrossRef]

Sample Availability: Samples of the compounds are not available from the authors.

MDPI
St. Alban-Anlage 66
4052 Basel
Switzerland
Tel. +41 61 683 77 34
Fax +41 61 302 89 18
www.mdpi.com

Molecules Editorial Office
E-mail: molecules@mdpi.com
www.mdpi.com/journal/molecules